D1569806

Measure, Integration, and Functional Analysis

Measure, Integration, and Functional Analysis

ROBERT B. ASH

University of Illinois

ACADEMIC PRESS

New York and London

ACADEMIC PRESS, INC.
111 Fifth Avenue, New York, New York 10003

United Kingdom Edition published by
ACADEMIC PRESS, INC. (LONDON) LTD.
24/28 Oval Road, London NW1 7DD

LIBRARY OF CONGRESS CATALOG CARD NUMBER: 76-159618

AMS (MOS) 1970 Subject Classification: 28-01

PRINTED IN THE UNITED STATES OF AMERICA

Contents

Preface

The subject matter of this book is fundamental to all areas of mathematical analysis and should be accessible to students who are in the early stages of their professional training. An undergraduate course in real variables is a prerequisite, and an acquaintance with complex analysis is desirable. However, since no use is made of the Cauchy theory, the exposure to complex variables need not be extensive. To fully appreciate Chapters 3 and 4 some background in elementary point-set topology is essential. The intended audience thus consists of mathematics majors, most likely seniors or beginning graduate students, plus students of engineering and physics who use measure theory or functional analysis in their work.

The book has been arranged so that it may be used in several ways. The first two chapters present the fundamentals of measure and integration theory, and can serve as the text for a short course in this subject. If time permits, material from Chapter 4, on the interplay between measure theory and topology, may be added. Chapter 4 is almost independent of Chapter 3; the only dependence occurs in Theorems 4.3.13 and 4.3.15. If the particular group of students has some background in measure theory, Chapter 3 may be used as a text for an introductory course in functional analysis. There is an Appendix on General Topology, in which those areas of topology that occur in the book are treated. Selections from the appendix and from Chapter 4 can be blended into a course covering aspects of topology that are of interest in analysis. Of course, the entire book may be covered, possibly in a leisurely one year course.

Problems are given at the end of each section. Fairly detailed solutions are given to many problems, and instructors may obtain solutions to those problems not worked out in the text by writing the publisher.

"Measure, Integration and Functional Analysis" is actually the first half of another book. The author's "Real Analysis and Probability" consists of the present text and four additional chapters on probability. Thus I have been careful to include those results that are of particular importance in probability. For example, the generalized product measure theorem (Section 2.6) is developed in such a way that it is immediately applicable to compound experiments in probability, where the probability of an event associated with step n of the experiment depends on the result of the first $n - 1$ steps. Also, measures on arbitrary product spaces are discussed, and the Kolmogorov extension theorem is proved in full generality.

Although the book has a probabilistic flavor and indeed its main purpose is to prepare probability students for later work, nevertheless considerable care has been taken to make the presentation suitable for the analysis student who is not necessarily a probability specialist. The basic training needed for work in probability theory is quite similar to the background required for the study of other areas of modern analysis. There are only a few sections in the book which can be regarded as specialized, and these sections may be omitted without loss of continuity. Specifically, Section 1.4 (Parts 6–10), Section 2.7, and Sections 4.4 and 4.5 may be skipped, although even the nonprobabilist may encounter this material in his later work.

It is a pleasure to thank Professors Melvin Gardner and Samuel Saslaw, who used the manuscript in their classes and made many helpful suggestions, Mrs. Dee Keel for her expert typing, and the staff at Academic Press for their encouragement and cooperation.

Summary of Notation

We indicate here the notational conventions to be used throughout the book. The numbering system is standard; for example, 2.7.4 means Chapter 2, Section 7, Part 4. In the Appendix on General Topology, the letter A is used; thus A6.3 means Section 6, Part 3 of the appendix.

The symbol ▌ will be used to mark the end of a proof.

1 Sets

If A and B are subsets of a set Ω, $A \cup B$ will denote the *union* of A and B, and $A \cap B$ the *intersection* of A and B. The union and intersection of a family of sets A_i are denoted by $\bigcup_i A_i$ and $\bigcap_i A_i$. The *complement* of A (relative to Ω) is denoted by A^c.

The statement "B is a *subset* of A" is denoted by $B \subset A$; the inclusion need not be proper, that is, we have $A \subset A$ for any set A. We also write $B \subset A$ as $A \supset B$, to be read "A is an *overset* (or *superset*) of B."

The notation $A - B$ will always mean, unless otherwise specified, the set of points that belong to A but not to B. It is referred to as the *difference* between A and B; a *proper difference* is a set $A - B$, where $B \subset A$.

The *symmetric difference* between A and B is by definition the union of $A - B$ and $B - A$; it is denoted by $A \triangle B$.

If $A_1 \subset A_2 \subset \cdots$ and $\bigcup_{n=1}^{\infty} A_n = A$, we say that the A_n form an *increasing* sequence of sets (increasing to A) and write $A_n \uparrow A$. Similarly, if $A_1 \supset A_2 \supset \cdots$ and $\bigcap_{n=1}^{\infty} A_n = A$, we say that the A_n form a *decreasing* sequence of sets (decreasing to A) and write $A_n \downarrow A$.

The word "includes" will always imply a subset relation, and the word "contains" a membership relation. Thus if $\mathscr{C}$ and $\mathscr{D}$ are collections of sets, "$\mathscr{C}$ includes $\mathscr{D}$" means that $\mathscr{D} \subset \mathscr{C}$. Equivalently, we may say that $\mathscr{C}$ contains all sets in $\mathscr{D}$, in other words, each $A \in \mathscr{D}$ is also a member of $\mathscr{C}$.

A *countable* set is one that is either finite or countably infinite.

2 Real Numbers

The set of real numbers will be denoted by R, and R^n will denote n-dimensional Euclidean space. In R, the interval $(a, b]$ is defined as $\{x \in R: a < x \le b\}$, and (a, ∞) as $\{x \in R: x > a\}$; other types of intervals are defined similarly. If $a = (a_1, \ldots, a_n)$ and $b = (b_1, \ldots, b_n)$ are points in R^n, $a \le b$ will mean $a_i \le b_i$ for all i. The interval $(a, b]$ is defined as $\{x \in R^n: a_i < x_i \le b_i, i = 1, \ldots, n\}$, and other types of intervals are defined similarly.

The set of *extended real numbers* is the two-point compactification $R \cup \{\infty\} \cup \{-\infty\}$, denoted by $\overline{R}$; the set of n-tuples $(x_1, \ldots, x_n)$, with each $x_i \in \overline{R}$, is denoted by $\overline{R}^n$. We adopt the following rules of arithmetic in $\overline{R}$.

$$a + \infty = \infty + a = \infty, \qquad a - \infty = -\infty + a = -\infty, \qquad a \in R,$$

$$\infty + \infty = \infty, \qquad -\infty - \infty = -\infty \qquad (\infty - \infty \quad \text{is not defined}),$$

$$b \cdot \infty = \infty \cdot b = \begin{cases} \infty & \text{if} \quad b \in \overline{R}, \quad b > 0, \\ -\infty & \text{if} \quad b \in \overline{R}, \quad b < 0, \end{cases}$$

$$\frac{a}{\infty} = \frac{a}{-\infty} = 0, \qquad a \in R \qquad \left(\frac{\infty}{\infty} \quad \text{is not defined}\right),$$

$$0 \cdot \infty = \infty \cdot 0 = 0.$$

The rules are convenient when developing the properties of the abstract Lebesgue integral, but it should be emphasized that $\overline{R}$ is not a field under these operations.

Unless otherwise specified (notably in the definition of a positive linear functional in Chapter 4), *positive* means (strictly) greater than zero, and *nonnegative* means greater than or equal to zero.

The set of *complex numbers* is denoted by C, and the set of n-tuples of complex numbers by C^n.

3 Functions

If f is a function from Ω to Ω' (written as $f: \Omega \to \Omega'$) and $B \subset \Omega'$, the *preimage* of B under f is given by $f^{-1}(B) = \{\omega \in \Omega: f(\omega) \in B\}$. It follows from the definition that $f^{-1}(\bigcup_i B_i) = \bigcup_i f^{-1}(B_i)$, $f^{-1}(\bigcap_i B_i) = \bigcap_i f^{-1}(B_i)$, $f^{-1}(A - B) = f^{-1}(A) - f^{-1}(B)$; hence $f^{-1}(A^c) = [f^{-1}(A)]^c$. If $\mathscr{C}$ is a class of sets, $f^{-1}(\mathscr{C})$ means the collection of sets $f^{-1}(B)$, $B \in \mathscr{C}$.

If $f: R \to R$, f is *increasing* iff $x < y$ implies $f(x) \le f(y)$; *decreasing* iff $x < y$ implies $f(x) \ge f(y)$. Thus, "increasing" and "decreasing" do not have the strict connotation. If $f_n: \Omega \to \overline{R}$, $n = 1, 2, \ldots$, the f_n are said to form an *increasing sequence* iff $f_n(\omega) \le f_{n+1}(\omega)$ for all n and ω; a *decreasing sequence* is defined similarly.

If f and g are functions from Ω to $\overline{R}$, statements such as $f \le g$ are always interpreted as holding pointwise, that is, $f(\omega) \le g(\omega)$ for all $\omega \in \Omega$. Similarly, if $f_i: \Omega \to \overline{R}$ for each $i \in I$, $\sup_i f_i$ is the function whose value at ω is $\sup\{f_i(\omega): i \in I\}$.

If $f_1, f_2, \ldots$ form an increasing sequence of functions with limit f [that is, $\lim_{n\to\infty} f_n(\omega) = f(\omega)$ for all ω], we write $f_n \uparrow f$. (Similarly, $f_n \downarrow f$ is used for a decreasing sequence.)

Sometimes, a set such as $\{\omega \in \Omega: f(\omega) \le g(\omega)\}$ is abbreviated as $\{f \le g\}$; similarly, the preimage $\{\omega \in \Omega: f(\omega) \in B\}$ is written as $\{f \in B\}$.

If $A \subset \Omega$, the *indicator* of A is the function defined by $I_A(\omega) = 1$ if $\omega \in A$ and by $I_A(\omega) = 0$ if $\omega \notin A$. The phrase "characteristic function" is often used in the literature, but we shall not adopt this term here.

If f is a function of two variables x and y, the symbol $f(x, \cdot)$ is used for the mapping $y \to f(x, y)$ with x fixed.

The *composition* of two functions $X: \Omega \to \Omega'$ and $f: \Omega' \to \Omega''$ is denoted by $f \circ X$ or $f(X)$.

If $f: \Omega \to \overline{R}$, the *positive* and *negative parts* of f are defined by $f^+ = \max(f, 0)$ and $f^- = \max(-f, 0)$, that is,

$$f^+(\omega) = \begin{cases} f(\omega) & \text{if } f(\omega) \ge 0, \\ 0 & \text{if } f(\omega) < 0, \end{cases}$$

$$f^-(\omega) = \begin{cases} -f(\omega) & \text{if } f(\omega) \le 0, \\ 0 & \text{if } f(\omega) > 0. \end{cases}$$

4 Topology

A *metric space* is a set Ω with a function d (called a *metric*) from $\Omega \times \Omega$ to the nonnegative reals, satisfying $d(x, y) \ge 0$, $d(x, y) = 0$ iff $x = y$, $d(x, y) =$

$d(y, x)$, and $d(x, z) \le d(x, y) + d(y, z)$. If $d(x, y)$ can be 0 for $x \ne y$, but d satisfies the remaining properties, d is called a *pseudometric* (the term *semi-metric* is also used in the literature).

A *ball* (or *open ball*) in a metric or pseudometric space is a set of the form $B(x, r) = \{y \in \Omega: d(x, y) < r\}$ where x, the *center* of the ball, is a point of Ω, and r, the *radius*, is a positive real number. A *closed ball* is a set of the form $\bar{B}(x, r) = \{y \in \Omega: d(x, y) \le r\}$.

A *topological space* is a set Ω with a collection $\mathscr{T}$ of subsets of Ω, called a *topology*, such that $\varnothing$ and Ω belong to $\mathscr{T}$ and $\mathscr{T}$ is closed under finite intersection and arbitrary union. The members of $\mathscr{T}$ are called *open sets*. A *base* for $\mathscr{T}$ is a collection of sets $\mathscr{B}$ such that each open set is a union of sets in $\mathscr{B}$.

A *neighborhood* of a point $x \in \Omega$ is an open set containing x; an *over-neighborhood* of x is an overset of a neighborhood of x. A *base for the neighborhood system at* x (or simply a *base at* x) is a collection $\mathscr{U}$ of neighborhoods of x such that for each neighborhood V of x there is a set $U \in \mathscr{U}$ with $U \subset V$. A *base for the overneighborhood system at* x is defined similarly, with neighborhood replaced by overneighborhood.

If Ω is a topological space, $C(\Omega)$ denotes the class of continuous real-valued functions, and $C_b(\Omega)$ the class of bounded, real-valued, continuous functions, on Ω. The phrase "lower semicontinuous" is abbreviated LSC, and "upper semicontinuous" is abbreviated USC. Sequences in Ω are denoted by $\{x_n, n = 1, 2, \ldots\}$ and nets by $\{x_n, n \in D\}$.

5 Vector Spaces

The terms "vector space" and "linear space" are synonymous. All vector spaces are over the real or complex field, and the complex field is assumed unless the phrase "real vector space" is used.

A *Hamel basis* for a vector space L is a maximal linearly independent subset B of L. (Linear independence means that if $x_1, \ldots, x_n \in B$, $n = 1, 2, \ldots$, and $c_1, \ldots, c_n$ are scalars, then $\sum_{i=1}^n c_i x_i = 0$ iff all $c_i = 0$.) Alternatively, a Hamel basis is a linearly independent subset B with the property that each $x \in L$ is a finite linear combination of elements in B. [An *orthonormal basis* for a Hilbert space (Chapter 3) is a different concept.]

The terms "subspace" and "linear manifold" are synonymous, each referring to a subset M of a vector space L that is itself a vector space under the operations of addition and scalar multiplication in L. If there is a topology on L and M is a closed set in the topology, then M is called a *closed subspace*.

If B is an arbitrary subset of L, the *linear manifold generated by* B, denoted by $L(B)$, is the smallest linear manifold containing all elements of B, that is, the collection of finite linear combinations of elements of B. Assuming a topology on L, the *space spanned by* B, denoted by $S(B)$, is the smallest closed subspace containing all elements of B. Explicitly, $S(B)$ is the closure of $L(B)$.

6 Zorn's Lemma

A *partial ordering* on a set S is a relation "$\leq$" that is

(1) *reflexive*: $a \leq a$,
(2) *antisymmetric*: if $a \leq b$ and $b \leq a$, then $a = b$, and
(3) *transitive*: if $a \leq b$ and $b \leq c$, then $a \leq c$.

(All elements a, b, c belong to S.)

If $C \subset S$, C is said to be *totally ordered* iff for all $a, b \in C$, either $a \leq b$ or $b \leq a$. A totally ordered subset of S is also called a *chain* in S.

The form of Zorn's lemma that will be used in the text is as follows:

Let S be a set with a partial ordering "$\leq$." Assume that every chain C in S has an upper bound; in other words, there is an element $x \in S$ such that $x \geq a$ for all $a \in C$. Then S has a maximal element, that is, an element m such that for each $a \in S$ it is not possible to have $m \leq a$ and $m \neq a$.

Measure, Integration, and Functional Analysis

1

Fundamentals of Measure and Integration Theory

In this chapter we give a self-contained presentation of the basic concepts of the theory of measure and integration. The principles discussed here and in Chapter 2 will serve as background for the study of probability as well as harmonic analysis, linear space theory, and other areas of mathematics.

1.1 Introduction

It will be convenient to start with a little practice in the algebra of sets. This will serve as a refresher and also as a way of collecting a few results that will often be useful.

Let $A_1, A_2, \ldots$ be subsets of a set Ω. If $A_1 \subset A_2 \subset \cdots$ and $\bigcup_{n=1}^{\infty} A_n = A$, we say that the A_n form an *increasing* sequence of sets with limit A, or that the A_n increase to A; we write $A_n \uparrow A$. If $A_1 \supset A_2 \supset \cdots$ and $\bigcap_{n=1}^{\infty} A_n = A$, we say that the A_n form a *decreasing* sequence of sets with limit A, or that the A_n decrease to A; we write $A_n \downarrow A$.

The *De Morgan laws,* namely, $(\bigcup_n A_n)^c = \bigcap_n A_n^c$, $(\bigcap_n A_n)^c = \bigcup_n A_n^c$, imply that

(1) if $A_n \uparrow A$, then $A_n^c \downarrow A^c$; if $A_n \downarrow A$, then $A_n^c \uparrow A^c$.

It is sometimes useful to write a union of sets as a disjoint union. This may be done as follows:

Let $A_1, A_2, \ldots$ be subsets of Ω. For each n we have

(2) $\bigcup_{i=1}^{n} A_i = A_1 \cup (A_1{}^c \cap A_2) \cup (A_1{}^c \cap A_2{}^c \cap A_3) \cup \cdots \cup (A_1{}^c \cap \cdots A_{n-1}^c \cap A_n)$.

Furthermore,

(3) $\bigcup_{n=1}^{\infty} A_n = \bigcup_{n=1}^{\infty} (A_1{}^c \cap \cdots \cap A_{n-1}^c \cap A_n)$.

In (2) and (3), the sets on the right are disjoint. If the A_n form an increasing sequence, the formulas become

(4) $\bigcup_{i=1}^{n} A_i = A_1 \cup (A_2 - A_1) \cup \cdots \cup (A_n - A_{n-1})$

and

(5) $\bigcup_{n=1}^{\infty} A_n = \bigcup_{n=1}^{\infty} (A_n - A_{n-1})$

(Take A_0 as the empty set).

The results (1)–(5) are proved using only the definitions of union, intersection, and complementation; see Problem 1.

The following set operation will be of particular interest. If $A_1, A_2, \ldots$ are subsets of Ω, we define

(6) $\limsup_n A_n = \bigcap_{n=1}^{\infty} \bigcup_{k=n}^{\infty} A_k$.

Thus $\omega \in \limsup_n A_n$ iff for every n, $\omega \in A_k$ for some $k \geq n$, in other words,

(7) $\omega \in \limsup_n A_n$ iff $\omega \in A_n$ for infinitely many n.

Also define

(8) $\liminf_n A_n = \bigcup_{n=1}^{\infty} \bigcap_{k=n}^{\infty} A_k$.

Thus $\omega \in \liminf_n A_n$ iff for some n, $\omega \in A_k$ for all $k \geq n$, in other words,

(9) $\omega \in \liminf_n A_n$ iff $\omega \in A_n$ eventually, that is, for all but finitely many n.

We call $\limsup_n A_n$ the *upper limit* of the sequence of sets A_n, and $\liminf_n A_n$ the *lower limit*. The terminology is, of course, suggested by the analogous concepts for sequences of real numbers

$$\limsup_n x_n = \inf_n \sup_{k \geq n} x_k,$$

$$\liminf_n x_n = \sup_n \inf_{k \geq n} x_k.$$

See Problem 4 for a further development of the analogy.

The following facts may be verified (Problem 5):

(10) $(\limsup_n A_n)^c = \liminf_n A_n{}^c$

(11) $(\liminf_n A_n)^c = \limsup_n A_n{}^c$

(12) $\liminf_n A_n \subset \limsup_n A_n$

(13) If $A_n \uparrow A$ or $A_n \downarrow A$, then $\liminf_n A_n = \limsup_n A_n = A$.

In general, if $\liminf_n A_n = \limsup_n A_n = A$, A is said to be the *limit* of the sequence $A_1, A_2, \ldots$; we write $A = \lim_n A_n$.

Problems

1. Establish formulas (1)–(5).
2. Define sets of real numbers as follows. Let $A_n = (-1/n, 1]$ if n is odd, and $A_n = (-1, 1/n]$ if n is even. Find $\limsup_n A_n$ and $\liminf_n A_n$.
3. Let $\Omega = R^2$, A_n the interior of the circle with center at $((-1)^n/n, 0)$ and radius 1. Find $\limsup_n A_n$ and $\liminf_n A_n$.
4. Let $\{x_n\}$ be a sequence of real numbers, and let $A_n = (-\infty, x_n)$. What is the connection between $\limsup_{n\to\infty} x_n$ and $\limsup_n A_n$ (similarly for lim inf)?
5. Establish formulas (10)–(13).

1.2 Fields, σ-Fields, and Measures

Length, area, and volume, as well as probability are instances of the measure concept that we are going to discuss. A measure is a *set function*, that is, an assignment of a number $\mu(A)$ to each set A in a certain class. Some structure must be imposed on the class of sets on which μ is defined, and probability considerations provide a good motivation for the type of structure required. If Ω is a set whose points correspond to the possible outcomes of a random experiment, certain subsets of Ω will be called "events" and assigned a probability. Intuitively, A is an event if the question "Does ω belong to A?" has a definite yes or no answer after the experiment is performed (and the outcome corresponds to the point $\omega \in \Omega$). Now if we can answer the question "Is $\omega \in A$?" we can certainly answer the question "Is $\omega \in A^c$?", and if, for each $i = 1, \ldots, n$, we can decide whether or not ω belongs to A_i, then we can determine whether or not ω belongs to $\bigcup_{i=1}^n A_i$ (and similarly for $\bigcap_{i=1}^n A_i$). Thus it is natural to require that the class of events be closed under complementation, finite union, and finite intersection; furthermore, since the answer to the question "Is $\omega \in \Omega$?" is always "yes," the entire space Ω should be an event. Closure under *countable* union and intersection is difficult to justify physically, and perhaps the most convincing reason for requiring it is that a richer mathematical theory is obtained. We shall have more to say about this point after we give the definition of a measure. First, we concentrate on the underlying class of sets.

1.2.1 ***Definitions.*** Let $\mathscr{F}$ be a collection of subsets of a set Ω. Then $\mathscr{F}$ is called a *field* (the term *algebra* is also used) iff $\Omega \in \mathscr{F}$ and $\mathscr{F}$ is closed under complementation and finite union, that is,

(a) $\Omega \in \mathscr{F}$.
(b) If $A \in \mathscr{F}$ then $A^c \in \mathscr{F}$.
(c) If $A_1, A_2, \ldots, A_n \in \mathscr{F}$ then $\bigcup_{i=1}^n A_i \in \mathscr{F}$.

It follows that $\mathscr{F}$ is closed under finite intersection. For if $A_1, \ldots, A_n \in \mathscr{F}$, then

$$\bigcap_{i=1}^n A_i = \left(\bigcup_{i=1}^n A_i^c\right)^c \in \mathscr{F}.$$

If (c) is replaced by closure under *countable* union, that is,

(d) If $A_1, A_2, \ldots \in \mathscr{F}$, then $\bigcup_{i=1}^\infty A_i \in \mathscr{F}$,

$\mathscr{F}$ is called a *σ-field* (the term *σ-algebra* is also used). Just as above, $\mathscr{F}$ is also closed under countable intersection.

If $\mathscr{F}$ is a field, a countable union of sets in $\mathscr{F}$ can be expressed as the limit of an increasing sequence of sets in $\mathscr{F}$, and conversely. To see this, note that if $A = \bigcup_{n=1}^\infty A_n$, then $\bigcup_{i=1}^n A_i \uparrow A$; conversely, if $A_n \uparrow A$, then $A = \bigcup_{n=1}^\infty A_n$. This shows that a σ-field is a field that is closed under limits of increasing sequences.

1.2.2 ***Examples.*** The largest σ-field of subsets of a fixed set Ω is the collection of all subsets of Ω. The smallest σ-field consists of the two sets $\varnothing$ and Ω.

Let A be a nonempty proper subset of Ω, and let $\mathscr{F} = \{\varnothing, \Omega, A, A^c\}$. Then $\mathscr{F}$ is the smallest σ-field containing A. For if $\mathscr{G}$ is a σ-field and $A \in \mathscr{G}$, then by definition of a σ-field, Ω, $\varnothing$, and A^c belong to $\mathscr{G}$, hence $\mathscr{F} \subset \mathscr{G}$. But $\mathscr{F}$ is a σ-field, for if we form complements or unions of sets in $\mathscr{F}$, we invariably obtain sets in $\mathscr{F}$. Thus $\mathscr{F}$ is a σ-field that is included in any σ-field containing A, and the result follows.

If $A_1, \ldots, A_n$ are arbitrary subsets of Ω, the smallest σ-field containing $A_1, \ldots, A_n$ may be described explicitly; see Problem 8.

If $\mathscr{S}$ is a class of sets, the smallest σ-field containing the sets of $\mathscr{S}$ will be written as $\sigma(\mathscr{S})$, and sometimes called the *minimal σ-field over $\mathscr{S}$*.

Let Ω be the set R of real numbers. Let $\mathscr{F}$ consist of all finite disjoint unions of right-semiclosed intervals. (A right-semiclosed interval is a set of the form $(a, b] = \{x: a < x \leq b\}, -\infty \leq a < b < \infty$; by convention we also count (a, ∞) as right-semiclosed for $-\infty \leq a < \infty$. The convention is necessary because $(-\infty, a]$ belongs to $\mathscr{F}$, and if $\mathscr{F}$ is to be a field, the complement (a, ∞) must also belong to $\mathscr{F}$.) It may be verified that conditions (a)–(c) of 1.2.1 hold; and thus $\mathscr{F}$ is a field. But $\mathscr{F}$ is not a σ-field; for example, $A_n = (0, 1 - (1/n)] \in \mathscr{F}, n = 1, 2, \ldots$, and $\bigcup_{n=1}^{\infty} A_n = (0, 1) \notin \mathscr{F}$.

If Ω is the set $\overline{R} = [-\infty, \infty]$ of extended real numbers, then just as above, the collection of finite disjoint unions of right-semiclosed intervals forms a field but not a σ-field. Here, the right-semiclosed intervals are sets of the form $(a, b] = \{x: a < x \leq b\}, \ -\infty \leq a < b \leq \infty$, and, by convention, the sets $[-\infty, b] = \{x: -\infty \leq x \leq b\}, \ -\infty \leq b \leq \infty$. (In this case the convention is necessary because $(b, \infty]$ must belong to $\mathscr{F}$, and therefore the complement $[-\infty, b]$ also belongs to $\mathscr{F}$.)

There is a type of reasoning that occurs so often in problems involving σ-fields that it deserves to be displayed explicitly, as in the following typical illustration.

If $\mathscr{C}$ is a class of subsets of Ω and $A \subset \Omega$, we denote by $\mathscr{C} \cap A$ the class $\{B \cap A: B \in \mathscr{C}\}$. If the minimal σ-field over $\mathscr{C}$ is $\sigma(\mathscr{C}) = \mathscr{F}$, let us show that

$$\sigma_A(\mathscr{C} \cap A) = \mathscr{F} \cap A,$$

where $\sigma_A(\mathscr{C} \cap A)$ is the minimal σ-field *of subsets of* A over $\mathscr{C} \cap A$. (In other words, A rather than Ω is regarded as the entire space.)

Now $\mathscr{C} \subset \mathscr{F}$, hence $\mathscr{C} \cap A \subset \mathscr{F} \cap A$, and it is not hard to verify that $\mathscr{F} \cap A$ is a σ-field of subsets of A. Therefore $\sigma_A(\mathscr{C} \cap A) \subset \mathscr{F} \cap A$.

To establish the reverse inclusion we must show that $B \cap A \in \sigma_A(\mathscr{C} \cap A)$ for all $B \in \mathscr{F}$. This is not obvious, so we resort to the following basic reasoning process, which might be called the *good sets principle*. Let $\mathscr{S}$ be the class of good sets, that is, let $\mathscr{S}$ consist of those sets $B \in \mathscr{F}$ such that

$$B \cap A \in \sigma_A(\mathscr{C} \cap A).$$

Since $\mathscr{F}$ and $\sigma_A(\mathscr{C} \cap A)$ are σ-fields, it follows quickly that $\mathscr{S}$ is a σ-field. But $\mathscr{C} \subset \mathscr{S}$, so that $\sigma(\mathscr{C}) \subset \mathscr{S}$, hence $\mathscr{F} = \mathscr{S}$ and the result follows. Briefly, every set in $\mathscr{C}$ is good and the class of good sets forms a σ-field; consequently, every set in $\sigma(\mathscr{C})$ is good.

One other comment: If $\mathscr{C}$ is closed under finite intersection and $A \in \mathscr{C}$, then $\mathscr{C} \cap A = \{C \in \mathscr{C}: C \subset A\}$. (Observe that if $C \subset A$, then $C = C \cap A$.)

1.2.3 ***Definitions and Comments.*** A *measure* on a σ-field $\mathscr{F}$ is a nonnegative, extended real-valued function μ on $\mathscr{F}$ such that whenever $A_1, A_2, \ldots$ form a finite or countably infinite collection of disjoint sets in $\mathscr{F}$, we have

$$\mu\left(\bigcup_n A_n\right) = \sum_n \mu(A_n).$$

If $\mu(\Omega) = 1$, μ is called a *probability measure.*

A *measure space* is a triple $(\Omega, \mathscr{F}, \mu)$ where Ω is a set, $\mathscr{F}$ is a σ-field of subsets of Ω, and μ is a measure on $\mathscr{F}$. If μ is a probability measure, $(\Omega, \mathscr{F}, \mu)$ is called a *probability space.*

It will be convenient to have a slight generalization of the notion of a measure on a σ-field. Let $\mathscr{F}$ be a *field,* μ a set function on $\mathscr{F}$ (a map from $\mathscr{F}$ to $\overline{R}$). We say that μ is *countably additive* on $\mathscr{F}$ iff whenever $A_1, A_2, \ldots$ form a finite or countably infinite collection of disjoint sets in $\mathscr{F}$ whose union also belongs to $\mathscr{F}$ (this will always be the case if $\mathscr{F}$ is a σ-field) we have

$$\mu\left(\bigcup_n A_n\right) = \sum_n \mu(A_n).$$

If this requirement holds only for finite collections of disjoint sets in $\mathscr{F}$, μ is said to be *finitely additive* on $\mathscr{F}$. To avoid the appearance of terms of the form $+\infty - \infty$ in the summation, we always assume that $+\infty$ and $-\infty$ cannot both belong to the range of μ.

If μ is countably additive and $\mu(A) \geq 0$ for all $A \in \mathscr{F}$, μ is called a *measure* on $\mathscr{F}$, a *probability measure* if $\mu(\Omega) = 1$.

Note that countable additivity actually implies finite additivity. For if $\mu(A) = +\infty$ for all $A \in \mathscr{F}$, or if $\mu(A) = -\infty$ for all $A \in \mathscr{F}$, the result is immediate; therefore assume $\mu(A)$ finite for some $A \in \mathscr{F}$. By considering the sequence $A, \varnothing, \varnothing, \ldots$, we find that $\mu(\varnothing) = 0$, and finite additivity is now established by considering the sequence $A_1, \ldots, A_n, \varnothing, \varnothing, \ldots$, where $A_1, \ldots, A_n$ are disjoint sets in $\mathscr{F}$.

Although the set function given by $\mu(A) = +\infty$ for all $A \in \mathscr{F}$ satisfies the definition of a measure, and similarly $\mu(A) = -\infty$ for all $A \in \mathscr{F}$ defines a countably additive set function, we shall from now on exclude these cases. Thus by the above discussion, we always have $\mu(\varnothing) = 0$.

It is possible to develop a theory of measure with the countable additivity requirement replaced by the weaker condition of finite additivity. The disadvantage of doing this is that the resulting mathematical equipment is much less powerful. However, a convincing physical justification of countable additivity has yet to be given. If the probability $P(A)$ of an event A is to represent the long run relative frequency of A in a sequence of performances of a random experiment, P must be a finitely additive set function; but only finitely many measurements can be made in a finite time interval, so countable additivity is not inevitable on physical grounds. Dubins and Savage (1965) have considered certain problems in stochastic processes using only finitely additive set functions, and they assert that for their purposes, finite additivity avoids some of the complications of countable additivity without sacrificing power or scope. On the other hand, at the present time almost all applications of measure theory in mathematics (and physics and engineering as well) use countable rather than finite additivity, and we shall follow this practice here.

1.2.4 ***Examples.*** Let Ω be any set, and let $\mathscr{F}$ consist of all subsets of Ω. Define $\mu(A)$ as the number of points of A. Thus if A has n members, $n = 0, 1, 2, \ldots$, then $\mu(A) = n$; if A is an infinite set, $\mu(A) = \infty$. The set function μ is a measure on $\mathscr{F}$, called *counting measure* on Ω.

A closely related measure is defined as follows. Let $\Omega = \{x_1, x_2, \ldots\}$ be a finite or countably infinite set, and let $p_1, p_2, \ldots$ be nonnegative numbers. Take $\mathscr{F}$ as all subsets of Ω, and define

$$\mu(A) = \sum_{x_i \in A} p_i .$$

Thus if $A = \{x_{i_1}, x_{i_2}, \ldots\}$, then $\mu(A) = p_{i_1} + p_{i_2} + \cdots$. The set function μ is a measure on $\mathscr{F}$ and $\mu\{x_i\} = p_i$, $i = 1, 2, \ldots$. A probability measure will be obtained iff $\sum_i p_i = 1$; if all $p_i = 1$, then μ is counting measure.

Now if A is a subset of R, we try to arrive at a definition of the *length* of A. If A is an interval (open, closed, or semiclosed) with endpoints a and b, it is reasonable to take the length of A to be $\mu(A) = b - a$. If A is a complicated set, we may not have any intuition about its length, but we shall see in Section 1.4 that the requirements that $\mu(a, b] = b - a$ for all $a, b \in R$, $a < b$, and that μ be a measure, determine μ on a large class of sets.

Specifically, μ is determined on the collection of *Borel sets* of R, denoted by $\mathscr{B}(R)$ and defined as the smallest σ-field of subsets of R containing all intervals $(a, b]$, $a, b \in R$.

Note that $\mathscr{B}(R)$ is guaranteed to exist; it may be described (admittedly in a rather ethereal way) as the intersection of all σ-fields containing the intervals $(a, b]$. Also, if a σ-field contains, say, all open intervals, it must contain all intervals $(a, b]$, and conversely. For

$$(a, b] = \bigcap_{n=1}^{\infty} \left(a, b + \frac{1}{n}\right) \qquad \text{and} \qquad (a, b) = \bigcup_{n=1}^{\infty} \left(a, b - \frac{1}{n}\right].$$

Thus $\mathscr{B}(R)$ is the smallest σ-field containing all open intervals. Similarly we may replace the intervals $(a, b]$ by other classes of intervals, for instance,

all closed intervals,
all intervals $[a, b)$, $a, b \in R$,
all intervals (a, ∞), $a \in R$,
all intervals $[a, \infty)$, $a \in R$,
all intervals $(-\infty, b)$, $b \in R$,
all intervals $(-\infty, b]$, $b \in R$.

Since a σ-field that contains all intervals of a given type contains all intervals of any other type, $\mathscr{B}(R)$ may be described as the smallest σ-field that contains the class of all intervals of R. Similarly, $\mathscr{B}(R)$ is the smallest σ-field containing all open sets of R. (To see this, recall that an open set is a countable union of open intervals.) Since a set is open iff its complement is closed, $\mathscr{B}(R)$ is the smallest σ-field containing all closed sets of R. Finally, if $\mathscr{F}_0$ is the field of finite disjoint unions of right-semiclosed intervals (see 1.2.2), then $\mathscr{B}(R)$ is the smallest σ-field containing the sets of $\mathscr{F}_0$.

Intuitively, we may think of generating the Borel sets by starting with the intervals and forming complements and countable unions and intersections in all possible ways. This idea is made precise in Problem 11.

The class of Borel sets of $\overline{R}$, denoted by $\mathscr{B}(\overline{R})$, is defined as the smallest σ-field of subsets of $\overline{R}$ containing all intervals $(a, b]$, $a, b \in \overline{R}$. The above discussion concerning the replacement of the right-semiclosed intervals by other classes of sets applies equally well to $\overline{R}$.

If $E \in \mathscr{B}(R)$, $\mathscr{B}(E)$ will denote $\{B \in \mathscr{B}(R): B \subset E\}$; this coincides with $\{A \cap E: A \in \mathscr{B}(R)\}$ (see 1.2.2).

We now begin to develop some properties of set functions:

1.2.5 ***Theorem.*** Let μ be a finitely additive set function on the field $\mathscr{F}$.

(a) $\mu(\varnothing) = 0$.
(b) $\mu(A \cup B) + \mu(A \cap B) = \mu(A) + \mu(B)$ for all $A, B \in \mathscr{F}$.
(c) If $A, B \in \mathscr{F}$ and $B \subset A$, then $\mu(A) = \mu(B) + \mu(A - B)$

(hence $\mu(A - B) = \mu(A) - \mu(B)$ if $\mu(B)$ is finite, and $\mu(B) \le \mu(A)$ if

$$\mu(A - B) \ge 0).$$

(d) If μ is nonnegative,

$$\mu\left(\bigcup_{i=1}^{n} A_i\right) \le \sum_{i=1}^{n} \mu(A_i) \qquad \text{for all} \qquad A_1, \ldots, A_n \in \mathscr{F}.$$

If μ is a measure,

$$\mu\left(\bigcup_{n=1}^{\infty} A_n\right) \le \sum_{n=1}^{\infty} \mu(A_n)$$

for all $A_1, A_2, \ldots \in \mathscr{F}$ such that $\bigcup_{n=1}^{\infty} A_n \in \mathscr{F}$.

PROOF. (a) Pick $A \in \mathscr{F}$ such that $\mu(A)$ is finite; then

$$\mu(A) = \mu(A \cup \varnothing) = \mu(A) + \mu(\varnothing).$$

(b) By finite additivity,

$$\mu(A) = \mu(A \cap B) + \mu(A - B),$$
$$\mu(B) = \mu(A \cap B) + \mu(B - A).$$

Add the above equations to obtain

$$\begin{aligned}\mu(A) + \mu(B) &= \mu(A \cap B) + [\mu(A - B) + \mu(B - A) + \mu(A \cap B)] \\ &= \mu(A \cap B) + \mu(A \cup B).\end{aligned}$$

(c) We may write $A = B \cup (A - B)$, hence $\mu(A) = \mu(B) + \mu(A - B)$.
(d) We have

$$\begin{aligned}\bigcup_{i=1}^{n} A_i = A_1 \cup (A_1{}^c \cap A_2) \cup (A_1{}^c \cap A_2{}^c \cap A_3) \cup \cdots \\ \cup (A_1{}^c \cap \cdots \cap A_{n-1}^c \cap A_n)\end{aligned}$$

[see Section 1.1, formula (2)]. The sets on the right are disjoint and

$$\mu(A_1{}^c \cap \cdots \cap A_{n-1}^c \cap A_n) \le \mu(A_n) \qquad \text{by (c).}$$

The case in which μ is a measure is handled using the identity (3) of Section 1.1. ▮

1.2.6 ***Definitions.*** A set function μ defined on $\mathscr{F}$ is said to be *finite* iff $\mu(A)$ is finite, that is, not $\pm\infty$, for each $A \in \mathscr{F}$. If μ is finitely additive, it is sufficient to require that $\mu(\Omega)$ be finite; for $\Omega = A \cup A^c$, and if $\mu(A)$ is, say, $+\infty$, so is $\mu(\Omega)$.

A nonnegative, finitely additive set function μ on the field $\mathscr{F}$ is said to be *σ-finite* on $\mathscr{F}$ iff Ω can be written as $\bigcup_{n=1}^{\infty} A_n$ where the A_n belong to $\mathscr{F}$ and $\mu(A_n) < \infty$ for all n. [By formula (3) of Section 1.1, the A_n may be assumed disjoint.] We shall see that many properties of finite measures can be extended quickly to σ-finite measures.

It follows from 1.2.5(c) that a nonnegative, finitely additive set function μ on a field $\mathscr{F}$ is finite iff it is bounded; that is, $\sup\{|\mu(A)| : A \in \mathscr{F}\} < \infty$.

This no longer holds if the nonnegativity assumption is dropped (see Problem 4). It is true, however, that a *countably* additive set function on a σ-field is finite iff it is bounded; this will be proved in 2.1.3.

Countably additive set functions have a basic continuity property, which we now describe.

1.2.7 ***Theorem.*** Let μ be a countably additive set function on the σ-field $\mathscr{F}$.

(a) If $A_1, A_2, \ldots \in \mathscr{F}$ and $A_n \uparrow A$, then $\mu(A_n) \to \mu(A)$ as $n \to \infty$.

(b) If $A_1, A_2, \ldots \in \mathscr{F}$, $A_n \downarrow A$, and $\mu(A_1)$ is finite [hence $\mu(A_n)$ is finite for all n since $\mu(A_1) = \mu(A_n) + \mu(A_1 - A_n)$], then $\mu(A_n) \to \mu(A)$ as $n \to \infty$.

The same results hold if $\mathscr{F}$ is only assumed to be a field, if we add the hypothesis that the limit sets A belong to $\mathscr{F}$. [If $A \notin \mathscr{F}$ and $\mu \geq 0$, 1.2.5(c) implies that $\mu(A_n)$ increases to a limit in part (a), and decreases to a limit in part (b), but we cannot identify the limit with $\mu(A)$.]

PROOF. (a) If $\mu(A_n) = \infty$ for some n, then $\mu(A) = \mu(A_n) + \mu(A - A_n) = \infty + \mu(A - A_n) = \infty$. Replacing A by A_k we find that $\mu(A_k) = \infty$ for all $k \geq n$, and we are finished. In the same way we eliminate the case in which $\mu(A_n) = -\infty$ for some n. Thus we may assume that all $\mu(A_n)$ are finite.

Since the A_n form an increasing sequence, we may use the identity (5) of Section 1.1:

$$A = A_1 \cup (A_2 - A_1) \cup \cdots \cup (A_n - A_{n-1}) \cup \cdots.$$

Therefore, by 1.2.5(c),

$$\begin{aligned}\mu(A) &= \mu(A_1) + \mu(A_2) - \mu(A_1) + \cdots + \mu(A_n) - \mu(A_{n-1}) + \cdots \\ &= \lim_{n \to \infty} \mu(A_n).\end{aligned}$$

(b) If $A_n \downarrow A$, then $A_1 - A_n \uparrow A_1 - A$, hence $\mu(A_1 - A_n) \to \mu(A_1 - A)$ by (a). The result now follows from 1.2.5(c). ∎

We shall frequently encounter situations in which finite additivity of a particular set function is easily established, but countable additivity is more difficult. It is useful to have the result that finite additivity plus continuity implies countable additivity.

1.2.8 ***Theorem.*** Let μ be a finitely additive set function on the field $\mathscr{F}$.

(a) Assume that μ is *continuous from below* at each $A \in \mathscr{F}$, that is, if $A_1, A_2, \ldots \in \mathscr{F}$, $A = \bigcup_{n=1}^{\infty} A_n \in \mathscr{F}$, and $A_n \uparrow A$, then $\mu(A_n) \to \mu(A)$. It follows that μ is countably additive on $\mathscr{F}$.

(b) Assume that μ is *continuous from above* at the empty set, that is, if $A_1, A_2, \ldots \in \mathscr{F}$ and $A_n \downarrow \emptyset$, then $\mu(A_n) \to 0$. It follows that μ is countably additive on $\mathscr{F}$.

PROOF. (a) Let $A_1, A_2, \ldots$ be disjoint sets in $\mathscr{F}$ whose union A belongs to $\mathscr{F}$. If $B_n = \bigcup_{i=1}^{n} A_i$ then $B_n \uparrow A$, hence $\mu(B_n) \to \mu(A)$ by hypothesis. But $\mu(B_n) = \sum_{i=1}^{n} \mu(A_i)$ by finite additivity, hence $\mu(A) = \lim_{n\to\infty} \sum_{i=1}^{n} \mu(A_i)$, the desired result.

(b) Let $A_1, A_2, \ldots$ be disjoint sets in $\mathscr{F}$ whose union A belongs to $\mathscr{F}$, and let $B_n = \bigcup_{i=1}^{n} A_i$. By 1.2.5(c), $\mu(A) = \mu(B_n) + \mu(A - B_n)$; but $A - B_n \downarrow \emptyset$, so by hypothesis, $\mu(A - B_n) \to 0$. Thus $\mu(B_n) \to \mu(A)$, and the result follows as in (a). ∎

If μ_1 and μ_2 are measures on the σ-field $\mathscr{F}$, then $\mu = \mu_1 - \mu_2$ is countably additive on $\mathscr{F}$, assuming either μ_1 or μ_2 is finite-valued. We shall see later (in 2.1.3) that any countably additive set function on a σ-field can be expressed as the difference of two measures.

For examples of finitely additive set functions that are not countably additive, see Problems 1, 3, and 4.

Problems

1. Let Ω be a countably infinite set, and let $\mathscr{F}$ consist of all subsets of Ω. Define $\mu(A) = 0$ if A is finite, $\mu(A) = \infty$ if A is infinite.
   (a) Show that μ is finitely additive but not countably additive.
   (b) Show that Ω is the limit of an increasing sequence of sets A_n with $\mu(A_n) = 0$ for all n, but $\mu(\Omega) = \infty$.
2. Let μ be counting measure on Ω, where Ω is an infinite set. Show that there is a sequence of sets $A_n \downarrow \emptyset$ with $\lim_{n\to\infty} \mu(A_n) \neq 0$.
3. Let Ω be a countably infinite set, and let $\mathscr{F}$ be the field consisting of all finite subsets of Ω and their complements. If A is finite, set $\mu(A) = 0$, and if A^c is finite, set $\mu(A) = 1$.
   (a) Show that μ is finitely additive but not countably additive on $\mathscr{F}$.
   (b) Show that Ω is the limit of an increasing sequence of sets $A_n \in \mathscr{F}$ with $\mu(A_n) = 0$ for all n, but $\mu(\Omega) = 1$.

4. Let $\mathscr{F}$ be the field of finite disjoint unions of right-semiclosed intervals of R, and define the set function μ on $\mathscr{F}$ as follows.

$$\mu(-\infty, a] = a, \qquad a \in R,$$

$$\mu(a, b] = b - a, \qquad a, b \in R, \qquad a < b,$$

$$\mu(b, \infty) = -b, \qquad b \in R,$$

$$\mu(R) = 0,$$

$$\mu\left(\bigcup_{i=1}^{n} I_i\right) = \sum_{i=1}^{n} \mu(I_i)$$

if $I_1, \ldots, I_n$ are disjoint right-semiclosed intervals.

(a) Show that μ is finitely additive but not countably additive on $\mathscr{F}$.
(b) Show that μ is finite but unbounded on $\mathscr{F}$.

5. Let μ be a nonnegative, finitely additive set function on the field $\mathscr{F}$. If $A_1, A_2, \ldots$ are disjoint sets in $\mathscr{F}$ and $\bigcup_{n=1}^{\infty} A_n \in \mathscr{F}$, show that

$$\mu\left(\bigcup_{n=1}^{\infty} A_n\right) \geq \sum_{n=1}^{\infty} \mu(A_n).$$

6. Let $f: \Omega \to \Omega'$, and let $\mathscr{C}$ be a class of subsets of Ω'. Show that

$$\sigma(f^{-1}(\mathscr{C})) = f^{-1}(\sigma(\mathscr{C})),$$

where $f^{-1}(\mathscr{C}) = \{f^{-1}(A): A \in \mathscr{C}\}$. (Use the good sets principle.)

7. If A is a Borel subset of R, show that the smallest σ-field of subsets of A containing the sets open in A (in the relative topology inherited from R) is $\{B \in \mathscr{B}(R): B \subset A\}$.

8. Let $A_1, \ldots, A_n$ be arbitrary subsets of a set Ω. Describe (explicitly) the smallest σ-field $\mathscr{F}$ containing $A_1, \ldots, A_n$. How many sets are there in $\mathscr{F}$? (Give an upper bound that is attainable under certain conditions.)

9. (a) Let $\mathscr{C}$ be an arbitrary class of subsets of Ω, and let $\mathscr{G}$ be the collection of all finite unions $\bigcup_{i=1}^{n} A_i$, $n = 1, 2, \ldots$, where each A_i is a finite intersection $\bigcap_{j=1}^{r} B_{ij}$, with B_{ij} a set in $\mathscr{C}$ or its complement.

   Show that $\mathscr{G}$ is the minimal field (not σ-field) over $\mathscr{C}$.

   (b) Show that the minimal field can also be described as the collection $\mathscr{D}$ of all finite *disjoint* unions $\bigcup_{i=1}^{n} A_i$, where the A_i are as above.

   (c) If $\mathscr{F}_1, \ldots, \mathscr{F}_n$ are fields of subsets of Ω, show that the smallest field including $\mathscr{F}_1, \ldots, \mathscr{F}_n$ consists of all finite (disjoint) unions of sets $A_1 \cap \cdots \cap A_n$ with $A_i \in \mathscr{F}_i$, $i = 1, \ldots, n$.

10. Let μ be a finite measure on the σ-field $\mathscr{F}$. If $A_n \in \mathscr{F}$, $n = 1, 2, \ldots$, and $A = \lim_n A_n$ (see Section 1.1), show that $\mu(A) = \lim_{n\to\infty} \mu(A_n)$.

11. Let $\mathscr{C}$ be any class of subsets of Ω, with $\varnothing, \Omega \in \mathscr{C}$. Define $\mathscr{C}_0 = \mathscr{C}$, and for any ordinal $\alpha > 0$ write, inductively,

$$\mathscr{C}_\alpha = \left(\bigcup \{\mathscr{C}_\beta : \beta < \alpha\}\right)',$$

where $\mathscr{D}'$ denotes the class of all countable unions of differences of sets in $\mathscr{D}$.

Let $\mathscr{S} = \bigcup\{\mathscr{C}_\alpha : \alpha < \beta_1\}$, where β_1 is the first uncountable ordinal, and let $\mathscr{F}$ be the minimal σ-field over $\mathscr{C}$. Since each $\mathscr{C}_\alpha \subset \mathscr{F}$, we have $\mathscr{S} \subset \mathscr{F}$. Also, the $\mathscr{C}_\alpha$ increase with α, and $\mathscr{C} \subset \mathscr{C}_\alpha$ for all α.

(a) Show that $\mathscr{S}$ is a σ-field (hence $\mathscr{S} = \mathscr{F}$ by minimality of $\mathscr{F}$).

(b) If the cardinality of $\mathscr{C}$ is at most c, the cardinality of the reals, show that card $\mathscr{F} \leq c$ also.

1.3 Extension of Measures

In Section 1.2.4, we discussed the concept of length of a subset of R. The problem was to extend the set function given on intervals by $\mu(a, b] = b - a$ to a larger class of sets. If $\mathscr{F}_0$ is the field of finite disjoint unions of right-semiclosed intervals, there is no problem extending μ to $\mathscr{F}_0$: if $A_1, \ldots, A_n$ are disjoint right-semiclosed intervals, we set $\mu(\bigcup_{i=1}^n A_i) = \sum_{i=1}^n \mu(A_i)$. The resulting set function on $\mathscr{F}_0$ is finitely additive, but countable additivity is not clear at this point. Even if we can prove countable additivity on $\mathscr{F}_0$, we still have the problem of extending μ to the minimal σ-field over $\mathscr{F}_0$, namely, the Borel sets.

We are going to consider a generalization of the above problem. Instead of working only with length, we shall examine set functions given by $\mu(a, b] = F(b) - F(a)$ where F is an increasing right-continuous function from R to R. The extension technique to be developed is not restricted to set functions defined on subsets of R; we shall prove a general result concerning the extension of a measure from a field $\mathscr{F}_0$ to the minimal σ-field over $\mathscr{F}_0$.

It will be convenient to consider finite measures at first, and nothing is lost if we normalize and work with probability measures:

1.3.1 ***Lemma.*** Let $\mathscr{F}_0$ be a field of subsets of a set Ω, and let P be a probability measure on $\mathscr{F}_0$. Suppose that the sets $A_1, A_2, \ldots$ belong to $\mathscr{F}_0$ and increase to a limit A, and that the sets $A_1', A_2', \ldots$ belong to $\mathscr{F}_0$ and increase to A'. (A and A' need not belong to $\mathscr{F}_0$.) If $A \subset A'$, then

$$\lim_{m\to\infty} P(A_m) \leq \lim_{n\to\infty} P(A_n').$$

PROOF. If m is fixed, $A_m \cap A_n' \uparrow A_m \cap A' = A_m$ as $n \to \infty$, hence

$$P(A_m \cap A_n') \to P(A_m)$$

by 1.2.7(a). But $P(A_m \cap A_n') \leq P(A_n')$ by 1.2.5(c), hence

$$P(A_m) = \lim_{n\to\infty} P(A_m \cap A_n') \leq \lim_{n\to\infty} P(A_n').$$

Let $m \to \infty$ to finish the proof. ∎

We are now ready for the first extension of P to a larger class of sets:

1.3.2 ***Lemma.*** Let P be a probability measure on the field $\mathscr{F}_0$. Let $\mathscr{G}$ be the collection of all limits of increasing sequences of sets in $\mathscr{F}_0$, that is, $A \in \mathscr{G}$ iff there are sets $A_n \in \mathscr{F}_0$, $n = 1, 2, \ldots$, such that $A_n \uparrow A$. (Note that $\mathscr{G}$ can also be described as the collection of all countable unions of sets in $\mathscr{F}_0$; see 1.2.1.)

Define μ on $\mathscr{G}$ as follows. If $A_n \in \mathscr{F}_0$, $n = 1, 2, \ldots$, $A_n \uparrow A$ $(\in \mathscr{G})$, set $\mu(A) = \lim_{n\to\infty} P(A_n)$; μ is well defined by 1.3.1, and $\mu = P$ on $\mathscr{F}_0$. Then:

(a) $\varnothing \in \mathscr{G}$ and $\mu(\varnothing) = 0$; $\Omega \in \mathscr{G}$ and $\mu(\Omega) = 1$; $0 \leq \mu(A) \leq 1$ for all $A \in \mathscr{G}$.

(b) If $G_1, G_2 \in \mathscr{G}$, then $G_1 \cup G_2$, $G_1 \cap G_2 \in \mathscr{G}$ and $\mu(G_1 \cup G_2) + \mu(G_1 \cap G_2) = \mu(G_1) + \mu(G_2)$.

(c) If $G_1, G_2 \in \mathscr{G}$ and $G_1 \subset G_2$, then $\mu(G_1) \leq \mu(G_2)$.

(d) If $G_n \in \mathscr{G}$, $n = 1, 2, \ldots$, and $G_n \uparrow G$, then $G \in \mathscr{G}$ and $\mu(G_n) \to \mu(G)$.

PROOF. (a) This is clear since $\mu = P$ on $\mathscr{F}_0$ and P is a probability measure.

(b) Let $A_{n1} \in \mathscr{F}_0$, $A_{n1} \uparrow G_1$; $A_{n2} \in \mathscr{F}_0$, $A_{n2} \uparrow G_2$. We have $P(A_{n1} \cup A_{n2}) + P(A_{n1} \cap A_{n2}) = P(A_{n1}) + P(A_{n2})$ by 1.2.5(b); let $n \to \infty$ to complete the argument.

(c) This follows from 1.3.1.

(d) Since G is a countable union of sets in $\mathscr{F}_0$, $G \in \mathscr{G}$. Now for each n we can find sets $A_{nm} \in \mathscr{F}_0$, $m = 1, 2, \ldots$, with $A_{nm} \uparrow G_n$ as $m \to \infty$. The situation may be represented schematically as follows:

$$\begin{array}{cccccc}
A_{11} & A_{12} & \cdots & A_{1m} & \cdots & \uparrow G_1 \\
A_{21} & A_{22} & \cdots & A_{2m} & \cdots & \uparrow G_2 \\
\vdots & \vdots & \vdots & \vdots & \vdots & \vdots \\
A_{n1} & A_{n2} & \cdots & A_{nm} & \cdots & \uparrow G_n \\
\vdots & \vdots & \vdots & \vdots & \vdots & \vdots
\end{array}$$

Let $D_m = A_{1m} \cup A_{2m} \cup \cdots \cup A_{mm}$ (the D_m form an increasing sequence). The key step in the proof is the observation that

$$A_{nm} \subset D_m \subset G_m \qquad \text{for} \quad n \le m \tag{1}$$

and, therefore,

$$P(A_{nm}) \le P(D_m) \le \mu(G_m) \qquad \text{for} \quad n \le m. \tag{2}$$

Let $m \to \infty$ in (1) to obtain $G_n \subset \bigcup_{m=1}^{\infty} D_m \subset G$; then let $n \to \infty$ to conclude that $D_m \uparrow G$, hence $P(D_m) \to \mu(G)$ by definition of μ. Now let $m \to \infty$ in (2) to obtain $\mu(G_n) \le \lim_{m\to\infty} P(D_m) \le \lim_{m\to\infty} \mu(G_m)$; then let $n \to \infty$ to conclude that $\lim_{n\to\infty} \mu(G_n) = \lim_{m\to\infty} P(D_m) = \mu(G)$. ∎

We now extend μ to the class of all subsets of Ω; however, the extension will not be countably additive on all subsets, but only on a smaller σ-field. The construction depends on properties (a)–(d) of 1.3.2, and not on the fact that μ was derived from a probability measure on a field. We express this explicitly as follows:

1.3.3 ***Lemma.*** Let $\mathscr{G}$ be a class of subsets of a set Ω, μ a nonnegative real-valued set function on $\mathscr{G}$ such that $\mathscr{G}$ and μ satisfy the four conditions (a)–(d) of 1.3.2. Define, for each $A \subset \Omega$,

$$\mu^*(A) = \inf\{\mu(G) \colon G \in \mathscr{G}, \quad G \supset A\}.$$

Then:

(a) $\mu^* = \mu$ on $\mathscr{G}$, $0 \le \mu^*(A) \le 1$ for all $A \subset \Omega$.

(b) $\mu^*(A \cup B) + \mu^*(A \cap B) \le \mu^*(A) + \mu^*(B)$; in particular, $\mu^*(A) + \mu^*(A^c) \ge \mu^*(\Omega) + \mu^*(\varnothing) = \mu(\Omega) + \mu(\varnothing) = 1$ by 1.3.2(a).

(c) If $A \subset B$, then $\mu^*(A) \le \mu^*(B)$.

(d) If $A_n \uparrow A$, then $\mu^*(A_n) \to \mu^*(A)$.

PROOF. (a) This is clear from the definition of μ^* and from 1.3.2(c).

(b) If $\varepsilon > 0$, choose $G_1, G_2 \in \mathscr{G}$, $G_1 \supset A$, $G_2 \supset B$, such that $\mu(G_1) \le \mu^*(A) + \varepsilon/2$, $\mu(G_2) \le \mu^*(B) + \varepsilon/2$. By 1.3.2(b),

$$\begin{aligned} \mu^*(A) + \mu^*(B) + \varepsilon &\ge \mu(G_1) + \mu(G_2) = \mu(G_1 \cup G_2) + \mu(G_1 \cap G_2) \\ &\ge \mu^*(A \cup B) + \mu^*(A \cap B). \end{aligned}$$

Since ε is arbitrary, the result follows.

(c) This follows from the definition of μ^*.

(d) By (c), $\mu^*(A) \ge \lim_{n\to\infty} \mu^*(A_n)$. If $\varepsilon > 0$, for each n we may choose $G_n \in \mathscr{G}$, $G_n \supset A_n$, such that

$$\mu(G_n) \le \mu^*(A_n) + \varepsilon 2^{-n}.$$

Now $A = \bigcup_{n=1}^{\infty} A_n \subset \bigcup_{n=1}^{\infty} G_n \in \mathscr{G}$; hence

$$\begin{aligned}
\mu^*(A) &\le \mu^*\left(\bigcup_{n=1}^{\infty} G_n\right) && \text{by (c)} \\
&= \mu\left(\bigcup_{n=1}^{\infty} G_n\right) && \text{by (a)} \\
&= \lim_{n\to\infty} \mu\left(\bigcup_{k=1}^{n} G_k\right) && \text{by 1.3.2(d).}
\end{aligned}$$

The proof will be accomplished if we prove that

$$\mu\left(\bigcup_{i=1}^{n} G_i\right) \le \mu^*(A_n) + \varepsilon \sum_{i=1}^{n} 2^{-i}, \qquad n = 1, 2, \ldots.$$

This is true for $n = 1$, by choice of G_1. If it holds for a given n, we apply 1.3.2(b) to the sets $\bigcup_{i=1}^{n} G_i$ and G_{n+1} to obtain

$$\mu\left(\bigcup_{i=1}^{n+1} G_i\right) = \mu\left(\bigcup_{i=1}^{n} G_i\right) + \mu(G_{n+1}) - \mu\left[\left(\bigcup_{i=1}^{n} G_i\right) \cap G_{n+1}\right].$$

Now $(\bigcup_{i=1}^{n} G_i) \cap G_{n+1} \supset G_n \cap G_{n+1} \supset A_n \cap A_{n+1} = A_n$, so that the induction hypothesis yields

$$\begin{aligned}
\mu\left(\bigcup_{i=1}^{n+1} G_i\right) &\le \mu^*(A_n) + \varepsilon \sum_{i=1}^{n} 2^{-i} + \mu^*(A_{n+1}) + \varepsilon 2^{-(n+1)} - \mu^*(A_n) \\
&\le \mu^*(A_{n+1}) + \varepsilon \sum_{i=1}^{n+1} 2^{-i}. \quad \blacksquare
\end{aligned}$$

Our aim in this section is to prove that a σ-finite measure on a field $\mathscr{F}_0$ has a unique extension to the minimal σ-field over $\mathscr{F}_0$. In fact an arbitrary measure μ on $\mathscr{F}_0$ can be extended to $\sigma(\mathscr{F}_0)$, but the extension is not necessarily unique. In proving this more general result (see Problem 3), the following concept plays a key role.

1.3.4 ***Definition.*** An *outer measure* on Ω is a nonnegative, extended real-valued set function λ on the class of all subsets of Ω, satisfying

(a) $\lambda(\varnothing) = 0$,
(b) $A \subset B$ implies $\lambda(A) \le \lambda(B)$ (monotonicity),
(c) $\lambda(\bigcup_{n=1}^{\infty} A_n) \le \sum_{n=1}^{\infty} \lambda(A_n)$ (countable subadditivity).

The set function μ^* of 1.3.3 is an outer measure on Ω. Parts 1.3.4(a) and (b) follow from 1.3.3(a), 1.3.2(a), and 1.3.3(c), and 1.3.4(c) is proved as follows:

$$\mu^*\left(\bigcup_{n=1}^{\infty} A_n\right) = \lim_{n\to\infty} \mu^*\left(\bigcup_{i=1}^{n} A_i\right) \qquad \text{by 1.3.3(d).}$$

$$\leq \lim_{n\to\infty} \sum_{i=1}^{n} \mu^*(A_i) \qquad \text{by 1.3.3(b),}$$

as desired.

We now identify a σ-field on which μ^* is countably additive:

1.3.5 ***Theorem.*** Under the hypothesis of 1.3.3, let

$$\mathscr{H} = \{H \subset \Omega \colon \mu^*(H) + \mu^*(H^c) = 1\}$$

[$\mathscr{G} \subset \mathscr{H}$ by 1.3.2(a) and (b) and $\mathscr{H} = \{H \subset \Omega \colon \mu^*(H) + \mu^*(H^c) \leq 1\}$ by 1.3.3(b)]. Then $\mathscr{H}$ is a σ-field and μ^* is a probability measure on $\mathscr{H}$.

PROOF. Clearly $\mathscr{H}$ is closed under complementation, and $\Omega \in \mathscr{H}$ by 1.3.3(a) and 1.3.2(a). If $H_1, H_2 \subset \Omega$, then by 1.3.3(b),

$$\mu^*(H_1 \cup H_2) + \mu^*(H_1 \cap H_2) \leq \mu^*(H_1) + \mu^*(H_2) \tag{1}$$

and since

$$(H_1 \cup H_2)^c = H_1{}^c \cap H_2{}^c, \qquad (H_1 \cap H_2)^c = H_1{}^c \cup H_2{}^c,$$

$$\mu^*(H_1 \cup H_2)^c + \mu^*(H_1 \cap H_2)^c \leq \mu^*(H_1{}^c) + \mu^*(H_2{}^c). \tag{2}$$

If $H_1, H_2 \in \mathscr{H}$, add (1) and (2); the sum of the left sides is at least 2 by 1.3.3(b), and the sum of the right sides is 2. Thus the sum of the left sides is 2 as well. If $a = \mu^*(H_1 \cup H_2) + \mu^*(H_1 \cup H_2)^c$, $b = \mu^*(H_1 \cap H_2) + \mu^*(H_1 \cap H_2)^c$, then $a + b = 2$, hence $a \leq 1$ or $b \leq 1$. If $a \leq 1$, then $a = 1$, so $b = 1$ also. Consequently $H_1 \cup H_2 \in \mathscr{H}$ and $H_1 \cap H_2 \in \mathscr{H}$. We have therefore shown that $\mathscr{H}$ is a field. Now equality holds in (1), for if not, the sum of the left sides of (1) and (2) would be less than the sum of the right sides, a contradiction. Thus μ^* is finitely additive on $\mathscr{H}$.

To show that $\mathscr{H}$ is a σ-field, let $H_n \in \mathscr{H}$, $n = 1, 2, \ldots$, $H_n \uparrow H$; $\mu^*(H) + \mu^*(H^c) \geq 1$ by 1.3.3(b). But $\mu^*(H) = \lim_{n\to\infty} \mu^*(H_n)$ by 1.3.3(d), hence for any $\varepsilon > 0$, $\mu^*(H) \leq \mu^*(H_n) + \varepsilon$ for large n. Since $\mu^*(H^c) \leq \mu^*(H_n{}^c)$ for all n by 1.3.3(c), and $H_n \in \mathscr{H}$, we have $\mu^*(H) + \mu^*(H^c) \leq 1 + \varepsilon$. Since ε is arbitrary, $H \in \mathscr{H}$, making $\mathscr{H}$ a σ-field.

Since $\mu^*(H_n) \to \mu^*(H)$, μ^* is countably additive by 1.2.8(a). ∎

We now have our first extension theorem:

1.3.6 ***Theorem.*** A finite measure on a field $\mathscr{F}_0$ can be extended to a measure on $\sigma(\mathscr{F}_0)$.

PROOF. Nothing is lost by considering a probability measure. The result then follows from 1.3.1–1.3.5 if we observe that $\mathscr{F}_0 \subset \mathscr{G} \subset \mathscr{H}$, hence $\sigma(\mathscr{F}_0) \subset \mathscr{H}$. Thus μ^* restricted to $\sigma(\mathscr{F}_0)$ is the desired extension. ∎

In fact there is very little difference between $\sigma(\mathscr{F}_0)$ and $\mathscr{H}$; if $B \in \mathscr{H}$, then B can be expressed as $A \cup N$, where $A \in \sigma(\mathscr{F}_0)$ and N is a subset of a set $M \in \sigma(\mathscr{F}_0)$ with $\mu^*(M) = 0$. To establish this, we introduce the idea of completion of a measure space.

1.3.7 ***Definitions.*** A measure μ on a σ-field $\mathscr{F}$ is said to be *complete* iff whenever $A \in \mathscr{F}$ and $\mu(A) = 0$ we have $B \in \mathscr{F}$ for all $B \subset A$.

In 1.3.5, μ^* on $\mathscr{H}$ is complete, for if $B \subset A \in \mathscr{H}$, $\mu^*(A) = 0$, then $\mu^*(B) + \mu^*(B^c) \le \mu^*(A) + \mu^*(B^c) = \mu^*(B^c) \le 1$; thus $B \in \mathscr{H}$.

The *completion* of a measure space $(\Omega, \mathscr{F}, \mu)$ is defined as follows. Let $\mathscr{F}_\mu$ be the class of sets $A \cup N$, where A ranges over $\mathscr{F}$ and N over all subsets of sets of measure 0 in $\mathscr{F}$.

Now $\mathscr{F}_\mu$ is a σ-field including $\mathscr{F}$, for it is clearly closed under countable union, and if $A \cup N \in \mathscr{F}$, $N \subset M \in \mathscr{F}$, $\mu(M) = 0$, then $(A \cup N)^c = A^c \cap N^c = (A^c \cap M^c) \cup (A^c \cap (N^c - M^c))$ and $A^c \cap (N^c - M^c) = A^c \cap (M - N) \subset M$, so $(A \cup N)^c \in \mathscr{F}_\mu$.

We extend μ to $\mathscr{F}_\mu$ by setting $\mu(A \cup N) = \mu(A)$. This is a valid definition, for if $A_1 \cup N_1 = A_2 \cup N_2 \in \mathscr{F}_\mu$, we have

$$\mu(A_1) = \mu(A_1 \cap A_2) + \mu(A_1 - A_2) = \mu(A_1 \cap A_2)$$

since $A_1 - A_2 \subset N_2$. Thus $\mu(A_1) \le \mu(A_2)$, and by symmetry, $\mu(A_1) = \mu(A_2)$. The measure space $(\Omega, \mathscr{F}_\mu, \mu)$ is called the *completion* of $(\Omega, \mathscr{F}, \mu)$, and $\mathscr{F}_\mu$ the completion of $\mathscr{F}$ relative to μ.

Note that the completion is in fact complete, for if $M \subset A \cup N \in \mathscr{F}_\mu$ where $A \in \mathscr{F}$, $\mu(A) = 0$, $N \subset B \in \mathscr{F}$, $\mu(B) = 0$, then $M \subset A \cup B \in \mathscr{F}$, $\mu(A \cup B) = 0$; hence $M \in \mathscr{F}_\mu$.

1.3.8 ***Theorem.*** In 1.3.6, $(\Omega, \mathscr{H}, \mu^*)$ is the completion of $(\Omega, \sigma(\mathscr{F}_0), \mu^*)$.

PROOF. We must show that $\mathscr{H} = \mathscr{F}_{\mu*}$ where $\mathscr{F} = \sigma(\mathscr{F}_0)$. If $A \in \mathscr{H}$, by definition of $\mu^*(A)$ and $\mu^*(A^c)$ we can find sets $G_n, G_n' \in \sigma(\mathscr{F}_0)$, $n = 1, 2, \ldots$, with $G_n \subset A \subset G_n'$ and $\mu^*(G_n) \to \mu^*(A)$, $\mu^*(G_n') \to \mu^*(A)$. Let $G = \bigcup_{n=1}^{\infty} G_n$, $G' = \bigcap_{n=1}^{\infty} G_n'$. Then $A = G \cup (A - G)$, $G \in \sigma(\mathscr{F}_0)$, $A - G \subset G' - G \in \sigma(\mathscr{F}_0)$, $\mu^*(G' - G) \leq \mu^*(G_n' - G_n) \to 0$, so that $\mu^*(G' - G) = 0$. Thus $A \in \mathscr{F}_{\mu*}$.

Conversely if $B \in \mathscr{F}_{\mu*}$, then $B = A \cup N$, $A \in \mathscr{F}$, $N \subset M \in \mathscr{F}$, $\mu^*(M) = 0$. Since $\mathscr{F} \subset \mathscr{H}$ we have $A \in \mathscr{H}$, and since $(\Omega, \mathscr{H}, \mu^*)$ is complete we have $N \in \mathscr{H}$. Thus $B \in \mathscr{H}$. ∎

To prove the uniqueness of the extension from $\mathscr{F}_0$ to $\mathscr{F}$, we need the following basic result:

1.3.9 ***Monotone Class Theorem.*** Let $\mathscr{F}_0$ be a field of subsets of Ω, and $\mathscr{C}$ a class of subsets of Ω that is *monotone* (if $A_n \in \mathscr{C}$ and $A_n \uparrow A$ or $A_n \downarrow A$, then $A \in \mathscr{C}$). If $\mathscr{C} \supset \mathscr{F}_0$, then $\mathscr{C} \supset \sigma(\mathscr{F}_0)$, the minimal σ-field over $\mathscr{F}_0$.

PROOF. The technique of the proof might be called "boot strapping." Let $\mathscr{F} = \sigma(\mathscr{F}_0)$ and let $\mathscr{M}$ be the smallest monotone class containing all sets of $\mathscr{F}_0$. We show that $\mathscr{M} = \mathscr{F}$, in other words, *the smallest monotone class and the smallest σ-field over a field coincide*. The proof is completed by observing that $\mathscr{M} \subset \mathscr{C}$.

Fix $A \in \mathscr{M}$ and let $\mathscr{M}_A = \{B \in \mathscr{M} \colon A \cap B,\ A \cap B^c \text{ and } A^c \cap B \in \mathscr{M}\}$; then $\mathscr{M}_A$ is a monotone class. In fact $\mathscr{M}_A = \mathscr{M}$; for if $A \in \mathscr{F}_0$, then $\mathscr{F}_0 \subset \mathscr{M}_A$ since $\mathscr{F}_0$ is a field, hence $\mathscr{M} \subset \mathscr{M}_A$ by minimality of $\mathscr{M}$; consequently $\mathscr{M}_A = \mathscr{M}$. But this shows that for any $B \in \mathscr{M}$ we have $A \cap B$, $A \cap B^c$, $A^c \cap B \in \mathscr{M}$ for any $A \in \mathscr{F}_0$, so that $\mathscr{M}_B \supset \mathscr{F}_0$. Again by minimality of $\mathscr{M}$, $\mathscr{M}_B = \mathscr{M}$.

Now $\mathscr{M}$ is a field (for if $A, B \in \mathscr{M} = \mathscr{M}_A$, then $A \cap B$, $A \cap B^c$, $A^c \cap B \in \mathscr{M}$) and *a monotone class that is also a field is a σ-field* (see 1.2.1), hence $\mathscr{M}$ is a σ-field. Thus $\mathscr{F} \subset \mathscr{M}$ by minimality of $\mathscr{F}$, and in fact $\mathscr{F} = \mathscr{M}$ because $\mathscr{F}$ is a monotone class including $\mathscr{F}_0$. ∎

We now prove the fundamental extension theorem.

1.3.10 ***Carathéodory Extension Theorem.*** Let μ be a measure on the field $\mathscr{F}_0$ of subsets of Ω, and assume that μ is σ-finite on $\mathscr{F}_0$, so that Ω can be decomposed as $\bigcup_{n=1}^{\infty} A_n$, where $A_n \in \mathscr{F}_0$ and $\mu(A_n) < \infty$ for all n. Then μ has a unique extension to a measure on the minimal σ-field $\mathscr{F}$ over $\mathscr{F}_0$.

PROOF. Since $\mathscr{F}_0$ is a field, the A_n may be taken as disjoint (replace A_n by $A_1^c \cap \cdots \cap A_{n-1}^c \cap A_n$, as in formula (3) of 1.1). Let $\mu_n(A) = \mu(A \cap A_n)$, $A \in \mathscr{F}_0$; then μ_n is a finite measure on $\mathscr{F}_0$, hence by 1.3.6 it has an extension $\mu_n{}^*$ to $\mathscr{F}$. Since $\mu = \sum_n \mu_n$, the set function $\mu^* = \sum_n \mu_n{}^*$ is an extension of μ, and it is a measure on $\mathscr{F}$ since the order of summation of any double series of nonnegative terms can be reversed.

Now suppose that λ is a measure on $\mathscr{F}$ and $\lambda = \mu$ on $\mathscr{F}_0$. Define $\lambda_n(A) = \lambda(A \cap A_n)$, $A \in \mathscr{F}$. Then λ_n is a finite measure on $\mathscr{F}$ and $\lambda_n = \mu_n = \mu_n{}^*$ on $\mathscr{F}_0$, and it follows that $\lambda_n = \mu_n{}^*$ on $\mathscr{F}$. For $\mathscr{C} = \{A \in \mathscr{F}: \lambda_n(A) = \mu_n{}^*(A)\}$ is a monotone class (by 1.2.7) that contains all sets of $\mathscr{F}_0$. hence $\mathscr{C} = \mathscr{F}$ by 1.3.9. But then $\lambda = \sum_n \lambda_n = \sum_n \mu_n{}^* = \mu^*$, proving uniqueness. ∎

The intuitive idea of constructing a minimal σ-field by forming complements and countable unions and intersections in all possible ways suggests that if $\mathscr{F}_0$ is a field and $\mathscr{F} = \sigma(\mathscr{F}_0)$, sets in $\mathscr{F}$ can be approximated in some sense by sets in $\mathscr{F}_0$. The following result formalizes this notion:

1.3.11 ***Approximation Theorem.*** Let $(\Omega, \mathscr{F}, \mu)$ be a measure space, and let $\mathscr{F}_0$ be a field of subsets of Ω such that $\sigma(\mathscr{F}_0) = \mathscr{F}$. Assume that μ is σ-finite on $\mathscr{F}_0$, and let $\varepsilon > 0$ be given. If $A \in \mathscr{F}$ and $\mu(A) < \infty$, there is a set $B \in \mathscr{F}_0$ such that $\mu(A \,\triangle\, B) < \varepsilon$.

PROOF. Let $\mathscr{G}$ be the class of all countable unions of sets of $\mathscr{F}_0$. The conclusion of 1.3.11 holds for any $A \in \mathscr{G}$, by 1.2.7(a). By 1.3.3, if μ is finite and $A \in \mathscr{F}$, A can be approximated arbitrarily closely (in the sense of 1.3.11) by a set in $\mathscr{G}$, and therefore 1.3.11 is proved for finite μ. In general, let Ω be the disjoint union of sets $A_n \in \mathscr{F}_0$ with $\mu(A_n) < \infty$, and let $\mu_n(C) = \mu(C \cap A_n)$, $C \in \mathscr{F}$.

Then μ_n is a finite measure on $\mathscr{F}$, hence if $A \in \mathscr{F}$, there is a set $B_n \in \mathscr{F}_0$ such that $\mu_n(A \,\triangle\, B_n) < \varepsilon 2^{-n}$. Since

$$\begin{aligned}\mu_n(A \,\triangle\, B_n) &= \mu((A \,\triangle\, B_n) \cap A_n) \\ &= \mu[(A \,\triangle\, (B_n \cap A_n)) \cap A_n] = \mu_n(A \,\triangle\, (B_n \cap A_n)),\end{aligned}$$

and $B_n \cap A_n \in \mathscr{F}_0$, we may assume that $B_n \subset A_n$. (The observation that $B_n \cap A_n \in \mathscr{F}_0$ is the point where we use the hypothesis that μ is σ-finite on $\mathscr{F}_0$, not merely on $\mathscr{F}$.) If $C = \bigcup_{n=1}^{\infty} B_n$, then $C \cap A_n = B_n$, so that

$$\mu_n(A \,\triangle\, C) = \mu((A \,\triangle\, C) \cap A_n) = \mu((A \,\triangle\, B_n) \cap A_n) = \mu_n(A \,\triangle\, B_n),$$

hence $\mu(A \,\triangle\, C) = \sum_{n=1}^{\infty} \mu_n(A \,\triangle\, C) < \varepsilon$. But $\bigcap_{k=1}^{N} B_k - A \uparrow C - A$ as $N \to \infty$, and $A - \bigcup_{k=1}^{N} B_k \downarrow A - C$. If $A \in \mathscr{F}$ and $\mu(A) < \infty$, it follows from 1.2.7 that $\mu(A \triangle \bigcup_{k=1}^{N} B_k) \to \mu(A \triangle C)$ as $N \to \infty$, hence is less than ε for large enough N. Set $B = \bigcup_{k=1}^{N} B_k \in \mathscr{F}_0$. ∎

1.3.12 ***Example.*** Let Ω be the rationals, $\mathscr{F}_0$ the field of finite disjoint unions of right-semiclosed intervals $(a, b] = \{\omega \in \Omega: a < \omega \le b\}$, a, b rational [counting (a, ∞) and Ω itself as right-semiclosed; see 1.2.2]. Let $\mathscr{F} = \sigma(\mathscr{F}_0)$. Then:

(a) $\mathscr{F}$ consists of all subsets of Ω.

(b) If $\mu(A)$ is the number of points in A (μ is counting measure), then μ is σ-finite on $\mathscr{F}$ but not on $\mathscr{F}_0$.

(c) There are sets $A \in \mathscr{F}$ of finite measure that cannot be approximated by sets in $\mathscr{F}_0$, that is, there is no sequence $A_n \in \mathscr{F}_0$ with $\mu(A \triangle A_n) \to 0$.

(d) If $\lambda = 2\mu$, then $\lambda = \mu$ on $\mathscr{F}_0$ but not on $\mathscr{F}$.

Thus both the approximation theorem and the Carathéodory extension theorem fail in this case.

PROOF. (a) We have $\{x\} = \bigcap_{n=1}^{\infty} (x - (1/n), x]$, and therefore all singletons are in $\mathscr{F}$. But then all sets are in $\mathscr{F}$ since Ω is countable.

(b) Since Ω is a countable union of singletons, μ is σ-finite on $\mathscr{F}$. But every nonempty set in $\mathscr{F}_0$ has infinite measure, so μ is not σ-finite on $\mathscr{F}_0$.

(c) If A is any finite nonempty subset of Ω, then $\mu(A \triangle B) = \infty$ for all nonempty $B \in \mathscr{F}_0$, since any nonempty set in $\mathscr{F}_0$ must contain infinitely many points not in A.

(d) Since $\lambda\{x\} = 2$ and $\mu\{x\} = 1$, $\lambda \ne \mu$ on $\mathscr{F}$. But $\lambda(A) = \mu(A) = \infty$, $A \in \mathscr{F}_0$ (except for $A = \varnothing$). ∎

Problems

1. Let $(\Omega, \mathscr{F}, \mu)$ be a measure space, and let $\mathscr{F}_\mu$ be the completion of $\mathscr{F}$ relative to μ. If $A \subset \Omega$, define

$$\mu_0(A) = \sup\{\mu(B) : B \in \mathscr{F},\ B \subset A\}, \quad \mu^0(A) = \inf\{\mu(B) : B \in \mathscr{F},\ B \supset A\}.$$

If $A \in \mathscr{F}_\mu$, show that $\mu_0(A) = \mu^0(A) = \mu(A)$. Conversely, if $\mu_0(A) = \mu^0(A) < \infty$, show that $A \in \mathscr{F}_\mu$.

2. Show that the monotone class theorem (1.3.9) fails if $\mathscr{F}_0$ is not assumed to be a field.

3. This problem deals with the extension of an arbitrary (not necessarily σ-finite) measure on a field.
   (a) Let λ be an outer measure on the set Ω (see 1.3.4). We say that the set E is *λ-measurable* iff

$$\lambda(A) = \lambda(A \cap E) + \lambda(A \cap E^c) \qquad \text{for all} \quad A \subset \Omega.$$

   (The equals sign may be replaced by "$\geq$" by subadditivity of λ.) If $\mathscr{M}$ is the class of all λ-measurable sets, show that $\mathscr{M}$ is a σ-field, and that if $E_1, E_2, \ldots$ are disjoint sets in $\mathscr{M}$ whose union is E, and $A \subset \Omega$, we have

$$\lambda(A \cap E) = \sum_n \lambda(A \cap E_n). \tag{1}$$

   In particular, λ is a measure on $\mathscr{M}$. [Use the definition of λ-measurability to show that $\mathscr{M}$ is a field and that (1) holds for finite sequences. If $E_1, E_2, \ldots$ are disjoint sets in $\mathscr{M}$ and $F_n = \bigcup_{i=1}^n E_i \uparrow E$, show that

$$\lambda(A) \geq \lambda(A \cap F_n) + \lambda(A \cap E^c) = \sum_{i=1}^n \lambda(A \cap E_i) + \lambda(A \cap E^c),$$

   and then let $n \to \infty$.]
   (b) Let μ be a measure on a field $\mathscr{F}_0$ of subsets of Ω. If $A \subset \Omega$, define

$$\mu^*(A) = \inf\left\{\sum_n \mu(E_n): A \subset \bigcup_n E_n, E_n \in \mathscr{F}_0\right\}.$$

   Show that μ^* is an outer measure on Ω and that $\mu^* = \mu$ on $\mathscr{F}_0$.
   (c) In (b), if $\mathscr{M}$ is the class of μ^*-measurable sets, show that $\mathscr{F}_0 \subset \mathscr{M}$. Thus by (a) and (b), μ may be extended to the minimal σ-field over $\mathscr{F}_0$.
   (d) In (b), if μ is σ-finite on $\mathscr{F}_0$, show that $(\Omega, \mathscr{M}, \mu^*)$ is the completion of $(\Omega, \sigma(\mathscr{F}_0), \mu^*)$.

1.4 Lebesgue–Stieltjes Measures and Distribution Functions

We are now in a position to construct a large class of measures on the Borel sets of R. If F is an increasing, right-continuous function from R to R, we set $\mu(a, b] = F(b) - F(a)$; we then extend μ to a finitely additive set function on the field $\mathscr{F}_0(R)$ of finite disjoint unions of right-semiclosed intervals. If we can show that μ is countably additive on $\mathscr{F}_0(R)$, the Carathéodory extension theorem extends μ to $\mathscr{B}(R)$.

1.4.1 ***Definitions.*** A *Lebesgue–Stieltjes measure* on R is a measure μ on $\mathscr{B}(R)$ such that $\mu(I) < \infty$ for each bounded interval I. A *distribution function* on R is a map $F\colon R \to R$ that is *increasing* [$a < b$ implies $F(a) \le F(b)$] and *right-continuous* [$\lim_{x \to x_0^+} F(x) = F(x_0)$]. We are going to show that the formula $\mu(a,b] = F(b) - F(a)$ sets up a one-to-one correspondence between Lebesgue–Stieltjes measures and distribution functions, where two distribution functions that differ by a constant are identified.

1.4.2 ***Theorem.*** Let μ be a Lebesgue–Stieltjes measure on R. Let $F\colon R \to R$ be defined, up to an additive constant, by $F(b) - F(a) = \mu(a, b]$. [For example, fix $F(0)$ arbitrarily and set $F(x) - F(0) = \mu(0, x]$, $x > 0$; $F(0) - F(x) = \mu(x, 0]$, $x < 0$.] Then F is a distribution function.

PROOF. If $a < b$, then $F(b) - F(a) = \mu(a, b] \ge 0$. If $\{x_n\}$ is a sequence of points such that $x_1 > x_2 > \cdots \to x$, then $F(x_n) - F(x) = \mu(x, x_n] \to 0$ by 1.2.7(b). ∎

Now let F be a distribution function on R. It will be convenient to work in the compact space $\overline{R}$, so we extend F to a map of $\overline{R}$ into $\overline{R}$ by defining $F(\infty) = \lim_{x \to \infty} F(x)$, $F(-\infty) = \lim_{x \to -\infty} F(x)$; the limits exist by monotonicity. Define $\mu(a, b] = F(b) - F(a)$, $a, b \in \overline{R}$, $a < b$, and let $\mu[-\infty, b] = F(b) - F(-\infty) = \mu(-\infty, b]$; then μ is defined on all right-semiclosed intervals of $\overline{R}$ (counting $[-\infty, b]$ as right-semiclosed; see 1.2.2).

If $I_1, \ldots, I_k$ are disjoint right-semiclosed intervals of $\overline{R}$, we define $\mu(\bigcup_{j=1}^k I_j) = \sum_{j=1}^k \mu(I_j)$. Thus μ is extended to the field $\mathscr{F}_0(\overline{R})$ of finite disjoint unions of right-semiclosed intervals of $\overline{R}$, and μ is finitely additive on $\mathscr{F}_0(\overline{R})$. To show that μ is in fact countably additive on $\mathscr{F}_0(\overline{R})$, we make use of 1.2.8(b), as follows:

1.4.3 ***Lemma.*** The set function μ is countably additive on $\mathscr{F}_0(\overline{R})$.

PROOF. First assume that $F(\infty) - F(-\infty) < \infty$, so that μ is finite. Let $A_1, A_2, \ldots$ be a sequence of sets in $\mathscr{F}_0(\overline{R})$ decreasing to $\varnothing$. If $(a, b]$ is one of the intervals of A_n, then by right continuity of F, $\mu(a', b] = F(b) - F(a') \to F(b) - F(a) = \mu(a, b]$ as $a' \to a$ from above.

Thus we can find sets $B_n \in \mathscr{F}_0(\overline{R})$ whose closures $\overline{B}_n$ (in $\overline{R}$) are included in A_n, with $\mu(B_n)$ approximating $\mu(A_n)$. If $\varepsilon > 0$ is given, the finiteness of μ allows us to choose the B_n so that $\mu(A_n) - \mu(B_n) < \varepsilon 2^{-n}$. Now $\bigcap_{n=1}^\infty \overline{B}_n = \varnothing$, and it follows that $\bigcap_{k=1}^n \overline{B}_k = \varnothing$ for sufficiently large n. (Perhaps the easiest way to see this is to note that the sets $\overline{R} - \overline{B}_n$ form an open covering of the compact set $\overline{R}$, hence there is a finite subcovering, so that $\bigcup_{k=1}^{n}(\overline{R} - \overline{B}_k) = \overline{R}$ for some n. Therefore $\bigcap_{k=1}^{n} \overline{B}_k = \varnothing$.) Now

$$\begin{aligned}
\mu(A_n) &= \mu\left(A_n - \bigcap_{k=1}^{n} B_k\right) + \mu\left(\bigcap_{k=1}^{n} B_k\right) \\
&= \mu\left(A_n - \bigcap_{k=1}^{n} B_k\right) \\
&\le \mu\left(\bigcup_{k=1}^{n} (A_k - B_k)\right) \qquad \text{since} \quad A_n \subset A_{n-1} \subset \cdots \subset A_1 \\
&\le \sum_{k=1}^{n} \mu(A_k - B_k) \qquad \text{by 1.2.5(d)} \\
&< \varepsilon.
\end{aligned}$$

Thus $\mu(A_n) \to 0$.

Now if $F(\infty) - F(-\infty) = \infty$, define $F_n(x) = F(x)$, $|x| \le n$; $F_n(x) = F(n)$, $x \ge n$; $F_n(x) = F(-n)$, $x \le -n$. If μ_n is the set function corresponding to F_n, then $\mu_n \le \mu$ and $\mu_n \to \mu$ on $\mathscr{F}_0(\overline{R})$. Let $A_1, A_2, \ldots$ be disjoint sets in $\mathscr{F}_0(\overline{R})$ such that $A = \bigcup_{n=1}^{\infty} A_n \in \mathscr{F}_0(\overline{R})$. Then $\mu(A) \ge \sum_{n=1}^{\infty} \mu(A_n)$ (Problem 5, Section 1.2) so if $\sum_{n=1}^{\infty} \mu(A_n) = \infty$, we are finished. If $\sum_{n=1}^{\infty} \mu(A_n) < \infty$, then

$$\begin{aligned}
\mu(A) &= \lim_{n\to\infty} \mu_n(A) \\
&= \lim_{n\to\infty} \sum_{k=1}^{\infty} \mu_n(A_k)
\end{aligned}$$

since the μ_n are finite. Now since $\sum_{k=1}^{\infty} \mu(A_k) < \infty$, we may write

$$\begin{aligned}
0 \le \mu(A) - \sum_{k=1}^{\infty} \mu(A_k) \\
= \lim_{n\to\infty} \sum_{k=1}^{\infty} [\mu_n(A_k) - \mu(A_k)] \\
\le 0 \qquad \text{since} \quad \mu_n \le \mu. \quad ∎
\end{aligned}$$

We now complete the construction of Lebesgue–Stieltjes measures:

1.4.4 ***Theorem.*** Let F be a distribution function on R, and let $\mu(a, b] = F(b) - F(a)$, $a < b$. There is a unique extension of μ to a Lebesgue–Stieltjes measure on R.

PROOF. Extend μ to a countably additive set function on $\mathscr{F}_0(\overline{R})$ as above. Let $\mathscr{F}_0(R)$ be the field of all finite disjoint unions of right-semiclosed intervals of R [counting (a, ∞) as right-semiclosed; see 1.2.2], and extend μ to $\mathscr{F}_0(R)$

as in the discussion after 1.4.2. [Take $\mu(a, \infty) = F(\infty) - F(a)$; $\mu(-\infty, b] = F(b) - F(-\infty)$, $a, b \in R$; $\mu(R) = F(\infty) - F(-\infty)$; note that there is no other possible choice for μ on these sets, by 1.2.7(a).] Now the map

$$\begin{aligned} &(a, b] \to (a, b] \quad &&\text{if } a, b \in R \quad \text{or if } b \in R, \quad a = -\infty, \\ &(a, \infty] \to (a, \infty] \quad &&\text{if } a \in R \quad \text{or if } a = -\infty. \end{aligned}$$

sets up a one-to-one, μ-preserving correspondence between a subset of $\mathscr{F}_0(\bar{R})$ (everything in $\mathscr{F}_0(\bar{R})$ except sets including intervals of the form $[-\infty, b]$) and $\mathscr{F}_0(R)$. It follows that μ is countably additive on $\mathscr{F}_0(R)$. Furthermore, μ is σ-finite on $\mathscr{F}_0(R)$ since $\mu(-n, n] < \infty$; note that μ need not be σ-finite on $\mathscr{F}_0(\bar{R})$ since the sets $(-n, n]$ do not cover $\bar{R}$. By the Carathéodory extension theorem, μ has a unique extension to $\mathscr{B}(R)$. The extension is a Lebesgue–Stieltjes measure because $\mu(a, b] = F(b) - F(a) < \infty$ for $a, b \in R$, $a < b$. ▌

1.4.5 ***Comments and Examples.*** If F is a distribution function and μ the corresponding Lebesgue–Stieltjes measure, we have seen that $\mu(a, b] = F(b) - F(a)$, $a < b$. The measure of any interval, right-semiclosed or not, may be expressed in terms of F. For if $F(x^-)$ denotes $\lim_{y \to x^-} F(y)$, then

(1) $\mu(a, b] = F(b) - F(a)$, (3) $\mu[a, b] = F(b) - F(a^-)$,

(2) $\mu(a, b) = F(b^-) - F(a)$, (4) $\mu[a, b) = F(b^-) - F(a^-)$.

(Thus if F is continuous at a and b, all four expressions are equal.) For example, to prove (2), observe that

$$\mu(a, b) = \lim_{n \to \infty} \mu\left(a, b - \frac{1}{n}\right] = \lim_{n \to \infty} \left[F\left(b - \frac{1}{n}\right) - F(a)\right] = F(b^-) - F(a).$$

Statement (3) follows because

$$\mu[a, b] = \lim_{n \to \infty} \mu\left(a - \frac{1}{n}, b\right] = \lim_{n \to \infty} \left[F(b) - F\left(a - \frac{1}{n}\right)\right] = F(b) - F(a^-);$$

(4) is proved similarly. The proof of (3) works even if $a = b$, so that $\mu\{x\} = F(x) - F(x^-)$. Thus

(5) F is continuous at x iff $\mu\{x\} = 0$; the magnitude of a discontinuity of F at x coincides with the measure of $\{x\}$.

The following formulas are obtained from (1)–(3) by allowing a to approach $-\infty$ or b to approach $+\infty$.

(6) $\mu(-\infty, x] = F(x) - F(-\infty)$, (9) $\mu[x, \infty) = F(\infty) - F(x^-)$,

(7) $\mu(-\infty, x) = F(x^-) - F(-\infty)$, (10) $\mu(R) = F(\infty) - F(-\infty)$.

(8) $\mu(x, \infty) = F(\infty) - F(x)$,

(The formulas (6), (8), and (10) have already been observed in the proof of 1.4.4.)

If μ is finite, then F is bounded; since F may always be adjusted by an additive constant, nothing is lost in this case if we set $F(-\infty) = 0$.

We may now generate a large number of measures on $\mathscr{B}(R)$. For example, if $f: R \to R, f \geq 0$, and f is integrable (Riemann for now) on any finite interval, then if we fix $F(0)$ arbitrarily and define

$$F(x) - F(0) = \int_0^x f(t)\,dt, \qquad x > 0;$$

$$F(0) - F(x) = \int_x^0 f(t)\,dt, \qquad x < 0,$$

then F is a (continuous) distribution function and thus gives rise to a Lebesgue–Stieltjes measure; specifically,

$$\mu(a, b] = \int_a^b f(x)\,dx.$$

In particular, we may take $f(x) = 1$ for all x, and $F(x) = x$; then $\mu(a, b] = b - a$. The set function μ is called the *Lebesgue measure* on $\mathscr{B}(R)$. The completion of $\mathscr{B}(R)$ relative to Lebesgue measure is called the class of *Lebesgue measurable sets*, written $\overline{\mathscr{B}}(R)$. Thus a Lebesgue measurable set is the union of a Borel set and a subset of a Borel set of Lebesgue measure 0. The extension of Lebesgue measure to $\overline{\mathscr{B}}(R)$ is called "Lebesgue measure" also.

Now let μ be a Lebesgue–Stieltjes measure that is concentrated on a countable set $S = \{x_1, x_2, \ldots\}$, that is, $\mu(R - S) = 0$. [In general if $(\Omega, \mathscr{F}, \mu)$ is a measure space and $B \in \mathscr{F}$, we say that μ is concentrated on B iff $\mu(\Omega - B) = 0$.] In the present case, such a measure is easily constructed: If $a_1, a_2, \ldots$ are nonnegative numbers and $A \subset R$, set $\mu(A) = \sum\{a_i : x_i \in A\}$; μ is a measure on all subsets of R, not merely on the Borel sets (see 1.2.4). If $\mu(I) < \infty$ for each bounded interval I, μ will be a Lebesgue–Stieltjes measure on $\mathscr{B}(R)$; if $\sum_i a_i < \infty$, μ will be a finite measure. The distribution function F corresponding to μ is continuous on $R - S$; if $\mu\{x_n\} = a_n > 0$, F has a jump at x_n of magnitude a_n. If $x, y \in S$ and no point of S lies between x and y, then F is constant on $[x, y)$. For if $x \leq b < y$, then $F(b) - F(x) = \mu(x, b] = 0$.

Now if we take S to be the rational numbers, the above discussion yields a monotone function F from R to R that is continuous at each irrational point and discontinuous at each rational point.

If F is an increasing, right-continuous, real-valued function defined on a closed bounded interval $[a, b]$, there is a corresponding finite measure μ on the Borel subsets of $[a, b]$; explicitly, μ is determined by the requirement that $\mu(a', b'] = F(b') - F(a')$, $a \leq a' < b' \leq b$. The easiest way to establish the correspondence is to extend F by defining $F(x) = F(b)$, $x \geq b$; $F(x) = F(a)$, $x \leq a$; then take μ as the Lebesgue–Stieltjes measure corresponding to F, restricted to $\mathscr{B}[a, b]$.

We are going to consider Lebesgue–Stieltjes measures and distribution functions in Euclidean n-space. First, some terminology:

1.4.6 ***Definitions and Comments.*** If $a = (a_1, \ldots, a_n)$, $b = (b_1, \ldots, b_n) \in R^n$, the interval $(a, b]$ is defined as $\{x = (x_1, \ldots, x_n) \in R^n: a_i < x_i \leq b_i \text{ for all } i = 1, \ldots, n\}$; (a, ∞) is defined as $\{x \in R^n: x_i > a_i \text{ for all } i = 1, \ldots, n\}$, $(-\infty, b]$ as $\{x \in R^n: x_i \leq b_i \text{ for all } i = 1, \ldots, n\}$; other types of intervals are defined similarly. The smallest σ-field containing all intervals $(a, b]$, $a, b \in R^n$, is called the class of *Borel sets* of R^n, written $\mathscr{B}(R^n)$. The Borel sets form the minimal σ-field over many other classes of sets, for example, the open sets, the intervals $[a, b)$, and so on, exactly as in the discussion of the one-dimensional case in 1.2.4. The class of Borel sets of $\bar{R}^n$, written $\mathscr{B}(\bar{R}^n)$, is defined similarly.

A *Lebesgue–Stieltjes measure* on R^n is a measure μ on $\mathscr{B}(R^n)$ such that $\mu(I) < \infty$ for each bounded interval I.

The notion of a distribution function on R^n, $n \geq 2$, is more complicated than in the one-dimensional case. To see why, assume for simplicity that $n = 3$, and let μ be a finite measure on $\mathscr{B}(R^3)$. Define

$$F(x_1, x_2, x_3) = \mu\{\omega \in R^3: \omega_1 \leq x_1, \quad \omega_2 \leq x_2, \quad \omega_3 \leq x_3\}, \qquad (x_1, x_2, x_3) \in R^3.$$

By analogy with the one-dimensional case, we expect that F is a distribution function corresponding to μ [see formula (6) of 1.4.5]. This will turn out to be correct, but the correspondence is no longer by means of the formula $\mu(a, b] = F(b) - F(a)$. To see this, we compute $\mu(a, b]$ in terms of F.

Introduce the *difference operator* $\triangle$ as follows: If $G: R^n \to R$, $\triangle_{b_i a_i} G(x_1, \ldots, x_n)$ is defined as

$$G(x_1, \ldots, x_{i-1}, b_i, x_{i+1}, \ldots, x_n) - G(x_1, \ldots, x_{i-1}, a_i, x_{i+1}, \ldots, x_n).$$

1.4.7. ***Lemma.*** If $a \leq b$, that is, $a_i \leq b_i$, $i = 1, 2, 3$, then

(a) $\mu(a, b] = \triangle_{b_1 a_1} \triangle_{b_2 a_2} \triangle_{b_3 a_3} F(x_1, x_2, x_3)$, where

(b) $$\begin{aligned} &\triangle_{b_1 a_1} \triangle_{b_2 a_2} \triangle_{b_3 a_3} F(x_1, x_2, x_3) \\ &\quad = F(b_1, b_2, b_3) - F(a_1, b_2, b_3) - F(b_1, a_2, b_3) - F(b_1, b_2, a_3) \\ &\quad\quad + F(a_1, a_2, b_3) + F(a_1, b_2, a_3) + F(b_1, a_2, a_3) \\ &\quad\quad - F(a_1, a_2, a_3). \end{aligned}$$

Thus $\mu(a, b]$ is not simply $F(b) - F(a)$.

PROOF.

(a) $$\begin{aligned}\triangle_{b_3a_3} F(x_1, x_2, x_3) &= F(x_1, x_2, b_3) - F(x_1, x_2, a_3)\\ &= \mu\{\omega: \omega_1 \le x_1, \quad \omega_2 \le x_2, \quad \omega_3 \le b_3\}\\ &\quad - \mu\{\omega: \omega_1 \le x_1, \quad \omega_2 \le x_2, \quad \omega_3 \le a_3\}\\ &= \mu\{\omega: \omega_1 \le x_1, \quad \omega_2 \le x_2, \quad a_3 < \omega_3 \le b_3\}\\ &\quad \text{since} \quad a_3 \le b_3.\end{aligned}$$

Similarly,

$$\triangle_{b_2a_2} \triangle_{b_3a_3} F(x_1, x_2, x_3) = \mu\{\omega: \omega_1 \le x_1, \quad a_2 < \omega_2 \le b_2, \quad a_3 < \omega_3 \le b_3\}$$

and

$$\triangle_{b_1a_1} \triangle_{b_2a_2} \triangle_{b_3a_3} F(x_1, x_2, x_3) = \mu\{\omega: a_1 < \omega_1 \le b_1, a_2 < \omega_2 \le b_2, a_3 < \omega_3 \le b_3\}.$$

(b) $$\begin{aligned}\triangle_{b_3a_3} F(x_1, x_2, x_3) &= F(x_1, x_2, b_3) - F(x_1, x_2, a_3),\\ \triangle_{b_2a_2} \triangle_{b_3a_3} F(x_1, x_2, x_3) &= F(x_1, b_2, b_3) - F(x_1, a_2, b_3)\\ &\quad - F(x_1, b_2, a_3) + F(x_1, a_2, a_3).\end{aligned}$$

Thus $\triangle_{b_1a_1} \triangle_{b_2a_2} \triangle_{b_3a_3} F(x_1, x_2, x_3)$ is the desired expression. ▮

The extension of 1.4.7 to n dimensions is clear.

1.4.8 ***Theorem.*** Let μ be a finite measure on $\mathscr{B}(R^n)$ and define

$$F(x) = \mu(-\infty, x] = \mu\{\omega: \omega_i \le x_i, i = 1, \dots, n\}.$$

If $a \le b$, then

(a) $\mu(a, b] = \triangle_{b_1a_1} \cdots \triangle_{b_na_n} F(x_1, \dots, x_n)$, where

(b) $\triangle_{b_1a_1} \cdots \triangle_{b_na_n} F(x_1, \dots, x_n) = F_0 - F_1 + F_2 - F_3 + \cdots + (-1)^n F_n$;

F_i is the sum of all $\binom{n}{i}$ terms of the form $F(c_1, \dots, c_n)$ with $c_k = a_k$ for exactly i integers in $\{1, 2, \dots, n\}$, and $c_k = b_k$ for the remaining $n - i$ integers.

PROOF. Apply the computations of 1.4.7. ▮

We know that a distribution function on R determines a corresponding Lebesgue–Stieltjes measure. This is true in n dimensions if we change the definition of increasing.

Let $F: R^n \to R$, and, for $a \le b$, let $F(a, b]$ denote

$$\triangle_{b_1 a_1} \cdots \triangle_{b_n a_n} F(x_1, \ldots, x_n).$$

The function F is said to be *increasing* iff $F(a, b] \ge 0$ whenever $a \le b$; F is *right-continuous* iff it is right-continuous in all variables together, in other words, for any sequence $x^1 \ge x^2 \ge \cdots \ge x^k \ge \cdots \to x$ we have $F(x^k) \to F(x)$.

An increasing right-continuous $F: R^n \to R$ is said to be a *distribution function* on R^n. (Note that if F arises from a measure μ as in 1.4.8, F is a distribution function.)

If F is a distribution function on R^n, we set $\mu(a, b] = F(a, b]$ (this reduces to $F(b) - F(a)$ if $n = 1$). We are going to show that μ has a unique extension to a Lebesgue–Stieltjes measure on R^n. The technique of the proof is the same in any dimension, but to avoid cumbersome notation and to capture the essential ideas, we sometimes specialize to the case $n = 2$. We break the argument into several steps:

(1) If $a \le a' \le b' \le b$, $I = (a, b]$ is the union of the nine disjoint intervals $I_1, \ldots, I_9$ formed by first constraining the first coordinate in one of the following three ways:

$$a_1 < x \le a_1', \qquad a_1' < x \le b_1', \qquad b_1' < x \le b_1,$$

and then constraining the second coordinate in one of the following three ways:

$$a_2 < y \le a_2', \qquad a_2' < y \le b_2', \qquad b_2' < y \le b_2.$$

For example, a typical set in the union is

$$\{(x, y): b_1' < x \le b_1, \quad a_2 < y \le a_2'\};$$

in n dimensions we would obtain 3^n such sets.

The result (1) may be verified by looking at Fig. 1.1.

(2) In (1), $F(I) = \sum_{j=1}^{9} F(I_j)$, hence $a \le a' \le b' \le b$ implies $F(a', b'] \le F(a, b]$.

This is verified by brute force, using 1.4.8.

Now a *right-semiclosed interval* $(a, b]$ in $\overline{R}^n$ is, by convention, a set of the form $\{(x_1, \ldots, x_n): a_i < x_i \le b_i,\ i = 1, \ldots, n\}$, $a, b \in \overline{R}^n$, with the proviso that $a_i < x_i \le b_i$ can be replaced by $a_i \le x_i \le b_i$ if $a_i = -\infty$. With this assumption, the set $\mathscr{F}_0(\overline{R}^n)$ of finite disjoint unions of right-semiclosed intervals is a field. (The corresponding convention in R^n is that $a_i < x_i \le b_i$ can

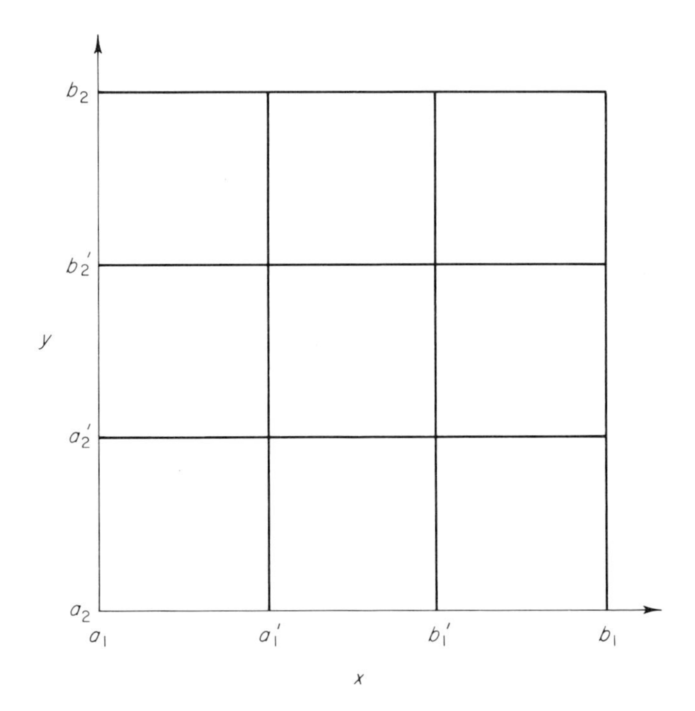

Figure 1.1.

be replaced by $a_i < x_i < b_i$ if $b_i = +\infty$. Both conventions are dictated by considerations similar to those of the one-dimensional case; see 1.2.2.)

(3) If a and b belong to $\overline{R}^n$ but not to R^n, we define $F(a, b]$ as the limit of $F(a,' b']$ where $a', b' \in R^n$, a' decreases to a, and b' increases to b. [The definition is sensible because of the monotonicity property in (2).] Similarly if $a \in R^n$, $b \in \overline{R}^n - R^n$, we take $F(a, b] = \lim_{b' \uparrow b} F(a, b']$; if $a \in \overline{R}^n - R^n$, $b \in R^n$, $F(a, b] = \lim_{a' \downarrow a} F(a', b]$.

Thus we define μ on right-semiclosed intervals of $\overline{R}^n$; μ extends to a finitely additive set function on $\mathscr{F}_0(\overline{R}^n)$, as in the discussion after 1.4.2. [There is a slight problem here; a given interval I may be expressible as a finite disjoint union of intervals $I_1, \ldots, I_r$, so that for the extension to be well defined we must have $F(I) = \sum_{j=1}^r F(I_j)$; but this follows just as in (2).]

(4) The set function μ is countably additive on $\mathscr{F}_0(\overline{R}^n)$.

First assume that $\mu(\overline{R}^n)$ is finite. If $a \in R^n$, $F(a', b] \to F(a, b]$ as a' decreases to a by the right-continuity of F and 1.4.8(b); if $a \in \overline{R}^n - R^n$, the same result holds by (3). The argument then proceeds word for word as in 1.4.3.

Now assume $\mu(\bar{R}^n) = \infty$. Then F, restricted to $C_k = \{x\colon -k < x_i \le k, i = 1, \ldots, n\}$, induces a finite-valued set function μ_k on $\mathscr{F}_0(\bar{R}^n)$ that is concentrated on C_k, so that $\mu_k(B) = \mu_k(B \cap C_k)$, $B \in \mathscr{F}_0(\bar{R}^n)$. Since $\mu_k \le \mu$ and $\mu_k \to \mu$ on $\mathscr{F}_0(\bar{R}^n)$, the proof of 1.4.3 applies verbatim.

1.4.9 ***Theorem.*** Let F be a distribution function on R^n, and let $\mu(a, b] = F(a, b]$, $a, b \in R^n$, $a \le b$. There is a unique extension of μ to a Lebesgue–Stieltjes measure on R^n.

PROOF. Repeat the proof of 1.4.4, with appropriate notational changes. For example, in extending μ to $\mathscr{F}_0(R^n)$, the field of finite disjoint unions of right-semiclosed intervals of R^n, we take (say for $n = 3$)

$$\mu\{(x, y, z)\colon a_1 < x \le b_1, \quad a_2 < y < \infty, \quad a_3 < z < \infty\} = \lim_{b_2, b_3 \to \infty} F(a, b].$$

The one-to-one μ-preserving correspondence is given by

$$(a, b] \to (a, b] \quad \text{if} \quad a, b \in R^n \quad \text{or if} \quad b \in R^n \text{ and at least one component of } a \text{ is } -\infty;$$

also, if the interval $\{(x_1, \ldots, x_n)\colon a_i < x_i \le b_i\colon i = 1, \ldots, n\}$ has some $b_i = \infty$, the corresponding interval in R^n has $a_i < x_i < \infty$. The remainder of the proof is as before. ∎

1.4.10 ***Examples.*** (a) Let $F_1, F_2, \ldots, F_n$ be distribution functions on R, and define $F(x_1, \ldots, x_n) = F_1(x_1)F_2(x_2) \cdots F_n(x_n)$. Then F is a distribution function on R^n since

$$F(a, b] = \prod_{i=1}^{n} [F_i(b_i) - F_i(a_i)].$$

In particular, if $F_i(x_i) = x_i$, $i = 1, \ldots, n$, then each F_i corresponds to Lebesgue measure on $\mathscr{B}(R)$. In this case we have $F(x_1, \ldots, x_n) = x_1 x_2 \cdots x_n$ and $\mu(a, b] = F(a, b] = \prod_{i=1}^{n} (b_i - a_i)$. Thus the measure of any rectangular box is its volume; μ is called *Lebesgue measure* on $\mathscr{B}(R^n)$. Just as in one dimension, the completion of $\mathscr{B}(R^n)$ relative to Lebesgue measure is called the class of *Lebesgue measurable sets* in R^n, written $\bar{\mathscr{B}}(R^n)$.

(b) Let f be a nonnegative function from R^n to R such that

$$\int_{-\infty}^{\infty} \cdots \int_{-\infty}^{\infty} f(x_1, \ldots, x_n)\, dx_1 \cdots dx_n < \infty.$$

(For now, we assume the integration is in the Riemann sense.) Define

$$F(x) = \int_{(-\infty, x]} f(t)\, dt,$$

that is,

$$F(x_1, \ldots, x_n) = \int_{-\infty}^{x_1} \cdots \int_{-\infty}^{x_n} f(t_1, \ldots, t_n)\, dt_1 \cdots dt_n.$$

Then

$$\triangle_{b_n a_n} F(x_1, \ldots, x_n) = \int_{-\infty}^{x_1} \cdots \int_{-\infty}^{x_{n-1}} \int_{a_n}^{b_n} f(t_1, \ldots, t_n)\, dt_1 \cdots dt_n,$$

and we find by repeating this computation that

$$F(a, b] = \int_{a_1}^{b_1} \cdots \int_{a_n}^{b_n} f(t_1, \ldots, t_n)\, dt_1 \cdots dt_n.$$

Thus F is a distribution function. If μ is the Lebesgue–Stieltjes measure determined by F, we have

$$\mu(a, b] = \int_{(a, b]} f(x)\; dx.$$

We have seen that if F is a distribution function on R^n, there is a unique Lebesgue–Stieltjes measure determined by $\mu(a, b] = F(a, b]$, $a \le b$. Also, if μ is a finite measure on $\mathscr{B}(R^n)$ and $F(x) = \mu(-\infty, x]$, $x \in R^n$, then F is a distribution function on R^n and $\mu(a, b] = F(a, b]$, $a \le b$. It is possible to associate a distribution function with an arbitrary Lebesgue–Stieltjes measure on R^n, and thus establish a one-to-one correspondence between Lebesgue–Stieltjes measures and distribution functions, provided distribution functions with the same increments $F(a, b]$, $a, b \in R^n$, $a \le b$, are identified. The result will not be needed, and the details are quite tedious and will be omitted.

Problems

1. Let F be the distribution function on R given by $F(x) = 0$, $x < -1$; $F(x) = 1 + x$, $-1 \le x < 0$; $F(x) = 2 + x^2$, $0 \le x < 2$; $F(x) = 9$, $x \ge 2$. If μ is the Lebesgue–Stieltjes measure corresponding to F, compute the measure of each of the following sets:

   (a) $\{2\}$,
   (b) $[-\frac{1}{2}, 3)$,
   (c) $(-1, 0] \cup (1, 2)$,
   (d) $[0, \frac{1}{2}) \cup (1, 2]$,
   (e) $\{x \colon |x| + 2x^2 > 1\}$.

2. Let μ be a Lebesgue–Stieltjes measure on R corresponding to a continuous distribution function.
   (a) If A is a countable subset of R, show that $\mu(A) = 0$.
   (b) If $\mu(A) > 0$, must A include an open interval?
   (c) If $\mu(A) > 0$ and $\mu(R - A) = 0$, must A be dense in R?
   (d) Do the answers to (b) or (c) change if μ is restricted to be Lebesgue measure?
3. If B is a Borel set in R^n and $a \in R^n$, show that $a + B = \{a + x: x \in B\}$ is a Borel set, and $-B = \{-x: x \in B\}$ is a Borel set. (Use the good sets principle.)
4. Show that if $B \in \overline{\mathscr{B}}(R^n)$, $a \in R^n$, then $a + B \in \overline{\mathscr{B}}(R^n)$ and $\mu(a + B) = \mu(B)$, where μ is Lebesgue measure. Thus Lebesgue measure is translation-invariant. (The good sets principle works here also, in conjunction with the monotone class theorem.)
5. Let μ be a measure on $\mathscr{B}(R^n)$ such that $\mu(a + I) = \mu(I)$ for all $a \in R^n$ and all (right-semiclosed) intervals I in R^n. In other words, μ is translation-invariant on intervals. Show that μ is a constant times Lebesgue measure.
6. (*A set that is not Lebesgue measurable*) Call two real numbers x and y equivalent iff $x - y$ is rational. Choose a member of each distinct equivalence class $B_x = \{y: y - x \text{ rational}\}$ to form a set A; assume that the representatives are chosen so that $A \subset [0, 1]$. Establish the following:
   (a) If r and s are distinct rational numbers, $(r + A) \cap (s + A) = \varnothing$; also $R = \bigcup\{r + A: r \text{ rational}\}$.
   (b) If A is Lebesgue measurable (so that $r + A$ is Lebesgue measurable by Problem 4), then $\mu(r + A) = 0$ for all rational r (μ is Lebesgue measure). Conclude that A cannot be Lebesgue measurable.

   The only properties of Lebesgue measure needed in this problem are translation-invariance and finiteness on bounded intervals. Therefore, the result implies that there is no translation-invariant measure λ (except $\lambda \equiv 0$) on the class of all subsets of R such that $\lambda(I) < \infty$ for each bounded interval I.
7. (*The Cantor ternary set*) Let E_1 be the middle third of the interval $[0, 1]$, that is, $E_1 = (\frac{1}{3}, \frac{2}{3})$; thus $x \in [0, 1] - E_1$ iff x can be written in ternary form using 0 or 2 in the first digit. Let E_2 be the union of the middle thirds of the two intervals that remain after E_1 is removed, that is, $E_2 = (\frac{1}{9}, \frac{2}{9}) \cup (\frac{7}{9}, \frac{8}{9})$; thus $x \in [0, 1] - (E_1 \cup E_2)$ iff x can be written in ternary form using 0 or 2 in the first two digits. Continue the construction; let E_n be the union of the middle thirds of the intervals that remain after $E_1, \ldots, E_{n-1}$ are removed. The Cantor ternary set C is defined as $[0, 1] - \bigcup_{n=1}^{\infty} E_n$; thus $x \in C$ iff x can be expressed in

ternary form using only digits 0 and 2. Various topological properties of C follow from the definition: C is closed, perfect (every point of C is a limit point of C), and nowhere dense.

Show that C is uncountable and has Lebesgue measure 0.

Comment. In the above construction, we have $m(E_n) = (\frac{1}{3})(\frac{2}{3})^{n-1}$, $n = 1, 2, \ldots$, where m is Lebesgue measure. If $0 < \alpha < 1$, the procedure may be altered slightly so that $m(E_n) = \alpha(\frac{1}{2})^n$. We then obtain a set $C(\alpha)$, homeomorphic to C, of measure $1 - \alpha$; such sets are called *Cantor sets of positive measure.*

8. Give an example of a function $F: R^2 \to R$ such that F is right-continuous and is increasing in each coordinate separately, but F is not a distribution function on R^2.
9. A distribution function on R is monotone and thus has only countably many points of discontinuity. Is this also true for a distribution function on R^n, $n > 1$?
10. (a) Let F and G be distribution functions on R^n. If $F(a, b] = G(a, b]$ for all $a, b \in R^n$, $a \leq b$, does it follow that F and G differ by a constant?
    (b) Must a distribution function on R^n be increasing in each coordinate separately?
11. If c is the cardinality of the reals, show that there are only c Borel subsets of R^n, but 2^c Lebesgue measurable sets.
12. The following result shows that under appropriate conditions, a Borel set can be approximated from below by a compact set, and from above by an open set. Problems of this type will be examined systematically in Chapter 4.

    If μ is a σ-finite measure on $\mathscr{B}(R^n)$, show that for each $B \in \mathscr{B}(R^n)$,

    (a) $\mu(B) = \sup\{\mu(K): K \subset B, K \text{ compact}\}$.

    If μ is in fact a Lebesgue–Stieltjes measure, show that

    (b) $\mu(B) = \inf\{\mu(V): V \supset B, B \text{ open}\}$.

    (c) Give an example of a σ-finite measure on $\mathscr{B}(R^n)$ that is not a Lebesgue–Stieltjes measure and for which (b) fails.

1.5 Measurable Functions and Integration

If f is a real-valued function defined on a bounded interval $[a, b]$ of reals, we can talk about the Riemann integral of f, at least if f is piecewise continuous. We are going to develop a much more general integration process, one that applies to functions from an arbitrary set to the extended reals, provided that certain "measurability" conditions are satisfied.

Probability considerations may again be used to motivate the concept of measurability. Suppose that $(\Omega, \mathscr{F}, P)$ is a probability space, and that h is a function from Ω to $\overline{R}$. Thus if the outcome of the experiment corresponds to the point $\omega \in \Omega$, we may compute the number $h(\omega)$. Suppose that we are interested in the probability that $a \leq h(\omega) \leq b$, in other words, we wish to compute $P\{\omega: h(\omega) \in B\}$ where $B = [a, b]$. For this to be possible, the set $\{\omega: h(\omega) \in B\} = h^{-1}(B)$ must belong to the σ-field $\mathscr{F}$. If $h^{-1}(B) \in \mathscr{F}$ for each interval B (and hence, as we shall see below, for each Borel set B), then h is a "measurable function," in other words, probabilities of events involving h can be computed. In the language of probability theory, h is a "random variable."

1.5.1 ***Definitions and Comments.*** If $h: \Omega_1 \to \Omega_2$, h is *measurable* relative to the σ-fields $\mathscr{F}_j$ of subsets of Ω_j, $j = 1, 2$, iff $h^{-1}(A) \in \mathscr{F}_1$ for each $A \in \mathscr{F}_2$.

It is sufficient that $h^{-1}(A) \in \mathscr{F}_1$ for each $A \in \mathscr{C}$, where $\mathscr{C}$ is a class of subsets of Ω_2 such that the minimal σ-field over $\mathscr{C}$ is $\mathscr{F}_2$. For $\{A \in \mathscr{F}_2 : h^{-1}(A) \in \mathscr{F}_1\}$ is a σ-field that contains all sets of $\mathscr{C}$, hence coincides with $\mathscr{F}_2$. This is another application of the good sets principle.

The notation $h: (\Omega_1, \mathscr{F}_1) \to (\Omega_2, \mathscr{F}_2)$ will mean that $h: \Omega_1 \to \Omega_2$, measurable relative to $\mathscr{F}_1$ and $\mathscr{F}_2$.

If $\mathscr{F}$ is a σ-field of subsets of Ω, $(\Omega, \mathscr{F})$ is sometimes called a *measurable space*, and the sets in $\mathscr{F}$ are sometimes called *measurable sets.*

Notice that measurability of h does not imply that $h(A) \in \mathscr{F}_2$ for each $A \in \mathscr{F}_1$. For example, if $\mathscr{F}_2 = \{\varnothing, \Omega_2\}$, then any $h: \Omega_1 \to \Omega_2$ is measurable, regardless of $\mathscr{F}_1$, but if $A \in \mathscr{F}_1$ and $h(A)$ is a nonempty proper subset of Ω_2, then $h(A) \notin \mathscr{F}_2$. Actually, in measure theory, the inverse image is a much more desirable object than the direct image since the basic set operations are preserved by inverse images but not in general by direct images. Specifically, we have $h^{-1}(\bigcup_i B_i) = \bigcup_i h^{-1}(B_i)$, $h^{-1}(\bigcap_i B_i) = \bigcap_i h^{-1}(B_i)$, and $h^{-1}(B^c) = [h^{-1}(B)]^c$. We also have $h(\bigcup_i A_i) = \bigcup_i h(A_i)$, but $h(\bigcap_i A_i) \subset \bigcap_i h(A_i)$, and the inclusion may be proper. Furthermore, $h(A^c)$ need not equal $[h(A)]^c$, in fact when h is a constant function the two sets are disjoint.

If $(\Omega, \mathscr{F})$ is a measurable space and $h: \Omega \to R^n$ (or $\overline{R}^n$), h is said to be *Borel measurable* [on $(\Omega, \mathscr{F})$] iff h is measurable relative to the σ-fields $\mathscr{F}$ and $\mathscr{B}$, the class of Borel sets. If Ω is a Borel subset of R^k (or $\overline{R}^k$) and we use the term "Borel measurable," we always assume that $\mathscr{F} = \mathscr{B}$.

A continuous map h from R^k to R^n is Borel measurable; for if $\mathscr{C}$ is the class of open subsets of R^n, then $h^{-1}(A)$ is open, hence belongs to $\mathscr{B}(R^k)$, for each $A \in \mathscr{C}$.

If A is a subset of R that is not a Borel set (Section 1.4, Problems 6 and 11) and I_A is the *indicator* of A, that is, $I_A(\omega) = 1$ for $\omega \in A$ and 0 for $\omega \notin A$, then I_A is not Borel measurable; for $\{\omega: I_A(\omega) = 1\} = A \notin \mathscr{B}(R)$.

To show that a function $h: \Omega \to R$ (or $\bar{R}$) is Borel measurable, it is sufficient to show that $\{\omega: h(\omega) > c\} \in \mathscr{F}$ for each real c. For if $\mathscr{C}$ is the class of sets $\{x: x > c\}$, $c \in R$, then $\sigma(\mathscr{C}) = \mathscr{B}(R)$. Similarly, $\{\omega: h(\omega) > c\}$ can be replaced by $\{\omega: h(\omega) \geq c\}$, $\{\omega: h(\omega) < c\}$ or $\{\omega: h(\omega) \leq c\}$, or equally well by $\{\omega: a \leq h(\omega) \leq b\}$ for all real a and b, and so on.

If $(\Omega, \mathscr{F}, \mu)$ is a measure space the terminology "h is Borel measurable on $(\Omega, \mathscr{F}, \mu)$" will mean that h is Borel measurable on $(\Omega, \mathscr{F})$ and μ is a measure on $\mathscr{F}$.

1.5.2 ***Definition.*** Let $(\Omega, \mathscr{F})$ be a measurable space, fixed throughout the discussion. If $h: \Omega \to \bar{R}$, h is said to be *simple* iff h is Borel measurable and takes on only finitely many distinct values. Equivalently, h is simple iff it can be written as a finite sum $\sum_{i=1}^{r} x_i I_{A_i}$ where the A_i are disjoint sets in $\mathscr{F}$ and I_{A_i} is the indicator of A_i; the x_i need not be distinct.

We assume the standard arithmetic of $\bar{R}$; if $a \in R$, $a + \infty = \infty$, $a - \infty = -\infty$, $a/\infty = a/-\infty = 0$, $a \cdot \infty = \infty$ if $a > 0$, $a \cdot \infty = -\infty$ if $a < 0$, $0 \cdot \infty = 0 \cdot (-\infty) = 0$, $\infty + \infty = \infty$, $-\infty - \infty = -\infty$, with commutativity of addition and multiplication. It is then easy to check that sums, differences, products, and quotients of simple functions are simple, as long as the operations are well-defined, in other words we do not try to add $+\infty$ and $-\infty$, divide by 0, or divide ∞ by ∞.

Let μ be a measure on $\mathscr{F}$, again fixed throughout the discussion. If $h: \Omega \to \bar{R}$ is Borel measurable we are going to define the *abstract Lebesgue integral* of h with respect to μ, written as $\int_\Omega h\, d\mu$, $\int_\Omega h(\omega)\mu(d\omega)$, or $\int_\Omega h(\omega)\, d\mu(\omega)$.

1.5.3 ***Definition of the Integral.*** First let h be simple, say $h = \sum_{i=1}^{r} x_i I_{A_i}$ where the A_i are disjoint sets in $\mathscr{F}$. We define

$$\int_\Omega h\, d\mu = \sum_{i=1}^{r} x_i \mu(A_i)$$

as long as $+\infty$ and $-\infty$ do not both appear in the sum; if they do, we say that the integral does not exist. Strictly speaking, it must be verified that if h has a different representation, say $\sum_{j=1}^{s} y_j I_{B_j}$, then

$$\sum_{i=1}^{r} x_i \mu(A_i) = \sum_{j=1}^{s} y_j \mu(B_j).$$

(For example, if $A = B \cup C$, where $B \cap C = \varnothing$, then $xI_A = xI_B + xI_C$.) The proof is based on the observation that

$$h = \sum_{i=1}^{r} \sum_{j=1}^{s} z_{ij} I_{A_i \cap B_j},$$

where $z_{ij} = x_i = y_j$. Thus

$$\begin{aligned}\sum_{i,j} z_{ij}\mu(A_i \cap B_j) &= \sum_i x_i \sum_j \mu(A \cap B_j)\\ &= \sum_i x_i \mu(A_i)\\ &= \sum_j y_j \mu(B_j) \qquad \text{by a symmetrical argument.}\end{aligned}$$

If h is nonnegative Borel measurable, define

$$\int_\Omega h\, d\mu = \sup\left\{\int_\Omega s\, d\mu : s \quad \text{simple}, \quad 0 \le s \le h\right\}.$$

This agrees with the previous definition if h is simple. Furthermore, we may if we like restrict s to be finite-valued.

Notice that according to the definition, the integral of a nonnegative Borel measurable function always exists; it may be $+\infty$.

Finally, if h is an arbitrary Borel measurable function, let $h^+ = \max(h, 0)$, $h^- = \max(-h, 0)$, that is,

$$\begin{aligned}h^+(\omega) &= h(\omega) \quad \text{if} \quad h(\omega) \ge 0; \qquad h^+(\omega) = 0 \quad \text{if} \quad h(\omega) < 0;\\ h^-(\omega) &= -h(\omega) \quad \text{if} \quad h(\omega) \le 0; \qquad h^-(\omega) = 0 \quad \text{if} \quad h(\omega) > 0.\end{aligned}$$

The function h^+ is called the *positive part* of h, h^- the *negative part*. We have $|h| = h^+ + h^-$, $h = h^+ - h^-$, and h^+ and h^- are Borel measurable. For example, $\{\omega: h^+(\omega) \in B\} = \{\omega: h(\omega) \ge 0, h(\omega) \in B\} \cup \{\omega: h(\omega) < 0, 0 \in B\}$. The first set is $h^{-1}[0, \infty] \cap h^{-1}(B) \in \mathscr{F}$; the second is $h^{-1}[-\infty, 0)$ if $0 \in B$, and $\varnothing$ if $0 \notin B$. Thus $(h^+)^{-1}(B) \in \mathscr{F}$ for each $B \in \mathscr{B}(\bar{R})$, and similarly for h^-. Alternatively, if h_1 and h_2 are Borel measurable, then $\max(h_1, h_2)$ and $\min(h_1, h_2)$ are Borel measurable; to see this, note that

$$\{\omega: \max(h_1(\omega), h_2(\omega)) \le c\} = \{\omega: h_1(\omega) \le c\} \cap \{\omega: h_2(\omega) \le c\}$$

and $\{\omega: \min(h_1(\omega), h_2(\omega) \le c\} = \{\omega: h_1(\omega) \le c\} \cup \{\omega: h_2(\omega) \le c\}$. It follows that if h is Borel measurable, so are h^+ and h^-.

We define

$$\int_\Omega h\, d\mu = \int_\Omega h^+\, d\mu - \int_\Omega h^-\, d\mu \qquad \text{if this is not of the form } +\infty - \infty;$$

if it is, we say that the integral does not exist. The function h is said to be *μ-integrable* (or simply *integrable* if μ is understood) iff $\int_\Omega h\, d\mu$ is finite, that is, iff $\int_\Omega h^+ d\mu$ and $\int_\Omega h^- d\mu$ are both finite.

If $A \in \mathscr{F}$, we define

$$\int_A h\, d\mu = \int_\Omega h I_A\, d\mu.$$

(The proof that hI_A is Borel measurable is similar to the first proof above that h^+ is Borel measurable.)

If h is a step function from R to R and μ is Lebesgue measure, $\int_R h\,d\mu$ agrees with the Riemann integral. However, the integral of h with respect to Lebesgue measure exists for many functions that are not Riemann integrable, as we shall see in Section 1.7.

Before examining the properties of the integral, we need to know more about Borel measurable functions. One of the basic reasons why such functions are useful in anaylsis is that a pointwise limit of Borel measurable functions is still Borel measurable.

1.5.4 ***Theorem.*** If $h_1, h_2, \ldots$ are Borel measurable functions from Ω to $\overline{R}$ and $h_n(\omega) \to h(\omega)$ for all $\omega \in \Omega$, then h is Borel measurable.

PROOF. It is sufficient to show that $\{\omega: h(\omega) > c\} \in \mathscr{F}$ for each real c. We have

$$\begin{aligned}
\{\omega: h(\omega) > c\} &= \left\{\omega: \lim_{n\to\infty} h_n(\omega) > c\right\} \\
&= \left\{\omega: h_n(\omega) \quad \text{is eventually} > c + \frac{1}{r} \text{ for some } r = 1, 2, \ldots\right\} \\
&= \bigcup_{r=1}^{\infty} \left\{\omega: h_n(\omega) > c + \frac{1}{r} \text{ for all but finitely many } n\right\} \\
&= \bigcup_{r=1}^{\infty} \liminf_n \left\{\omega: h_n(\omega) > c + \frac{1}{r}\right\} \\
&= \bigcup_{r=1}^{\infty} \bigcup_{n=1}^{\infty} \bigcap_{k=n}^{\infty} \left\{\omega: h_k(\omega) > c + \frac{1}{r}\right\} \in \mathscr{F}. \quad \blacksquare
\end{aligned}$$

To show that the class of Borel measurable functions is closed under algebraic operations, we need the following basic approximation theorem.

1.5.5 ***Theorem.*** (a) A nonnegative Borel measurable function h is the limit of an increasing sequence of nonnegative, finite-valued, simple functions h_n.

(b) An arbitrary Borel measurable function f is the limit of a sequence of finite-valued simple functions f_n, with $|f_n| \le |f|$ for all n.

PROOF. (a) Define

$$h_n(\omega) = \frac{k-1}{2^n} \qquad \text{if} \quad \frac{k-1}{2^n} \le h(\omega) < \frac{k}{2^n}, \quad k = 1, 2, \ldots, n2^n,$$

and let $h_n(\omega) = n$ if $h(\omega) \geq n$. [Or equally well, $h_n(\omega) = (k-1)/2^n$ if $(k-1)/2^n < h(\omega) \leq k/2^n$, $k = 1, 2, \ldots, n2^n$; $h_n(\omega) = n$ if $h(\omega) > n$; $h_n(\omega) = 0$ if $h(\omega) = 0$.] The h_n have the desired properties (Problem 1).

(b) Let g_n and h_n be nonnegative, finite-valued, simple functions with $g_n \uparrow f^+$ and $h_n \uparrow f^-$; take $f_n = g_n - h_n$. ▮

1.5.6 ***Theorem.*** If h_1 and h_2 are Borel measurable functions from Ω to $\overline{R}$, so are $h_1 + h_2$, $h_1 - h_2$, $h_1 h_2$, and h_1/h_2 [assuming these are well-defined, in other words, $h_1(\omega) + h_2(\omega)$ is never of the form $+\infty - \infty$ and $h_1(\omega)/h_2(\omega)$ is never of the form ∞/∞ or $a/0$].

PROOF. As in 1.5.5, let s_{1n}, s_{2n} be finite-valued simple functions with $s_{1n} \to h_1$, $s_{2n} \to h_2$. Then $s_{1n} + s_{2n} \to h_1 + h_2$,

$$s_{1n} s_{2n} I_{\{h_1 \neq 0\}} I_{\{h_2 \neq 0\}} \to h_1 h_2,$$

and

$$\frac{s_{1n}}{s_{2n} + (1/n) I_{\{s_{2n} = 0\}}} \to \frac{h_1}{h_2}.$$

Since

$$s_{1n} \pm s_{2n}, \qquad s_{1n} s_{2n} I_{\{h_1 \neq 0\}} I_{\{h_2 \neq 0\}}, \qquad s_{1n}\left(s_{2n} + \frac{1}{n} I_{\{s_{2n} = 0\}}\right)^{-1}$$

are simple, the result follows from 1.5.4. ▮

We are going to extend 1.5.4 and part of 1.5.6 to Borel measurable functions from Ω to $\overline{R}^n$; to do this, we need the following useful result.

1.5.7 ***Lemma.*** A composition of measurable functions is measurable; specifically, if $g: (\Omega_1, \mathscr{F}_1) \to (\Omega_2, \mathscr{F}_2)$ and $h: (\Omega_2, \mathscr{F}_2) \to (\Omega_3, \mathscr{F}_3)$, then $h \circ g: (\Omega_1, \mathscr{F}_1) \to (\Omega_3, \mathscr{F}_3)$.

PROOF. If $B \in \mathscr{F}_3$, then $(h \circ g)^{-1}(B) = g^{-1}(h^{-1}(B)) \in \mathscr{F}_1$. ▮

Since some books contain the statement "A composition of measurable functions need not be measurable," some explanation is called for. If $h: R \to R$, some authors call h "measurable" iff the preimage of a Borel set is a Lebesgue measurable set. We shall call such a function *Lebesgue measurable.* Note that every Borel measurable function is Lebesgue measurable, but not conversely. (Consider the indicator of a Lebesgue measurable set that is not a

Borel set; see Section 1.4, Problem 11.) If g and h are Lebesgue measurable, the composition $h \circ g$ need not be Lebesgue measurable. For let $\mathscr{B}$ be the Borel sets, and $\overline{\mathscr{B}}$ the Lebesgue measurable sets. If $B \in \mathscr{B}$ then $h^{-1}(B) \in \overline{\mathscr{B}}$; but $g^{-1}(h^{-1}(B))$ is known to belong to $\overline{\mathscr{B}}$ only when $h^{-1}(B) \in \mathscr{B}$, so we cannot conclude that $(h \circ g)^{-1}(B) \in \overline{\mathscr{B}}$. For an explicit example, see Royden (1968, p. 70). If $g^{-1}(A) \in \overline{\mathscr{B}}$ for all $A \in \overline{\mathscr{B}}$, not just for all $A \in \mathscr{B}$, then we are in the situation described in Lemma 1.5.7, and $h \circ g$ is Lebesgue measurable; similarly, if h is Borel measurable (and g is Lebesgue measurable), then $h \circ g$ is Lebesgue measurable.

It is rarely necessary to replace Borel measurability of functions from R to R (or R^k to R^n) by the slightly more general concept of Lebesgue measurability; in this book, the only instance is in Section 1.7. The integration theory that we are developing works for extended real-valued functions on an arbitrary measure space $(\Omega, \mathscr{F}, \mu)$. Thus there is no problem in integrating Lebesgue measurable functions; set $\Omega = R$, $\mathscr{F} = \overline{\mathscr{B}}$.

We may now assert that if $h_1, h_2, \ldots$ are Borel measurable functions from Ω to $\overline{R}^n$ and h_n converges pointwise to h, then h is Borel measurable; furthermore, if h_1 and h_2 are Borel measurable functions from Ω to $\overline{R}^n$, so are $h_1 + h_2$ and $h_1 - h_2$, assuming these are well-defined. The reason is that if $h(\omega) = (h_1(\omega), \ldots, h_n(\omega))$ describes a map from Ω to $\overline{R}^n$, Borel measurability of h is equivalent to Borel measurability of all the component functions h_i.

1.5.8 ***Theorem.*** Let $h: \Omega \to \overline{R}^n$; if p_i is the projection map of $\overline{R}^n$ onto $\overline{R}$, taking $(x_1, \ldots, x_n)$ to x_i, set $h_i = p_i \circ h$, $i = 1, \ldots, n$. Then h is Borel measurable iff h_i is Borel measurable for all $i = 1, \ldots, n$.

PROOF. Assume h Borel measurable. Since

$$p_i^{-1}\{x_i : a_i \le x_i \le b_i\} = \{x \in \overline{R}^n : a_i \le x_i \le b_i, \quad -\infty \le x_j \le \infty, \quad j \ne i\},$$

which is an interval of $\overline{R}^n$, p_i is Borel measurable. Thus

$$h: (\Omega, \mathscr{F}) \to (\overline{R}^n, \mathscr{B}(\overline{R}^n)), \qquad p_i : (\overline{R}^n, \mathscr{B}(\overline{R}^n)) \to (\overline{R}, \mathscr{B}(\overline{R})),$$

and therefore by 1.5.7, $h_i : (\Omega, \mathscr{F}) \to (\overline{R}, \mathscr{B}(\overline{R}))$.

Conversely, assume each h_i to be Borel measurable. Then

$$h^{-1}\{x \in \overline{R}^n : a_i \le x_i \le b_i, \quad i = 1, \ldots, n\} = \bigcap_{i=1}^{n} \{\omega \in \Omega : a_i \le h_i(\omega) \le b_i\} \in \mathscr{F},$$

and the result follows. ▌

We now proceed to some properties of the integral. In the following result, all functions are assumed Borel measurable from Ω to $\overline{R}$.

1.5.9 ***Theorem.*** (a) If $\int_\Omega h\,d\mu$ exists and $c \in R$, then $\int_\Omega ch\,d\mu$ exists and equals $c\int_\Omega h\,d\mu$.

(b) If $g(\omega) \geq h(\omega)$ for all ω, then $\int_\Omega g\,d\mu \geq \int_\Omega h\,d\mu$ in the sense that if $\int_\Omega h\,d\mu$ exists and is greater than $-\infty$, then $\int_\Omega g\,d\mu$ exists and $\int_\Omega g\,d\mu \geq \int_\Omega h\,d\mu$; if $\int_\Omega g\,d\mu$ exists and is less than $+\infty$, then $\int_\Omega h\,d\mu$ exists and $\int_\Omega h\,d\mu \leq \int_\Omega g\,d\mu$. Thus if both integrals exist, $\int_\Omega g\,d\mu \geq \int_\Omega h\,d\mu$, whether or not the integrals are finite.

(c) If $\int_\Omega h\,d\mu$ exists, then $|\int_\Omega h\,d\mu| \leq \int_\Omega |h|\,d\mu$.

(d) If $h \geq 0$ and $B \in \mathscr{F}$, then $\int_B h\,d\mu = \sup\{\int_B s\,d\mu\colon 0 \leq s \leq h,\ s \text{ simple}\}$.

(e) If $\int_\Omega h\,d\mu$ exists, so does $\int_A h\,d\mu$ for each $A \in \mathscr{F}$; if $\int_\Omega h\,d\mu$ is finite, then $\int_A h\,d\mu$ is also finite for each $A \in \mathscr{F}$.

PROOF. (a) It is immediate that this holds when h is simple. If h is nonnegative and $c > 0$, then

$$\int_\Omega ch\,d\mu = \sup\left\{\int_\Omega s\,d\mu;\quad 0 \leq s \leq ch,\quad s \quad \text{simple}\right\}$$

$$= c\sup\left\{\int_\Omega \frac{s}{c}\,d\mu;\quad 0 \leq \frac{s}{c} \leq h,\quad \frac{s}{c} \quad \text{simple}\right\} = c\int_\Omega h\,d\mu.$$

In general, if $h = h^+ - h^-$ and $c > 0$, then $(ch)^+ = ch^+$, $(ch)^- = ch^-$; hence $\int_\Omega ch\,d\mu = c\int_\Omega h^+\,d\mu - c\int_\Omega h^-\,d\mu$ by what we have just proved, so that $\int_\Omega ch\,d\mu = c\int_\Omega h\,d\mu$. If $c < 0$, then

$$(ch)^+ = -ch^-, \qquad (ch)^- = -ch^+,$$

so

$$\int_\Omega ch\,d\mu = -c\int_\Omega h^-\,d\mu + c\int_\Omega h^+\,d\mu = c\int_\Omega h\,d\mu.$$

(b) If g and h are nonnegative and $0 \leq s \leq h$, s simple, then $0 \leq s \leq g$; hence $\int_\Omega h\,d\mu \leq \int_\Omega g\,d\mu$. In general, $h \leq g$ implies $h^+ \leq g^+$, $h^- \geq g^-$. If $\int_\Omega h\,d\mu > -\infty$, we have $\int_\Omega g^-\,d\mu \leq \int_\Omega h^-\,d\mu < \infty$; hence $\int_\Omega g\,d\mu$ exists and equals

$$\int_\Omega g^+\,d\mu - \int_\Omega g^-\,d\mu \geq \int_\Omega h^+\,d\mu - \int_\Omega h^-\,d\mu = \int_\Omega h\,d\mu.$$

The case in which $\int_\Omega g\,d\mu < \infty$ is handled similarly.

(c) We have $-|h| \leq h \leq |h|$ so by (a) and (b), $-\int_\Omega |h|\,d\mu \leq \int_\Omega h\,d\mu \leq \int_\Omega |h|\,d\mu$ and the result follows. (Note that $|h|$ is Borel measurable by 1.5.6 since $h = h^+ + h^-$.)

(d) If $0 \leq s \leq h$, then $\int_B s\,d\mu \leq \int_B h\,d\mu$ by (b); hence

$$\int_B h\,d\mu \geq \sup\left\{\int_B s\,d\mu\colon 0 \leq s \leq h\right\}.$$

If $0 \le t \le hI_B$, t simple, then $t = tI_B \le h$ so $\int_\Omega t\,d\mu \le \sup\{\int_\Omega sI_B\,d\mu: 0 \le s \le h,\ s \text{ simple}\}$. Take the sup over t to obtain $\int_B h\,d\mu \le \sup\{\int_B s\,d\mu: 0 \le s \le h,\ s \text{ simple}\}$.

(e) This follows from (b) and the fact that $(hI_A)^+ = h^+I_A \le h^+$, $(hI_A)^- = h^-I_A \le h^-$. ∎

Problems

1. Show that the functions proposed in the proof of 1.5.5(a) have the desired properties. Show also that if h is bounded, the approximating sequence converges to h uniformly on Ω.
2. Let f and g be extended real-valued Borel measurable functions on $(\Omega, \mathscr{F})$, and define
$$h(\omega) = f(\omega) \quad \text{if } \omega \in A,$$
$$= g(\omega) \quad \text{if } \omega \in A^c,$$
where A is a set in $\mathscr{F}$. Show that h is Borel measurable.
3. If $f_1, f_2, \ldots$ are extended real-valued Borel measurable functions on $(\Omega, \mathscr{F})$, $n = 1, 2, \ldots$, show that $\sup_n f_n$ and $\inf_n f_n$ are Borel measurable (hence $\limsup_{n\to\infty} f_n$ and $\liminf_{n\to\infty} f_n$ are Borel measurable).
4. Let $(\Omega, \mathscr{F}, \mu)$ be a complete measure space. If $f: (\Omega, \mathscr{F}) \to (\Omega', \mathscr{F}')$ and $g: \Omega \to \Omega'$, $g = f$ except on a subset of a set $A \in \mathscr{F}$ with $\mu(A) = 0$, show that g is measurable (relative to $\mathscr{F}$ and $\mathscr{F}'$).
5. (a) Let f be a function from R^k to R^m, not necessarily Borel measurable. Show that $\{x: f \text{ is discontinuous at } x\}$ is an F_σ (a countable union of closed subsets of R^k), and hence is a Borel set. Does this result hold in spaces more general than the Euclidean space R^n?
   (b) Show that there is no function from R to R whose discontinuity set is the irrationals. (In 1.4.5 we constructed a distribution function whose discontinuity set was the rationals.)
6. How many Borel measurable functions are there from R^n to R^k?
7. We have seen that a pointwise limit of measurable functions is measurable. We may also show that under certain conditions, a pointwise limit of measures is a measure. The following result, known as Steinhaus' lemma, will be needed in the problem: If $\{a_{nk}\}$ is a double sequence of real numbers satisfying

   (i) $\sum_{k=1}^\infty a_{nk} = 1$ for all n,
   (ii) $\sum_{k=1}^\infty |a_{nk}| \le c < \infty$ for all n, and
   (iii) $a_{nk} \to 0$ as $n \to \infty$ for all k,

   there is a sequence $\{x_n\}$, with $x_n = 0$ or 1 for all n, such that $t_n = \sum_{k=1}^\infty a_{nk} x_k$ fails to converge to a finite or infinite limit.

To prove this, choose positive integers n_1 and k_1 arbitrarily; having chosen $n_1, \ldots, n_r$, $k_1, \ldots, k_r$, choose $n_{r+1} > n_r$ such that $\sum_{k \le k_r} |a_{n_{r+1}k}| < \frac{1}{8}$; this is possible by (iii). Then choose $k_{r+1} > k_r$ such that $\sum_{k > k_{r+1}} |a_{n_{r+1}k}| < \frac{1}{8}$; this is possible by (ii). Set $x_k = 0$, $k_{2s-1} < k \le k_{2s}$, $x_k = 1$, $k_{2s} < k \le k_{2s+1}$, $s = 1, 2, \ldots$. We may write $t_{n_{r+1}}$ as $h_1 + h_2 + h_3$, where h_1 is the sum of $a_{n_{r+1}k} x_k$ for $k \le k_r$, h_2 corresponds to $k_r < k \le k_{r+1}$, and h_3 to $k > k_{r+1}$. If r is odd, then $x_k = 0$, $k_r < k \le k_{r+1}$; hence $|t_{n_{r+1}}| < \frac{1}{4}$. If r is even, then $h_2 = \sum_{k_r < k \le k_{r+1}} a_{n_{r+1}k}$; hence by (i),

$$h_2 = 1 - \sum_{k \le k_r} a_{n_{r+1}k} - \sum_{k > k_{r+1}} a_{n_{r+1}k} > \tfrac{3}{4}.$$

Thus $t_{n_{r+1}} > \frac{3}{4} - |h_1| - |h_3| > \frac{1}{2}$, so $\{t_n\}$ cannot converge.

(a) *Vitali–Hahn–Saks Theorem.* Let $(\Omega, \mathscr{F})$ be a measurable space, and let P_n, $n = 1, 2, \ldots$, be probability measures on $\mathscr{F}$. If $P_n(A) \to P(A)$ for all $A \in \mathscr{F}$, then P is a probability measure on $\mathscr{F}$; furthermore, if $\{B_k\}$ is a sequence of sets in $\mathscr{F}$ decreasing to $\varnothing$, then $\sup_n P_n(B_k) \downarrow 0$ as $k \to \infty$. [Let A be the disjoint union of sets $A_k \in \mathscr{F}$; without loss of generality, assume $A = \Omega$ (otherwise add A^c to both sides). It is immediate that P is finitely additive, so by Problem 5, Section 1.2, $\alpha = \sum_k P(A_k) \le P(\Omega) = 1$. If $\alpha < 1$, set $a_{nk} = (1 - \alpha)^{-1}[P_n(A_k) - P(A_k)]$ and apply Steinhaus' lemma.]

(b) Extend the Vitali–Hahn–Saks theorem to the case where the P_n are not necessarily probability measures, but $P_n(\Omega) \le c < \infty$ for all n. [For further extensions, see Dunford and Schwartz (1958).]

1.6 Basic Integration Theorems

We are now ready to present the main properties of the integral. The results in this section will be used many times in the text. As above, $(\Omega, \mathscr{F}, \mu)$ is a fixed measure space, and all functions to be considered map Ω to $\overline{R}$.

1.6.1 ***Theorem.*** Let h be a Borel measurable function such that $\int_\Omega h\, d\mu$ exists. Define $\lambda(B) = \int_B h\, d\mu$, $B \in \mathscr{F}$. Then λ is countably additive on $\mathscr{F}$; thus if $h \ge 0$, λ is a measure.

PROOF. Let h be a nonnegative simple function $\sum_{i=1}^n x_i I_{A_i}$. Then $\lambda(B) = \int_B h\, d\mu = \sum_{i=1}^n x_i \mu(B \cap A_i)$; since μ is countably additive, so is λ.

Now let h be nonnegative Borel measurable, and let $B = \bigcup_{n=1}^{\infty} B_n$, the B_n disjoint sets in $\mathscr{F}$. If s is simple and $0 \le s \le h$, then

$$\int_B s\,d\mu = \sum_{n=1}^{\infty} \int_{B_n} s\,d\mu$$

by what we have proved for nonnegative simple functions

$$\le \sum_{n=1}^{\infty} \int_{B_n} h\,d\mu$$

by 1.5.9(b) (or the definition of the integral).

Take the sup over s to obtain, by 1.5.9(d), $\lambda(B) \le \sum_{n=1}^{\infty} \lambda(B_n)$.

Now $B_n \subset B$, hence $I_{B_n} \le I_B$, so by 1.5.9(b), $\lambda(B_n) \le \lambda(B)$. If $\lambda(B_n) = \infty$ for some n, we are finished, so assume all $\lambda(B_n)$ finite. Fix n and let $\varepsilon > 0$. It follows from 1.5.9(d), 1.5.9(b), and the fact that the maximum of a finite number of simple functions is simple that we can find a simple function s, $0 \le s \le h$, such that

$$\int_{B_i} s\,d\mu \ge \int_{B_i} h\,d\mu - \frac{\varepsilon}{n}, \qquad i = 1, 2, \ldots, n.$$

Now

$$\lambda(B_1 \cup \cdots \cup B_n) = \int_{\bigcup_{i=1}^{n} B_i} h\,d\mu \ge \int_{\bigcup_{i=1}^{n} B_i} s\,d\mu = \sum_{i=1}^{n} \int_{B_i} s\,d\mu$$

by what we have proved for nonnegative simple functions, hence

$$\lambda(B_1 \cup \cdots \cup B_n) \ge \sum_{i=1}^{n} \int_{B_i} h\,d\mu - \varepsilon = \sum_{i=1}^{n} \lambda(B_i) - \varepsilon.$$

Since $\lambda(B) \ge \lambda\left(\bigcup_{i=1}^{n} B_i\right)$ and ε is arbitrary, we have

$$\lambda(B) \ge \sum_{i=1}^{\infty} \lambda(B_i).$$

Finally let $h = h^+ - h^-$ be an arbitrary Borel measurable function. Then $\lambda(B) = \int_B h^+\,d\mu - \int_B h^-\,d\mu$. Since $\int_\Omega h^+\,d\mu < \infty$ or $\int_\Omega h^-\,d\mu < \infty$, the result follows. ▌

The proof of 1.6.1 shows that λ is the difference of two measures λ^+ and λ^-, where $\lambda^+(B) = \int_B h^+\,d\mu$, $\lambda^- = \int_B h^-\,d\mu$; at least one of the measures λ^+ and λ^- must be finite.

1.6.2 ***Monotone Convergence Theorem.*** Let $h_1, h_2, \ldots$ form an increasing sequence of nonnegative Borel measurable functions, and let $h(\omega) = \lim_{n\to\infty} h_n(\omega)$, $\omega \in \Omega$. Then $\int_\Omega h_n\,d\mu \to \int_\Omega h\,d\mu$. (Note that $\int_\Omega h_n\,d\mu$ increases with n by 1.5.9(b); for short, $0 \le h_n \uparrow h$ implies $\int_\Omega h_n\,d\mu \uparrow \int_\Omega h\,d\mu$.)

PROOF. By 1.5.9(b), $\int_\Omega h_n\,d\mu \le \int_\Omega h\,d\mu$ for all n, hence $k = \lim_{n\to\infty} \int_\Omega h_n\,d\mu \le \int_\Omega h\,d\mu$. Let $0 < b < 1$, and let s be a nonnegative, finite-valued, simple function with $s \le h$. Let $B_n = \{\omega\colon h_n(\omega) \ge bs(\omega)\}$. Then $B_n \uparrow \Omega$ since $h_n \uparrow h$ and s is finite-valued. Now $k \ge \int_\Omega h_n\,d\mu \ge \int_{B_n} h_n\,d\mu$ by 1.5.9(b), and $\int_{B_n} h_n\,d\mu \ge b\int_{B_n} s\,d\mu$ by 1.5.9(a) and (b). By 1.6.1 and 1.2.7, $\int_{B_n} s\,d\mu \to \int_\Omega s\,d\mu$, hence (let $b \to 1$) $k \ge \int_\Omega s\,d\mu$. Take the sup over s to obtain $k \ge \int_\Omega h\,d\mu$. ∎

1.6.3 ***Additivity Theorem.*** Let f and g be Borel measurable, and assume that $f + g$ is well-defined. If $\int_\Omega f\,d\mu$ and $\int_\Omega g\,d\mu$ exist and $\int_\Omega f\,d\mu + \int_\Omega g\,d\mu$ is well-defined (not of the form $+\infty \ -\infty$ or $-\infty \ +\infty$), then

$$\int_\Omega (f + g)\,d\mu = \int_\Omega f\,d\mu + \int_\Omega g\,d\mu.$$

In particular, if f and g are integrable, so is $f + g$.

PROOF. If f and g are nonnegative simple functions, this is immediate from the definition of the integral. Assume f and g are nonnegative Borel measurable, and let t_n, u_n be nonnegative simple functions increasing to f and g, respectively. Then $0 \le s_n = t_n + u_n \uparrow f + g$. Now $\int_\Omega s_n\,d\mu = \int_\Omega t_n\,d\mu + \int_\Omega u_n\,d\mu$ by what we have proved for nonnegative simple functions; hence by 1.6.2, $\int_\Omega (f+g)\,d\mu = \int_\Omega f\,d\mu + \int_\Omega g\,d\mu$.

Now if $f \ge 0$, $g \le 0$, $h = f + g \ge 0$ (so g must be finite), we have $f = h + (-g)$; hence $\int_\Omega f\,d\mu = \int_\Omega h\,d\mu - \int_\Omega g\,d\mu$. If $\int_\Omega g\,d\mu$ is finite, then $\int_\Omega h\,d\mu = \int_\Omega f\,d\mu + \int_\Omega g\,d\mu$, and if $\int_\Omega g\,d\mu = -\infty$, then since $h \ge 0$,

$$\int_\Omega f\,d\mu \ge -\int_\Omega g\,d\mu = \infty,$$

contradicting the hypothesis that $\int_\Omega f\,d\mu + \int_\Omega g\,d\mu$ is well-defined. Similarly if $f \ge 0$, $g \le 0$, $h \le 0$, we obtain $\int_\Omega h\,d\mu = \int_\Omega f\,d\mu + \int_\Omega g\,d\mu$ by replacing all functions by their negatives. (Explicitly, $-g \ge 0$, $-f \le 0$, $-h = -f - g \ge 0$, and the above argument applies.)

Let

$$\begin{aligned}
E_1 &= \{\omega\colon f(\omega) \ge 0, && g(\omega) \ge 0\},\\
E_2 &= \{\omega\colon f(\omega) \ge 0, && g(\omega) < 0, \quad h(\omega) \ge 0\},\\
E_3 &= \{\omega\colon f(\omega) \ge 0, && g(\omega) < 0, \quad h(\omega) < 0\},\\
E_4 &= \{\omega\colon f(\omega) < 0, && g(\omega) \ge 0, \quad h(\omega) \ge 0\},\\
E_5 &= \{\omega\colon f(\omega) < 0, && g(\omega) \ge 0, \quad h(\omega) < 0\},\\
E_6 &= \{\omega\colon f(\omega) < 0, && g(\omega) < 0\}.
\end{aligned}$$

The above argument shows that $\int_{E_i} h\,d\mu = \int_{E_i} f\,d\mu + \int_{E_i} g\,d\mu$. Now $\int_\Omega f\,d\mu = \sum_{i=1}^6 \int_{E_i} f\,d\mu$, $\int_\Omega g\,d\mu = \sum_{i=1}^6 \int_{E_i} g\,d\mu$ by 1.6.1, so that $\int_\Omega f\,d\mu + \int_\Omega g\,d\mu = \sum_{i=1}^6 \int_{E_i} h\,d\mu$, and this equals $\int_\Omega h\,d\mu$ by 1.6.1, if we can show that $\int_\Omega h\,d\mu$ exists; that is, $\int_\Omega h^+\,d\mu$ and $\int_\Omega h^-\,d\mu$ are not both infinite.

If this is the case, $\int_{E_i} h^+\,d\mu = \int_{E_j} h^-\,d\mu = \infty$ for some i, j (1.6.1 again), so that $\int_{E_i} h\,d\mu = \infty$, $\int_{E_j} h\,d\mu = -\infty$. But then $\int_{E_i} f\,d\mu$ or $\int_{E_i} g\,d\mu = \infty$; hence $\int_\Omega f\,d\mu$ or $\int_\Omega g\,d\mu = \infty$. (Note that $\int_\Omega f^+\,d\mu \geq \int_{E_i} f^+\,d\mu$.) Similarly $\int_\Omega f\,d\mu$ or $\int_\Omega g\,d\mu = -\infty$, and this is a contradiction. ∎

1.6.4 ***Corollaries.*** (a) If $h_1, h_2, \ldots$ are nonnegative Borel measurable,

$$\int_\Omega \left(\sum_{n=1}^\infty h_n\right) d\mu = \sum_{n=1}^\infty \int_\Omega h_n\,d\mu.$$

Thus any series of nonnegative Borel measurable functions may be integrated term by term.

(b) If h is Borel measurable, h is integrable iff $|h|$ is integrable.

(c) If g and h are Borel measurable with $|g| \leq h$, h integrable, then g is integrable.

PROOF. (a) $\sum_{k=1}^n h_k \uparrow \sum_{k=1}^\infty h_k$, and the result follows from 1.6.2 and 1.6.3.

(b) Since $|h| = h^+ + h^-$, this follows from the definition of the integral and 1.6.3.

(c) By 1.5.9(b), $|g|$ is integrable, and the result follows from (b) above. ∎

A condition is said to hold *almost everywhere* with respect to the measure μ (written a.e. [μ] or simply a.e. if μ is understood) iff there is a set $B \in \mathscr{F}$ of μ-measure 0 such that the condition holds outside of B. From the point of view of integration theory, functions that differ only on a set of measure 0 may be identified. This is established by the following result.

1.6.5 ***Theorem.*** Let f, g, and h be Borel measurable functions.

(a) If $f = 0$ a.e. [μ], then $\int_\Omega f\,d\mu = 0$.

(b) If $g = h$ a.e. [μ] and $\int_\Omega g\,d\mu$ exists, then so does $\int_\Omega h\,d\mu$, and $\int_\Omega g\,d\mu = \int_\Omega h\,d\mu$.

PROOF. (a) If $f = \sum_{i=1}^n x_i I_{A_i}$ is simple, then $x_i \neq 0$ implies $\mu(A_i) = 0$ by hypothesis, hence $\int_\Omega f\,d\mu = 0$. If $f \geq 0$ and $0 \leq s \leq f$, s simple, then $s = 0$ a.e. [μ], hence $\int_\Omega s\,d\mu = 0$; thus $\int_\Omega f\,d\mu = 0$. If $f = f^+ - f^-$, then f^+ and f^-, being less than or equal to $|f|$, are 0 a.e. [μ], and the result follows.

(b) Let $A = \{\omega\colon g(\omega) = h(\omega)\}$, $B = A^c$. Then $g = gI_A + gI_B$, $h = hI_A + hI_B = gI_A + hI_B$. Since $gI_B = hI_B = 0$ except on B, a set of measure 0, the result follows from part (a) and 1.6.3. ∎

Thus in any integration theorem, we may freely use the phrase "almost everywhere." For example, if $\{h_n\}$ is an increasing sequence of nonnegative Borel measurable functions converging a.e. to the Borel measurable function h, then $\int_\Omega h_n\, d\mu \to \int_\Omega h\, d\mu$.

Another example: If g and h are Borel measurable and $g \geq h$ a.e., then $\int_\Omega g\, d\mu \geq \int_\Omega h\, d\mu$ [in the sense of 1.5.9(b)].

1.6.6 ***Theorem.*** Let h be Borel measurable.

(a) If h is integrable, then h is finite a.e.
(b) If $h \geq 0$ and $\int_\Omega h\, d\mu = 0$, then $h = 0$ a.e.

PROOF. (a) Let $A = \{\omega\colon |h(\omega)| = \infty\}$. If $\mu(A) > 0$, then $\int_\Omega |h|\, d\mu \geq \int_A |h|\, d\mu = \infty\mu(A) = \infty$, a contradiction.

(b) Let $B = \{\omega\colon h(\omega) > 0\}$, $B_n = \{\omega\colon h(\omega) \geq 1/n\} \uparrow B$. We have $0 \leq hI_{B_n} \leq hI_B = h$; hence by 1.5.9(b), $\int_{B_n} h\, d\mu = 0$. But $\int_{B_n} h\, d\mu \geq (1/n)\mu(B_n)$, so that $\mu(B_n) = 0$ for all n, and thus $\mu(B) = 0$. ∎

The monotone convergence theorem was proved under the hypothesis that all functions were nonnegative. This assumption can be relaxed considerably, as we now prove.

1.6.7 ***Extended Monotone Convergence Theorem.*** Let $g_1, g_2, \ldots, g, h$ be Borel measurable.

(a) If $g_n \geq h$ for all n, where $\int_\Omega h\, d\mu > -\infty$, and $g_n \uparrow g$, then

$$\int_\Omega g_n\, d\mu \uparrow \int_\Omega g\, d\mu.$$

(b) If $g_n \leq h$ for all n, where $\int_\Omega h\, d\mu < \infty$, and $g_n \downarrow g$, then

$$\int_\Omega g_n\, d\mu \downarrow \int_\Omega h\, d\mu.$$

PROOF. (a) If $\int_\Omega h\, d\mu = \infty$, then by 1.5.9(b), $\int_\Omega g_n\, d\mu = \infty$ for all n, and $\int_\Omega g\, d\mu = \infty$. Thus assume $\int_\Omega h\, d\mu < \infty$, so that by 1.6.6(a), h is a.e. finite; change h to 0 on the set where it is infinite. Then $0 \leq g_n - h \uparrow g - h$ a.e., hence by 1.6.2, $\int_\Omega (g_n - h)\, d\mu \uparrow \int_\Omega (g - h)\, d\mu$. The result follows from 1.6.3.

(We must check that the additivity theorem actually applies. Since $\int_\Omega h\,d\mu > -\infty$, $\int_\Omega g_n\,d\mu$ and $\int_\Omega g\,d\mu$ exist and are greater than $-\infty$ by 1.5.9(b). Also, $\int_\Omega h\,d\mu$ is finite, so that $\int_\Omega g_n\,d\mu - \int_\Omega h\,d\mu$ and $\int_\Omega g\,d\mu - \int_\Omega h\,d\mu$ are well-defined.)

(b) $-g_n \geq -h$, $\int_\Omega -h\,d\mu > -\infty$, and $-g_n \uparrow -g$. By part (a), $-\int_\Omega g_n\,d\mu \uparrow -\int_\Omega g\,d\mu$, so $\int_\Omega g_n\,d\mu \downarrow \int_\Omega g\,d\mu$. ∎

The extended monotone convergence theorem asserts that under appropriate conditions, the limit of the integrals of a sequence of functions is the integral of the limit function. More general theorems of this type can be obtained if we replace limits by upper or lower limits. If $f_1, f_2, \ldots$ are functions from Ω to $\overline{R}$, $\liminf_{n\to\infty} f_n$ and $\limsup_{n\to\infty} f_n$ are defined pointwise, that is,

$$\left(\liminf_{n\to\infty} f_n\right)(\omega) = \sup_n \inf_{k\geq n} f_k(\omega),$$

$$\left(\limsup_{n\to\infty} f_n\right)(\omega) = \inf_n \sup_{k\geq n} f_k(\omega).$$

1.6.8 ***Fatou's Lemma.*** Let $f_1, f_2, \ldots, f$ be Borel measurable.

(a) If $f_n \geq f$ for all n, where $\int_\Omega f\,d\mu > -\infty$, then

$$\liminf_{n\to\infty} \int_\Omega f_n\,d\mu \geq \int_\Omega \left(\liminf_{n\to\infty} f_n\right) d\mu.$$

(b) If $f_n \leq f$ for all n, where $\int_\Omega f\,d\mu < \infty$, then

$$\limsup_{n\to\infty} \int_\Omega f_n\,d\mu \leq \int_\Omega \left(\limsup_{n\to\infty} f_n\right) d\mu.$$

PROOF. (a) Let $g_n = \inf_{k\geq n} f_k$, $g = \liminf f_n$. Then $g_n \geq f$ for all n, $\int_\Omega f\,d\mu > -\infty$, and $g_n \uparrow g$. By 1.6.7, $\int_\Omega g_n\,d\mu \uparrow \int_\Omega (\liminf_{n\to\infty} f_n)\,d\mu$. But $g_n \leq f_n$, so

$$\lim_{n\to\infty} \int_\Omega g_n\,d\mu = \liminf_{n\to\infty} \int_\Omega g_n\,d\mu \leq \liminf_{n\to\infty} \int_\Omega f_n\,d\mu.$$

(b) We may write

$$\begin{aligned}\int_\Omega \left(\limsup_{n\to\infty} f_n\right) d\mu &= -\int_\Omega \liminf_{n\to\infty}(-f_n)\,d\mu \\ &\geq -\liminf_{n\to\infty} \int_\Omega (-f_n)\,d\mu \qquad \text{by (a)} \\ &= \limsup_{n\to\infty} \int_\Omega f_n\,d\mu. \quad ∎\end{aligned}$$

The following result is one of the "bread and butter" theorems of analysis; it will be used quite often in later chapters.

1.6.9 ***Dominated Convergence Theorem.*** If $f_1, f_2, \ldots, f, g$ are Borel measurable, $|f_n| \le g$ for all n, where g is μ-integrable, and $f_n \to f$ a.e. $[\mu]$, then f is μ-integrable and $\int_\Omega f_n \, d\mu \to \int_\Omega f \, d\mu$.

PROOF. We have $|f| \le g$ a.e.; hence f is integrable by 1.6.4(c). By 1.6.8,

$$\int_\Omega \left(\liminf_{n\to\infty} f_n\right) d\mu \le \liminf_{n\to\infty} \int_\Omega f_n \, d\mu \le \limsup_{n\to\infty} \int_\Omega f_n \, d\mu$$

$$\le \int_\Omega \left(\limsup_{n\to\infty} f_n\right) d\mu.$$

By hypothesis, $\liminf_{n\to\infty} f_n = \limsup_{n\to\infty} f_n = f$ a.e., so all terms of the above inequality are equal to $\int_\Omega f \, d\mu$. ∎

1.6.10 ***Corollary.*** If $f_1, f_2, \ldots, f, g$ are Borel measurable, $|f_n| \le g$ for all n, where $|g|^p$ is μ-integrable ($p > 0$, fixed), and $f_n \to f$ a.e. $[\mu]$, then $|f|^p$ is μ-integrable and $\int_\Omega |f_n - f|^p \, d\mu \to 0$ as $n \to \infty$.

PROOF. We have $|f_n|^p \le |g|^p$ for all n; so $|f|^p \le |g|^p$, and therefore $|f|^p$ is integrable. Also $|f_n - f|^p \le (|f_n| + |f|)^p \le (2|g|)^p$, which is integrable, and the result follows from 1.6.9. ∎

We have seen in 1.5.9(b) that $g \le h$ implies $\int_\Omega g \, d\mu \le \int_\Omega h \, d\mu$, and in fact $\int_A g \, d\mu \le \int_A h \, d\mu$ for all $A \in \mathscr{F}$. There is a converse to this result.

1.6.11 ***Theorem.*** If μ is σ-finite on $\mathscr{F}$, g and h are Borel measurable, $\int_\Omega g \, d\mu$ and $\int_\Omega h \, d\mu$ exist, and $\int_A g \, d\mu \le \int_A h \, d\mu$ for all $A \in \mathscr{F}$, then $g \le h$ a.e. $[\mu]$.

PROOF. It is sufficient to prove this when μ is finite. Let

$$A_n = \left\{\omega \colon g(\omega) \ge h(\omega) + \frac{1}{n}, \quad |h(\omega)| \le n\right\}.$$

Then

$$\int_{A_n} h \, d\mu \ge \int_{A_n} g \, d\mu \ge \int_{A_n} h \, d\mu + \frac{1}{n} \mu(A_n).$$

But

$$\left|\int_{A_n} h \, d\mu\right| \le \int_{A_n} |h| \, d\mu \le n\mu(A_n) < \infty,$$

and thus we may subtract $\int_{A_n} h\,d\mu$ to obtain $(1/n)\mu(A_n) \le 0$, hence $\mu(A_n) = 0$. Therefore $\mu(\bigcup_{n=1}^{\infty} A_n) = 0$; hence $\mu\{\omega: g(\omega) > h(\omega),\ h(\omega) \text{ finite}\} = 0$. Consequently $g \le h$ a.e. on $\{\omega: h(\omega) \text{ finite}\}$. Clearly, $g \le h$ everywhere on $\{\omega: h(\omega) = \infty\}$, and by taking $C_n = \{\omega: h(\omega) = -\infty, g(\omega) \ge -n\}$ we obtain

$$-\infty\mu(C_n) = \int_{C_n} h\,d\mu \ge \int_{C_n} g\,d\mu \ge -n\mu(C_n);$$

hence $\mu(C_n) = 0$. Thus $\mu(\bigcup_{n=1}^{\infty} C_n) = 0$, so that

$$\mu\{\omega: g(\omega) > h(\omega), h(\omega) = -\infty\} = 0.$$

Therefore $g \le h$ a.e. on $\{\omega: h(\omega) = -\infty\}$. ∎

If g and h are integrable, the proof is simpler. Let $B = \{\omega: g(\omega) > h(\omega)\}$. Then $\int_B g\,d\mu \le \int_B h\,d\mu \le \int_B g\,d\mu$; hence all three integrals are equal. Thus by 1.6.3, $0 = \int_B (g - h)\,d\mu = \int_\Omega (g - h)I_B\,d\mu$, with $(g - h)I_B \ge 0$. By 1.6.6(b), $(g - h)I_B = 0$ a.e., so that $g = h$ a.e. on B. But $g \le h$ on B^c, and the result follows.

The reader may have noticed that several integration theorems in this section were proved by starting with nonnegative simple functions and working up to nonnegative measurable functions and finally to arbitrary measurable functions. This technique is quite basic and will often be useful. A good illustration of the method is the following result, which introduces the notion of a measure-preserving transformation, a key concept in ergodic theory. In fact it is convenient here to start with indicators before proceeding to nonnegative simple functions.

1.6.12 ***Theorem.*** Let $T: (\Omega, \mathscr{F}) \to (\Omega_0, \mathscr{F}_0)$ be a measurable mapping, and let μ be a measure on $\mathscr{F}$. Define a measure $\mu_0 = \mu T^{-1}$ on $\mathscr{F}_0$ by

$$\mu_0(A) = \mu(T^{-1}(A)), \qquad A \in \mathscr{F}_0.$$

If $\Omega_0 = \Omega$, $\mathscr{F}_0 = \mathscr{F}$, and $\mu_0 = \mu$, T is said to *preserve* the measure μ.

If $f: (\Omega_0, \mathscr{F}_0) \to (\overline{R}, \mathscr{B}(\overline{R}))$ and $A \in \mathscr{F}_0$, then

$$\int_{T^{-1}A} f(T(\omega))\,d\mu(\omega) = \int_A f(\omega)\,d\mu_0(\omega)$$

in the sense that if one of the integrals exists, so does the other, and the two integrals are equal.

PROOF. If f is an indicator I_B, the desired formula states that

$$\mu(T^{-1}A \cap T^{-1}B) = \mu_0(A \cap B),$$

which is true by definition of μ_0. If f is a nonnegative simple function $\sum_{i=1}^{n} x_i I_{B_i}$, then

$$\int_{T^{-1}A} f(T(\omega))\, d\mu(\omega) = \sum_{i=1}^{n} x_i \int_{T^{-1}A} I_{B_i}(T(\omega))\, d\mu(\omega) \qquad \text{by 1.6.3}$$

$$= \sum_{i=1}^{n} x_i \int_A I_{B_i}(\omega)\, d\mu_0(\omega)$$

by what we have proved for indicators

$$= \int_A f(\omega)\, d\mu_0(\omega) \qquad \text{by 1.6.3.}$$

If f is a nonnegative Borel measurable function, let $f_1, f_2, \ldots$ be nonnegative simple functions increasing to f. Then $\int_{T^{-1}A} f_n(T(\omega))\, d\mu(\omega) = \int_A f_n(\omega)\, d\mu_0(\omega)$ by what we have proved for simple functions, and the monotone convergence theorem yields the desired result for f.

Finally, if $f = f^+ - f^-$ is an arbitrary Borel measurable function, we have proved that the result holds for f^+ and f^-. If, say, $\int_A f^+(\omega)\, d\mu_0(\omega) < \infty$, then $\int_{T^{-1}A} f^+(T(\omega))\, d\mu(\omega) < \infty$, and it follows that if one of the integrals exists, so does the other, and the two integrals are equal. ∎

If one is having difficulty proving a theorem about measurable functions or integration, it is often helpful to start with indicators and work upward. In fact it is possible to suspect that almost anything can be proved this way, but of course there are exceptions. For example, you will run into trouble trying to prove the proposition "All functions are indicators."

We shall adopt the following terminology: If μ is Lebesgue measure and A is a interval $[a, b]$, $\int_A f\, d\mu$, if it exists, will often be denoted by $\int_b^a f(x)\, dx$ (or $\int_{a_1}^{b_1} \cdots \int_{a_n}^{b_n} f(x_1, \ldots, x_n)\, dx_1 \cdots dx_n$ if we are integrating functions on R^n). The endpoints may be deleted from the interval without changing the integral, since the Lebesgue measure of a single point is 0. If f is integrable with respect to μ, then we say that f is *Lebesgue integrable*. A different notation, such as $r_{ab}(f)$, will be used for the Riemann integral of f on $[a, b]$.

Problems

The first three problems give conditions under which some of the most commonly occurring operations in real analysis may be performed: taking a limit under the integral sign, integrating an infinite series term by term, and differentiating under the integral sign.

1. Let $f = f(x, y)$ be a real-valued function of two real variables, defined for $a < y < b$, $c < x < d$. Assume that for each x, $f(x, \cdot)$ is a Borel measurable function of y, and that there is a Borel measurable g: $(a, b) \to R$ such that $|f(x, y)| \le g(y)$ for all x, y, and $\int_a^b g(y)\, dy < \infty$. If $x_0 \in (c, d)$ and $\lim_{x \to x_0} f(x, y)$ exists for all $y \in (a, b)$, show that

$$\lim_{x \to x_0} \int_a^b f(x, y)\, dy = \int_a^b \left[\lim_{x \to x_0} f(x, y) \right] dy.$$

2. Let $f_1, f_2, \ldots$ be Borel measurable functions on $(\Omega, \mathscr{F}, \mu)$. If

$$\sum_{n=1}^{\infty} \int_{\Omega} |f_n|\, d\mu < \infty,$$

show that $\sum_{n=1}^{\infty} f_n$ converges a.e. $[\mu]$ to a finite-valued function, and $\int_{\Omega} (\sum_{n=1}^{\infty} f_n)\, d\mu = \sum_{n=1}^{\infty} \int_{\Omega} f_n\, d\mu$.

3. Let $f = f(x, y)$ be a real-valued function of two real variables, defined for $a < y < b$, $c < x < d$, such that f is a Borel measurable function of y for each fixed x. Assume that for each x, $f(x, \cdot)$ is integrable over (a, b) (with respect to Lebesgue measure). Suppose that the partial derivative $f_1(x, y)$ of f with respect to x exists for all (x, y), and suppose there is a Borel measurable h: $(a, b) \to R$ such that $|f_1(x, y)| \le h(y)$ for all x, y, where $\int_a^b h(y)\, dy < \infty$.

   Show that $d[\int_a^b f(x, y)\, dy]/dx$ exists for all $x \in (c, d)$, and equals $\int_a^b f_1(x, y)\, dy$. (It must be verified that $f_1(x, \cdot)$ is Borel measurable for each x.)

4. If μ is a measure on $(\Omega, \mathscr{F})$ and $A_1, A_2, \ldots$ is a sequence of sets in $\mathscr{F}$, use Fatou's lemma to show that

$$\mu\left(\liminf_n A_n\right) \le \liminf_{n \to \infty} \mu(A_n).$$

   If μ is finite, show that

$$\mu\left(\limsup_n A_n\right) \ge \limsup_{n \to \infty} \mu(A_n).$$

   Thus if μ is finite and $A = \lim_n A_n$, then $\mu(A) = \lim_{n \to \infty} \mu(A_n)$. (For another proof of this, see Section 1.2, Problem 10.)

5. Give an example of a sequence of Lebesgue integrable functions f_n converging everywhere to a Lebesgue integrable function f, such that

$$\lim_{n \to \infty} \int_{-\infty}^{\infty} f_n(x)\, dx < \int_{-\infty}^{\infty} f(x)\, dx.$$

   Thus the hypotheses of the dominated convergence theorem and Fatou's lemma cannot be dropped.

6. (a) Show that $\int_1^\infty e^{-t} \ln t \, dt = \lim_{n\to\infty} \int_1^n [1 - (t/n)]^n \ln t \, dt$.
   (b) Show that $\int_0^1 e^{-t} \ln t \, dt = \lim_{n\to\infty} \int_0^1 [1 - (t/n)]^n \ln t \, dt$.
7. If $(\Omega, \mathscr{F}, \mu)$ is the completion of $(\Omega, \mathscr{F}_0, \mu)$ and f is a Borel measurable function on $(\Omega, \mathscr{F})$, show that there is a Borel measurable function g on $(\Omega, \mathscr{F}_0)$ such that $f = g$, except on a subset of a set in $\mathscr{F}_0$ of measure 0. (Start with indicators.)
8. If f is a Borel measurable function from R to R and $a \in R$, show that

$$\int_{-\infty}^{\infty} f(x) \, dx = \int_{-\infty}^{\infty} f(x - a) \, dx$$

in the sense that if one integral exists, so does the other, and the two are equal. (Start with indicators.)

1.7 Comparison of Lebesgue and Riemann Integrals

In this section we show that integration with respect to Lebesgue measure is more general than Riemann integration, and we obtain a precise criterion for Riemann integrability.

Let $[a, b]$ be a bounded closed interval of reals, and let f be a bounded real-valued function on $[a, b]$, fixed throughout the discussion. If P: $a = x_0 < x_1 < \cdots < x_n = b$ is a partition of $[a, b]$, we may construct the upper and lower sums of f relative to P as follows.

Let

$$M_i = \sup\{f(y) : x_{i-1} < y \le x_i\}, \qquad i = 1, \ldots, n,$$

$$m_i = \inf\{f(y) : x_{i-1} < y \le x_i\}, \qquad i = 1, \ldots, n,$$

and define step functions α and β, called the *upper* and *lower* functions corresponding to P, by

$$\alpha(x) = M_i \quad \text{if} \quad x_{i-1} < x \le x_i, \quad i = 1, \ldots, n,$$

$$\beta(x) = m_i \quad \text{if} \quad x_{i-1} < x < x_i, \quad i = 1, \ldots, n$$

[$\alpha(a)$ and $\beta(a)$ may be chosen arbitrarily]. The upper and lower sums are given by

$$\mathrm{U}(P) = \sum_{i=1}^{n} M_i(x_i - x_{i-1}),$$

$$\mathrm{L}(P) = \sum_{i=1}^{n} m_i(x_i - x_{i-1}).$$

Now we take as a measure space $\Omega = [a, b]$, $\mathscr{F} = \overline{\mathscr{B}}[a, b]$, the *Lebesgue measurable* subsets of $[a, b]$, $\mu =$ Lebesgue measure. Since α and β are simple functions, we have

$$\mathrm{U}(P) = \int_a^b \alpha \, d\mu, \qquad \mathrm{L}(P) = \int_a^b \beta \, d\mu.$$

Now let $P_1, P_2, \ldots$ be a sequence of partitions of $[a, b]$ such that P_{k+1} is a refinement of P_k for each k, and such that $|P_k|$ (the length of the largest subinterval of P_k) approaches 0 as $k \to \infty$. If α_k and β_k are the upper and lower functions corresponding to P_k, then

$$\alpha_1 \geq \alpha_2 \geq \cdots \geq f \geq \cdots \geq \beta_2 \geq \beta_1.$$

Thus α_k and β_k approach limit functions α and β. If $|f|$ is bounded by M, then all $|\alpha_k|$ and $|\beta_k|$ are bounded by M as well, and the function that is constant at M is integrable on $[a, b]$ with respect to μ, since

$$\mu[a, b] = b - a < \infty.$$

By the dominated convergence theorem,

$$\lim_{k \to \infty} \mathrm{U}(P_k) = \lim_{k \to \infty} \int_a^b \alpha_k \, d\mu = \int_a^b \alpha \, d\mu,$$

and

$$\lim_{k \to \infty} \mathrm{L}(P_k) = \lim_{k \to \infty} \int_a^b \beta_k \, d\mu = \int_a^b \beta \, d\mu.$$

We shall need one other fact, namely that if x is not an endpoint of any of the subintervals of the P_k,

$$f \text{ is continuous at } x \quad \text{iff} \quad \alpha(x) = f(x) = \beta(x).$$

This follows by a standard ε–δ argument.

If $\lim_{k \to \infty} \mathrm{U}(P_k) = \lim_{k \to \infty} \mathrm{L}(P_k) =$ a finite number r, independent of the particular sequence of partitions, f is said to be *Riemann integrable* on $[a, b]$, and $r = r_{ab}(f)$ is said to be the (value of the) *Riemann integral* of f on $[a, b]$. The above argument shows that f is Riemann integrable iff

$$\int_a^b \alpha \, d\mu = \int_a^b \beta \, d\mu = r,$$

independent of the particular sequence of partitions. If f is Riemann integrable,

$$r_{ab}(f) = \int_a^b \alpha \, d\mu = \int_a^b \beta \, d\mu.$$

We are now ready for the main results.

1.7.1 ***Theorem.*** Let f be a bounded real-valued function on $[a, b]$.

(a) The function f is Riemann integrable on $[a, b]$ iff f is continuous almost everywhere on $[a, b]$ (with respect to Lebesgue measure).

(b) If f is Riemann integrable on $[a, b]$, then f is integrable with respect to Lebesgue measure on $[a, b]$, and the two integrals are equal.

PROOF. (a) If f is Riemann integrable,

$$r_{ab}(f) = \int_a^b \alpha \, d\mu = \int_a^b \beta \, d\mu.$$

Since $\beta \le f \le \alpha$, 1.6.6(b) applied to $\alpha - \beta$ yields $\alpha = f = \beta$ a.e.; hence f is continuous a.e. Conversely, assume f continuous a.e.; then $\alpha = f = \beta$ a.e. Now α and β are limits of simple functions, and hence are Borel measurable. Thus f differs from a measurable function on a subset of a set of measure 0, and therefore f is measurable because of the completeness of the measure space. (See Section 1.5, Problem 4.) Since f is bounded, it is integrable with respect to μ, and since $\alpha = f = \beta$ a.e., we have

$$\int_a^b \alpha \, d\mu = \int_a^b \beta \, d\mu = \int_a^b f \, d\mu, \tag{1}$$

independent of the particular sequence of partitions. Therefore f is Riemann integrable.

(b) If f is Riemann integrable, then f is continuous a.e. by part (a). But then Eq. (1) yields $r_{ab}(f) = \int_a^b f \, d\mu$, as desired. ∎

Theorem 1.7.1 holds equally well in n dimensions, with $[a, b]$ replaced by a closed bounded interval of R^n; the proof is essentially the same.

A somewhat more complicated situation arises with *improper integrals*; here the interval of integration is infinite or the function f is unbounded. Some results are given in Problem 3.

We have seen that convenient conditions exist that allow the interchange of limit operations on Lebesgue integrable functions. (For example, see Problems 1–3 of Section 1.6.) The corresponding results for Riemann integrable functions are more complicated, basically because the limit of a sequence of Riemann integrable functions need not be Riemann integrable, even if the entire sequence is uniformly bounded (see Problem 4). Thus Riemann integrability of the limit function must be added as a hypothesis, and this is a serious limitation on the scope of the results.

Problems

1. The function defined on $[0, 1]$ by $f(x) = 1$ if x is irrational, and $f(x) = 0$ if x is rational, is the standard example of a function that is Lebesgue integrable (it is 1 a.e.) but not Riemann integrable. But what is wrong with the following reasoning:

   If we consider the behavior of f on the irrationals, f assumes the constant value 1 and is therefore continuous. Since the rationals have Lebesgue measure 0, f is therefore continuous almost everywhere and hence is Riemann integrable.
2. Let f be a bounded real-valued function on the bounded closed interval $[a, b]$. Let F be an increasing right-continuous function on $[a, b]$ with corresponding Lebesgue–Stieltjes measure μ (defined on the Borel subsets of $[a, b]$).

   Define M_i, m_i, α, and β as in Section 1.7, and take

$$\mathrm{U}(P) = \sum_{i=1}^{n} M_i(F(x_i) - F(x_{i-1})) = \int_a^b \alpha \, d\mu,$$

$$\mathrm{L}(P) = \sum_{i=1}^{n} m_i(F(x_i) - F(x_{i-1})) = \int_a^b \beta \, d\mu,$$

where $\int_a^b$ indicates that the integration is over $(a, b]$. If $\{P_k\}$ is a sequence of partitions with $|P_k| \to 0$ and P_{k+1} refining P_k, with α_k and β_k the upper and lower functions corresponding to P_k,

$$\lim_{k\to\infty} \mathrm{U}(P_k) = \int_a^b \alpha \, d\mu,$$

$$\lim_{k\to\infty} \mathrm{L}(P_k) = \int_a^b \beta \, d\mu,$$

where $\alpha = \lim_{k\to\infty} \alpha_k$, $\beta = \lim_{k\to\infty} \beta_k$. If $\mathrm{U}(P_k)$ and $\mathrm{L}(P_k)$ approach the same limit $r_{ab}(f; F)$ (independent of the particular sequence of partitions), this number is called the *Riemann–Stieltjes integral* of f with respect to F on $[a, b]$, and f is said to be *Riemann–Stieltjes integrable* with respect to F on $[a, b]$.

(a) Show that f is Riemann–Stieltjes integrable iff f is continuous a.e. $[\mu]$ on $[a, b]$.

(b) Show that if f is Riemann–Stieltjes integrable, then f is integrable with respect to the completion of the measure μ, and the two integrals are equal.

3. If $f: R \to R$, the improper Riemann integral of f may be defined as

$$r(f) = \lim_{\substack{a \to -\infty \\ b \to \infty}} r_{ab}(f)$$

   if the limit exists and is finite.
   (a) Show that if f has an improper Riemann integral, it is continuous a.e. [Lebesgue measure] on R, but not conversely.
   (b) If f is nonnegative and has an improper Riemann integral, show that f is integrable with respect to the completion of Lebesgue measure, and the two integrals are equal. Give a counterexample to this result if the nonnegativity hypothesis is dropped.
4. Give an example of a sequence of functions f_n on $[a, b]$ such that each f_n is Riemann integrable, $|f_n| \leq 1$ for all n, $f_n \to f$ everywhere, but f is not Riemann integrable.

Note: References on measure and integration will be given at the end of Chapter 2.

2

Further Results in Measure and Integration Theory

2.1 Introduction

This chapter consists of a variety of applications of the basic integration theory developed in Chapter 1. Perhaps the most important result is the Radon–Nikodym theorem, which is fundamental in modern probability theory and other parts of analysis. It will be instructive to consider a special case of this result before proceeding to the general theory. Suppose that F is a distribution function on R, and assume that F has a jump of magnitude a_k at the point x_k, $k = 1, 2, \ldots$. Let us subtract out the discontinuities of F; specifically let μ_1 be a measure concentrated on $\{x_1, x_2, \ldots\}$, with $\mu_1\{x_k\} = a_k$ for all k, and let F_1 be a distribution function corresponding to μ_1. Then $G = F - F_1$ is a continuous distribution function, so that the corresponding Lebesgue–Stieltjes measure λ satisfies $\lambda\{x\} = 0$ for all x. Now in any "practical" case, we can write $G(x) = \int_{-\infty}^{x} f(t)\, dt$, $x \in R$, for some nonnegative Borel measurable function f (the way to find f is to differentiate G). It follows that $\lambda(B) = \int_B f(x)\, dx$ for all $B \in \mathscr{B}(R)$. To see this, observe that if $\lambda'(B) = \int_B f(x)\, dx$, then λ' is a measure on $\mathscr{B}(R)$ and $\lambda'(a, b] = G(b) - G(a)$; thus λ' is the Lebesgue–Stieltjes measure determined by G; in other words, $\lambda' = \lambda$.

It is natural to conjecture that if λ is a measure on $\mathscr{B}(R)$ and $\lambda\{x\} = 0$ for all x, then we can write $\lambda(B) = \int_B f(x)\, dx$, $B \in \mathscr{B}(R)$, for some nonnegative

Borel measurable f. However, as found by Lebesgue, the conjecture is false unless the hypothesis is strengthened. Not only must we assume that λ assigns measure 0 to singletons, but in fact we must assume that λ is *absolutely continuous* with respect to Lebesgue measure μ, that is, if $\mu(B) = 0$, then $\lambda(B) = 0$. In general, λ may be represented as the sum of two measures λ_1 and λ_2, where λ_1 is absolutely continuous with respect to μ and λ_2 is *singular* with respect to μ, which means that λ_2 is concentrated on a set of Lebesgue measure 0. A simple example of a measure singular with respect to μ is one that is concentrated on a countable set; however, as we shall see, more complicated examples exist.

The first step in the development of the general Radon–Nikodym theorem is the Jordan–Hahn decomposition, which represents a countably additive set function as the difference of two measures.

Let $(\Omega, \mathscr{F}, \mu)$ be a measure space, h a Borel measurable function such that $\int_\Omega h\, d\mu$ exists. If $\lambda(A) = \int_A h\, d\mu$, $A \in \mathscr{F}$, then by 1.6.1, λ is a countably additive set function on $\mathscr{F}$. We call λ the *indefinite integral* of h (with respect to μ). If μ is Lebesgue measure and $A = [a, x]$, then $\lambda(A) = \int_a^x h(y)\, dy$, the familiar indefinite integral of calculus. As we noted after the proof of 1.6.1, λ is the difference of two measures, at least one of which is finite. We are going to show that any countably additive set function can be represented in this way. First, a preliminary result.

2.1.1 ***Theorem.*** Let λ be a countably additive extended real-valued set function on the σ-field $\mathscr{F}$ of subsets of Ω. Then λ assumes a maximum and a minimum value, that is, there are sets $C, D \in \mathscr{F}$ such that

$$\lambda(C) = \sup\{\lambda(A) : A \in \mathscr{F}\} \qquad \text{and} \qquad \lambda(D) = \inf\{\lambda(A) : A \in \mathscr{F}\}.$$

Before giving the proof, let us look at some special cases. If λ is a measure, the result is trivial: take $C = \Omega$, $D = \varnothing$. If λ is the indefinite integral of h with respect to μ, we may write

$$\lambda(A) = \int_A h\, d\mu = \int_{A \cap \{\omega:\, h(\omega) \geq 0\}} h\, d\mu + \int_{A \cap \{\omega:\, h(\omega) < 0\}} h\, d\mu.$$

Thus

$$\int_{\{\omega:\, h(\omega) < 0\}} h\, d\mu \leq \lambda(A) \leq \int_{\{\omega:\, h(\omega) \geq 0\}} h\, d\mu.$$

Therefore we may take $C = \{\omega : h(\omega) \geq 0\}$, $D = \{\omega : h(\omega) < 0\}$.

PROOF. First consider the sup. We may assume that $\lambda < \infty$, for if $\lambda(A_0) = \infty$ we take $C = A_0$. Let $A_n \in \mathscr{F}$ with $\lambda(A_n) \to \sup \lambda$, and let $A = \bigcup_{n=1}^\infty A_n \in \mathscr{F}$.

For each n, we may partition A into 2^n disjoint subsets A_{nm}, where each A_{nm} is of the form $A_1^* \cap A_2^* \cap \cdots \cap A_n^*$, with A_j^* either A_j or $A - A_j$. For example, if $n = 3$, we have (with intersections written as products)

$$A = A_1 A_2 A_3 \cup A_1 A_2 A_3' \cup A_1 A_2' A_3 \cup A_1 A_2' A_3' \cup A_1' A_2 A_3$$
$$\cup A_1' A_2 A_3' \cup A_1' A_2' A_3 \cup A_1' A_2' A_3', \qquad \text{where} \qquad A_j' = A - A_j.$$

Let $B_n = \bigcup_m \{A_{nm} : \lambda(A_{nm}) \geq 0\}$; set $B_n = \varnothing$ if $\lambda(A_{nm}) < 0$ for all m. Now each A_n is a finite union of some of the A_{nm}, hence $\lambda(A_n) \leq \lambda(B_n)$ by definition of B_n. Also, if $n' > n$, each $A_{n'm'}$ is either a subset of a given A_{nm} or disjoint from it (for example, $A_1 A_2' A_3' \subset A_1 A_2'$, and $A_1 A_2' A_3'$ is disjoint from $A_1 A_2$, $A_1' A_2$, and $A_1' A_2'$). Thus $\bigcup_{k=n}^r B_k$ can be expressed as a union of B_n and sets E disjoint from B_n such that $\lambda(E) \geq 0$. [Note that, for example, if $\lambda(A_1 A_2') = \lambda(A_1 A_2' A_3) + \lambda(A_1 A_2' A_3') \geq 0$, it may happen that $\lambda(A_1 A_2' A_3) < 0$, $\lambda(A_1 A_2' A_3') > 0$, so the sequence $\{B_n\}$ need not be monotone.] Consequently,

$$\lambda(A_n) \leq \lambda(B_n) \leq \lambda\left(\bigcup_{k=n}^{r} B_k\right) \to \lambda\left(\bigcup_{k=n}^{\infty} B_k\right) \qquad \text{as} \quad r \to \infty \qquad \text{by 1.2.7(a).}$$

Let $C = \lim_n \sup B_n = \bigcap_{n=1}^\infty \bigcup_{k=n}^\infty B_k$. Now $\bigcup_{k=n}^\infty B_k \downarrow C$, and $0 \leq \lambda(\bigcup_{k=n}^\infty B_k) < \infty$ for all n. By 1.2.7(b), $\lambda(\bigcup_{k=n}^\infty B_k) \to \lambda(C)$ as $n \to \infty$. Thus

$$\sup \lambda = \lim_{n\to\infty} \lambda(A_n) \leq \lim_{n\to\infty} \lambda\left(\bigcup_{k=n}^{\infty} B_k\right) = \lambda(C) \leq \sup \lambda;$$

hence $\lambda(C) = \sup \lambda$. The above argument applied to $-\lambda$ yields $D \in \mathscr{F}$ with $\lambda(D) = \inf \lambda$. ∎

We now prove the main theorem of this section.

2.1.2 ***Jordan–Hahn Decomposition Theorem.*** Let λ be a countably additive extended real-valued set function on the σ-field $\mathscr{F}$. Define

$$\lambda^+(A) = \sup\{\lambda(B) : B \in \mathscr{F}, \quad B \subset A\},$$
$$\lambda^-(A) = -\inf\{\lambda(B) : B \in \mathscr{F}, \quad B \subset A\}.$$

Then λ^+ and λ^- are measures on $\mathscr{F}$ and $\lambda = \lambda^+ - \lambda^-$.

PROOF. We may assume λ never takes on the value $-\infty$. For if $-\infty$ belongs to the range of λ, $+\infty$ does not, by definition of a countably additive set function. Thus $-\lambda$ never takes on the value $-\infty$. But $(-\lambda)^+ = \lambda^-$ and $(-\lambda)^- = \lambda^+$, so that if the theorem is proved for $-\lambda$ it holds for λ as well.

Let D be a set on which λ attains its minimum, as in 2.1.1. Since $\lambda(\varnothing) = 0$, we have $-\infty < \lambda(D) \leq 0$. We claim that

$$\lambda(A \cap D) \leq 0, \qquad \lambda(A \cap D^c) \geq 0 \qquad \text{for all} \quad A \in \mathscr{F}. \tag{1}$$

For if $\lambda(A \cap D) > 0$, then $\lambda(D) = \lambda(A \cap D) + \lambda(A^c \cap D)$. Since $\lambda(D)$ is finite, so are $\lambda(A \cap D)$ and $\lambda(A^c \cap D)$; hence $\lambda(A^c \cap D) = \lambda(D) - \lambda(A \cap D) < \lambda(D)$, contradicting the fact that $\lambda(D) = \inf \lambda$. If $\lambda(A \cap D^c) < 0$, then $\lambda(D \cup (A \cap D^c)) = \lambda(D) + \lambda(A \cap D^c) < \lambda(D)$, a contradiction.

We now show that

$$\lambda^+(A) = \lambda(A \cap D^c), \qquad \lambda^-(A) = -\lambda(A \cap D). \tag{2}$$

The theorem will follow from this. We have, for $B \in \mathscr{F}$, $B \subset A$,

$$\begin{aligned} \lambda(B) &= \lambda(B \cap D) + \lambda(B \cap D^c) \\ &\leq \lambda(B \cap D^c) \qquad \text{by (1)} \\ &\leq \lambda(B \cap D^c) + \lambda((A - B) \cap D^c) \\ &= \lambda(A \cap D^c). \end{aligned}$$

Thus $\lambda^+(A) \leq \lambda(A \cap D^c)$. But $\lambda(A \cap D^c) \leq \lambda^+(A)$ by definition of λ^+, proving the first assertion. Similarly,

$$\begin{aligned} \lambda(B) &= \lambda(B \cap D) + \lambda(B \cap D^c) \\ &\geq \lambda(B \cap D) \\ &\geq \lambda(B \cap D) + \lambda((A - B) \cap D) \\ &= \lambda(A \cap D). \end{aligned}$$

Hence $-\lambda^-(A) \geq \lambda(A \cap D)$. But $\lambda(A \cap D) \geq -\lambda^-(A)$ by definition of λ^-, completing the proof. ▌

2.1.3 ***Corollaries.*** Let λ be a countably additive extended real-valued set function on the σ-field $\mathscr{F}$.

(a) The set function λ is the difference of two measures, at least one of which is finite.

(b) If λ is finite ($\lambda(A)$ is never $\pm\infty$ for any $A \in \mathscr{F}$), then λ is bounded.

(c) There is a set $D \in \mathscr{F}$ such that $\lambda(A \cap D) \leq 0$ and $\lambda(A \cap D^c) \geq 0$ for all $A \in \mathscr{F}$.

(d) If D is any set in $\mathscr{F}$ such that $\lambda(A \cap D) \leq 0$ and $\lambda(A \cap D^c) \geq 0$ for all $A \in \mathscr{F}$, then $\lambda^+(A) = \lambda(A \cap D^c)$ and $\lambda^-(A) = -\lambda(A \cap D)$ for all $A \in \mathscr{F}$.

(e) If E is another set in $\mathscr{F}$ such that $\lambda(A \cap E) \leq 0$ and $\lambda(A \cap E^c) \geq 0$ for all $A \in \mathscr{F}$, then $|\lambda|(D \,\triangle\, E) = 0$, where $|\lambda| = \lambda^+ + \lambda^-$.

PROOF. (a) If $\lambda > -\infty$, then in 2.1.2, λ^- is finite; if $\lambda < +\infty$, λ^+ is finite [see Eq. (2)].

(b) In 2.1.2, λ^+ and λ^- are both finite; hence for any $A \in \mathscr{F}$, $|\lambda(A)| \leq \lambda^+(\Omega) + \lambda^-(\Omega) < \infty$.

(c) This follows from (1) of 2.1.2.

(d) Repeat the part of the proof of 2.1.2 after Eq. (2).

(e) By (d), $\lambda^+(A) = \lambda(A \cap D^c)$, $A \in \mathscr{F}$; take $A = D \cap E^c$ to obtain $\lambda^+(D \cap E^c) = 0$. Also by (d), $\lambda^+(A) = \lambda(A \cap E^c)$, $A \in \mathscr{F}$; take $A = D^c \cap E$ to obtain $\lambda^+(D^c \cap E) = 0$. Therefore $\lambda^+(D \triangle E) = 0$. The same argument using $\lambda^-(A) = -\lambda(A \cap D) = -\lambda(A \cap E)$ shows that $\lambda^-(D \triangle E) = 0$. The result follows. ▮

Corollary 2.1.3(d) is often useful in finding the Jordan–Hahn decomposition of a particular set function (see Problems 1 and 2).

2.1.4 ***Terminology.*** We call λ^+ the *upper variation* or *positive part* of λ, λ^- the *lower variation* or *negative part*, $|\lambda| = \lambda^+ + \lambda^-$ the *total variation*. Since $\lambda = \lambda^+ - \lambda^-$, it follows that $|\lambda(A)| \leq |\lambda|(A)$, $A \in \mathscr{F}$. For a sharper result, see Problem 4.

Note that if $A \in \mathscr{F}$, then $|\lambda|(A) = 0$ iff $\lambda(B) = 0$ for all $B \in \mathscr{F}$, $B \subset A$.

The phrase *signed measure* is sometimes used for the difference of two measures. By 2.1.3(a), this is synonomous (on a σ-field) with *countably additive set function.*

Problems

1. Let P be an arbitrary probability measure on $\mathscr{B}(R)$, and let Q be point mass at 0, that is, $Q(B) = 1$ if $0 \in B$, $Q(B) = 0$ if $0 \notin B$. Find the Jordan–Hahn decomposition of the signed measure $\lambda = P - Q$.
2. Let $\lambda(A) = \int_A f\,d\mu$, A in the σ-field $\mathscr{F}$, where $\int_\Omega f\,d\mu$ exists; thus λ is a signed measure on $\mathscr{F}$. Show that

$$\lambda^+(A) = \int_A f^+\,d\mu, \qquad \lambda^-(A) = \int_A f^-\,d\mu, \qquad |\lambda|(A) = \int_A |f|\,d\mu.$$

3. If a signed measure λ on the σ-field $\mathscr{F}$ is the difference of two measures λ_1 and λ_2, show that $\lambda_1 \geq \lambda^+$, $\lambda_2 \geq \lambda^-$.
4. Let λ be a signed measure on the σ-field $\mathscr{F}$. Show that $|\lambda|(A) = \sup\{\sum_{i=1}^n |\lambda(E_i)| : E_1, E_2, \ldots, E_n$ disjoint measurable subsets of A, $n = 1, 2, \ldots\}$. Consequently, if λ_1 and λ_2 are signed measures on $\mathscr{F}$, then $|\lambda_1 + \lambda_2| \leq |\lambda_1| + |\lambda_2|$.

2.2 Radon–Nikodym Theorem and Related Results

If $(\Omega, \mathscr{F}, \mu)$ is a measure space, then $\lambda(A) = \int_A g\, d\mu$, $A \in \mathscr{F}$, defines a signed measure if $\int_\Omega g\, d\mu$ exists. Furthermore, if $A \in \mathscr{F}$ and $\mu(A) = 0$, then $\lambda(A) = 0$. For $gI_A = 0$ on A^c, so that $gI_A = 0$ a.e. $[\mu]$, and the result follows from 1.6.5(a).

If μ is a measure on the σ-field $\mathscr{F}$, and λ is a signed measure on $\mathscr{F}$, we say that λ is *absolutely continuous* with respect to μ (notation $\lambda \ll \mu$) iff $\mu(A) = 0$ implies $\lambda(A) = 0$ $(A \in \mathscr{F})$. Thus if λ is an indefinite integral with respect to μ, then $\lambda \ll \mu$. The Radon–Nikodym theorem is an assertion in the converse direction; if $\lambda \ll \mu$ (and μ is σ-finite on $\mathscr{F}$), then λ is an indefinite integral with respect to μ. As we shall see, large areas of analysis are based on this theorem.

2.2.1 ***Radon–Nikodym Theorem.*** Let μ be a σ-finite measure and λ a signed measure on the σ-field $\mathscr{F}$ of subsets of Ω. Assume that λ is absolutely continuous with respect to μ. Then there is a Borel measurable function $g : \Omega \to \overline{R}$ such that

$$\lambda(A) = \int_A g\, d\mu \qquad \text{for all} \quad A \in \mathscr{F}.$$

If h is another such function, then $g = h$ a.e. $[\mu]$.

PROOF. The uniqueness statement follows from 1.6.11. We break the existence proof into several parts.

(a) Assume λ and μ are finite measures.

Let $\mathscr{S}$ be the set of all nonnegative μ-integrable functions f such that $\int_A f\, d\mu \le \lambda(A)$ for all $A \in \mathscr{F}$. Partially order $\mathscr{S}$ by calling $f \le g$ iff $f \le g$ a.e.$[\mu]$. Let $\mathscr{C}$ be a chain (totally ordered subset) of $\mathscr{S}$, and let $s = \sup\{\int_\Omega f\, d\mu : f \in \mathscr{C}\}$.

We can find functions $f_n \in \mathscr{C}$ with $\int_\Omega f_n\, d\mu \uparrow s$, and since $\mathscr{C}$ is totally ordered, $f_n \le f_{n+1}$ a.e. for all n. [If $f_{n+1} \le f_n$ a.e., then $\int_\Omega f_{n+1}\, d\mu \le \int_\Omega f_n\, d\mu \le \int_\Omega f_{n+1}\, d\mu$; hence by 1.6.6(b), $f_n = f_{n+1}$ a.e.]

Thus f_n increases to a limit f a.e., and by the monotone convergence theorem, $\int_\Omega f\, d\mu = s$.

We show that f is an upper bound of $\mathscr{C}$. Let h be an element of $\mathscr{C}$. If $h \le f_n$ a.e. for some n, then $h \le f$ a.e. If $h \ge f_n$ a.e. for all n, then $h \ge f$ a.e.; hence $\int_\Omega h\, d\mu = \int_\Omega f\, d\mu = s$. But then $h = f$ a.e. and therefore f is an upper bound of $\mathscr{C}$. (We have $f \in \mathscr{S}$ by definition of $\mathscr{S}$.)

By Zorn's lemma, there is a maximal element $g \in \mathscr{S}$. We must show that $\lambda(A) = \int_A g\, d\mu$, $A \in \mathscr{F}$: Let $\lambda_1(A) = \lambda(A) - \int_A g\, d\mu$, $A \in \mathscr{F}$. Then λ_1 is a measure, $\lambda_1 \ll \mu$, and $\lambda_1(\Omega) < \infty$. If λ_1 is not identically 0, then $\lambda_1(\Omega) > 0$, hence

$$\mu(\Omega) - k\lambda_1(\Omega) < 0 \qquad \text{for some} \quad k > 0. \tag{1}$$

Apply 2.1.3(c) to the signed measure $\mu - k\lambda_1$ to obtain $D \in \mathscr{F}$ such that for all $A \in \mathscr{F}$,

$$\mu(A \cap D) - k\lambda_1(A \cap D) \le 0, \tag{2}$$

and

$$\mu(A \cap D^c) - k\lambda_1(A \cap D^c) \ge 0. \tag{3}$$

We claim that $\mu(D) > 0$. For if $\mu(D) = 0$, then $\lambda(D) = 0$ by absolute continuity, and therefore $\lambda_1(D) = 0$ by definition of λ_1. Take $A = \Omega$ in (3) to obtain

$$\begin{aligned} 0 &\le \mu(D^c) - k\lambda_1(D^c) \\ &= \mu(\Omega) - k\lambda_1(\Omega) \qquad \text{since} \quad \mu(D) = \lambda_1(D) = 0 \\ &< 0 \qquad \text{by (1)}, \end{aligned}$$

a contradiction. Define $h(\omega) = 1/k$, $\omega \in D$; $h(\omega) = 0$, $\omega \notin D$. If $A \in \mathscr{F}$,

$$\begin{aligned} \int_A h\,d\mu = \frac{1}{k}\,\mu(A \cap D) &\le \lambda_1(A \cap D) \qquad \text{by (2)} \\ &\le \lambda_1(A) = \lambda(A) - \int_A g\,d\mu. \end{aligned}$$

Thus $\int_A (h + g)\,d\mu \le \lambda(A)$. But $h + g > g$ on the set D, with $\mu(D) > 0$, contradicting the maximality of g. Thus $\lambda_1 \equiv 0$, and the result follows.

(b) Assume μ is a finite measure, λ *a* σ-finite measure.

Let Ω be the disjoint union of sets A_n with $\lambda(A_n) < \infty$, and let $\lambda_n(A) = \lambda(A \cap A_n)$, $A \in \mathscr{F}$, $n = 1, 2, \ldots$. By part (a) we find a nonnegative Borel measurable g_n with $\lambda_n(A) = \int_A g_n\,d\mu$, $A \in \mathscr{F}$. Thus $\lambda(A) = \int_A g\,d\mu$, where $g = \sum_n g_n$.

(c) Assume μ is a finite measure, λ an arbitrary measure.

Let $\mathscr{C}$ be the class of sets $C \in \mathscr{F}$ such that λ on C (that is, λ restricted to $\mathscr{F}_C = \{A \cap C : A \in \mathscr{F}\}$) is σ-finite; note that $\varnothing \in \mathscr{C}$, so $\mathscr{C}$ is not empty. Let $s = \sup\{\mu(A) : A \in \mathscr{C}\}$ and pick $C_n \in \mathscr{C}$ with $\mu(C_n) \to s$. If $C = \bigcup_{n=1}^{\infty} C_n$, then $C \in \mathscr{C}$ by definition of $\mathscr{C}$, and $s \ge \mu(C) \ge \mu(C_n) \to s$; hence $\mu(C) = s$.

By part (b), there is a nonnegative $g': C \to \bar{R}$, measurable relative to $\mathscr{F}_C$ and $\mathscr{B}(\bar{R})$, such that

$$\lambda(A \cap C) = \int_{A \cap C} g' \, d\mu \qquad \text{for all} \quad A \in \mathscr{F}.$$

Now consider an arbitrary set $A \in \mathscr{F}$.

Case 1: Let $\mu(A \cap C^c) > 0$. Then $\lambda(A \cap C^c) = \infty$, for if $\lambda(A \cap C^c) < \infty$, then $C \cup (A \cap C^c) \in \mathscr{C}$; hence

$$s \geq \mu(C \cup (A \cup C^c)) = \mu(C) + \mu(A \cap C^c) > \mu(C) = s,$$

a contradiction.

Case 2: Let $\mu(A \cap C^c) = 0$. Then $\lambda(A \cap C^c) = 0$ by absolute continuity.

Thus in either case, $\lambda(A \cap C^c) = \int_{A \cap C^c} \infty \, d\mu$. It follows that

$$\lambda(A) = \lambda(A \cap C) + \lambda(A \cap C^c) = \int_A g \, d\mu,$$

where $g = g'$ on C, $g = \infty$ on C^c.

(d) Assume μ is a σ-finite measure, λ an arbitrary measure.

Let Ω be the union of disjoint sets A_n with $\mu(A_n) < \infty$. By part (c), there is a nonnegative function $g_n : A_n \to \bar{R}$, measurable with respect to $\mathscr{F}_{A_n}$ and $\mathscr{B}(\bar{R})$, such that $\lambda(A \cap A_n) = \int_{A \cap A_n} g_n \, d\mu$, $A \in \mathscr{F}$. We may write this as $\lambda(A \cap A_n) = \int_A g_n \, d\mu$ where $g_n(\omega)$ is taken as 0 for $\omega \notin A_n$. Thus $\lambda(A) = \sum_n \lambda(A \cap A_n) = \sum_n \int_A g_n \, d\mu = \int_A g \, d\mu$, where $g = \sum_n g_n$.

(e) Assume μ is a σ-finite measure, λ an arbitrary signed measure.

Write $\lambda = \lambda^+ - \lambda^-$ where, say, λ^- is finite. By part (d), there are nonnegative Borel measurable functions g_1 and g_2 such that

$$\lambda^+(A) = \int_A g_1 \, d\mu, \qquad \lambda^-(A) = \int_A g_2 \, d\mu, \qquad A \in \mathscr{F}.$$

Since λ^- is finite, g_2 is integrable; hence by 1.6.3 and 1.6.6(a), $\lambda(A) = \int_A (g_1 - g_2) \, d\mu$. ∎

2.2.2 ***Corollaries.*** Under the hypothesis of 2.2.1,

(a) If λ is finite, then g is μ-integrable, hence finite a.e. $[\mu]$.

(b) If $|\lambda|$ is σ-finite, so that Ω can be expressed as a countable union of sets A_n such that $|\lambda|(A_n)$ is finite (equivalently $\lambda(A_n)$ is finite), then g is finite a.e. $[\mu]$.

(c) If λ is a measure, then $g \geq 0$ a.e. $[\mu]$.

PROOF. All results may be obtained by examining the proof of 2.2.1. Alternatively, we may proceed as follows:

(a) Observe that $\lambda(\Omega) = \int_\Omega g\, d\mu$, finite by hypothesis.

(b) By (a), g is finite a.e. $[\mu]$ on each A_n, hence finite a.e. $[\mu]$ on Ω.

(c) Let $A = \{\omega: g(\omega) < 0\}$; then $0 \le \lambda(A) = \int_A g\, d\mu \le 0$. Thus $-gI_A$ is a nonnegative function whose integral is 0, so that $gI_A = 0$ a.e. $[\mu]$ by 1.6.6(b). Since $gI_A < 0$ on A, we must have $\mu(A) = 0$. ∎

If $\lambda(A) = \int_A g\, d\mu$ for each $A \in \mathscr{F}$, g is called the *Radon–Nikodym derivative* or *density* of λ with respect to μ, written $d\lambda/d\mu$. If μ is Lebesgue measure, then g is often called simply the density of λ.

There are converse assertions to 2.2.2(a) and (c). Suppose that

$$\lambda(A) = \int_A g\, d\mu, \qquad A \in \mathscr{F},$$

where $\int_\Omega g\, d\mu$ is assumed to exist. If g is μ-integrable, then λ is finite; if $g \ge 0$ a.e. $[\mu]$, then $\lambda \ge 0$, so that λ is a measure. (Note that σ-finiteness of μ is not assumed.) However, the converse to 2.2.2.(b) is false; if g is finite a.e. $[\mu]$, $|\lambda|$ need not be σ-finite (see Problem 1).

We now consider a property that is in a sense opposite to absolute continuity.

2.2.3 ***Definitions.*** Let μ_1 and μ_2 be measures on the σ-field $\mathscr{F}$. We say that μ_1 is *singular* with respect to μ_2 (written $\mu_1 \perp \mu_2$) iff there is a set $A \in \mathscr{F}$ such that $\mu_1(A) = 0$ and $\mu_2(A^c) = 0$; note μ_1 is singular with respect to μ_2 iff μ_2 is singular with respect to μ_1, so we may say that μ_1 and μ_2 are *mutually singular*. If λ_1 and λ_2 are signed measures on $\mathscr{F}$, we say that λ_1 and λ_2 are mutually singular iff $|\lambda_1| \perp |\lambda_2|$.

If $\mu_1 \perp \mu_2$, with $\mu_1(A) = \mu_2(A^c) = 0$, then μ_2 only assigns positive measure to subsets of A. Thus μ_2 concentrates its total effect on a set of μ_1-measure 0; on the other hand, if $\mu_2 \ll \mu_1$, μ_2 can have no effect on sets of μ_1-measure 0.

If λ is a signed measure with positive part λ^+ and negative part λ^-, we have $\lambda^+ \perp \lambda^-$ by 2.1.3(c) and (d).

Before establishing some facts about absolute continuity and singularity, we need the following lemma. Although the proof is quite simple, the result is applied very often in analysis, especially in probability theory.

2.2.4 ***Borel–Cantelli Lemma.*** If $A_1, A_2, \ldots \in \mathscr{F}$ and $\sum_{n=1}^\infty \mu(A_n) < \infty$, then $\mu(\limsup_n A_n) = 0$.

PROOF. Recall that $\limsup_n A_n = \bigcap_{n=1}^{\infty} \bigcup_{k=n}^{\infty} A_k$; hence

$$\mu(\limsup_n A_n) \le \mu\left(\bigcup_{k=n}^{\infty} A_k\right) \qquad \text{for all} \quad n$$

$$\le \sum_{k=n}^{\infty} \mu(A_k) \to 0 \qquad \text{as} \quad n \to \infty. \blacksquare$$

2.2.5 ***Lemma.*** Let μ be a measure, and λ_1 and λ_2 signed measures, on the σ-field $\mathscr{F}$.

(a) If $\lambda_1 \perp \mu$ and $\lambda_2 \perp \mu$, then $\lambda_1 + \lambda_2 \perp \mu$.
(b) If $\lambda_1 \ll \mu$, then $|\lambda_1| \ll \mu$, and conversely.
(c) If $\lambda_1 \ll \mu$ and $\lambda_2 \perp \mu$, then $\lambda_1 \perp \lambda_2$.
(d) If $\lambda_1 \ll \mu$ and $\lambda_1 \perp \mu$, then $\lambda_1 \equiv 0$.
(e) If λ_1 is finite, then $\lambda_1 \ll \mu$ iff $\lim_{\mu(A)\to 0} \lambda_1(A) = 0$.

PROOF. (a) Let $\mu(A) = \mu(B) = 0$, $|\lambda_1|(A^c) = |\lambda_2|(B^c) = 0$. Then $\mu(A \cup B) = 0$ and $\lambda_1(C) = \lambda_2(C) = 0$ for every $C \in \mathscr{F}$ with $C \subset A^c \cap B^c$; hence $|\lambda_1 + \lambda_2|\,[(A \cup B)^c] = 0$.

(b) Let $\mu(A) = 0$. If $\lambda_1^+(A) > 0$, then (see 2.1.2) $\lambda_1(B) > 0$ for some $B \subset A$; since $\mu(B) = 0$, this is a contradiction. It follows that λ_1^+, and similarly λ_1^-, is absolutely continuous with respect to μ; hence $|\lambda_1| \ll \mu$. (This may also be proved using Section 2.1, Problem 4.) The converse is clear.

(c) Let $\mu(A) = 0$, $|\lambda_2|(A^c) = 0$. By (b), $|\lambda_1|(A) = 0$, so $|\lambda_1| \perp |\lambda_2|$.

(d) By (c), $\lambda_1 \perp \lambda_1$; hence for some $A \in \mathscr{F}$, $|\lambda_1|(A) = |\lambda_1|(A^c) = 0$. Thus $|\lambda_1|(\Omega) = 0$.

(e) If $\mu(A_n) \to 0$ implies $\lambda_1(A_n) \to 0$, and $\mu(A) = 0$, set $A_n \equiv A$ to conclude that $\lambda_1(A) = 0$, so $\lambda \ll \mu$.

Conversely, let $\lambda_1 \ll \mu$.

If $\lim_{\mu(A)\to 0} |\lambda_1|(A) \ne 0$ we can find, for some $\varepsilon > 0$, sets $A_n \in \mathscr{F}$ with $\mu(A_n) < 2^{-n}$ and $|\lambda_1|(A_n) \ge \varepsilon$ for all n. Let $A = \lim_n \sup A_n$; by 2.2.4, $\mu(A) = 0$. But $|\lambda_1|(\bigcup_{k=n}^{\infty} A_k) \ge |\lambda_1|(A_n) \ge \varepsilon$ for all n; hence by 1.2.7(b), $|\lambda_1|(A) \ge \varepsilon$, contradicting (b). Thus $\lim_{\mu(A)\to 0} |\lambda_1|(A) = 0$, and the result follows since $|\lambda_1(A)| \le |\lambda_1|(A)$. $\blacksquare$

If λ_1 is an indefinite integral with respect to μ (hence $\lambda_1 \ll \mu$), then 2.2.5(e) has an easier proof. If $\lambda_1(A) = \int_A f\, d\mu$, $A \in \mathscr{F}$, then

$$\int_A |f|\, d\mu = \int_{A \cap \{|f| \le n\}} |f|\, d\mu + \int_{A \cap \{|f| > n\}} |f|\, d\mu$$

$$\le n\mu(A) + \int_{\{|f| > n\}} |f|\, d\mu.$$

By 1.6.1 and 1.2.7(b), $\int_{\{|f|>n\}} |f| \, d\mu$ may be made less than $\varepsilon/2$ for large n, say $n \geq N$. Fix $n = N$ and take $\mu(A) < \varepsilon/2N$, so that $\int_A |f| \, d\mu < \varepsilon$.

If μ is a σ-finite measure and λ a signed measure on the σ-field $\mathscr{F}$, λ may be neither absolutely continuous nor singular with respect to μ. However, if $|\lambda|$ is σ-finite, the two concepts of absolute continuity and singularity are adequate to describe the relation between λ and μ, in the sense that λ can be written as the sum of two signed measures, one absolutely continuous and the other singular with respect to μ.

2.2.6 ***Lebesgue Decomposition Theorem.*** Let μ be a σ-finite measure on the σ-field $\mathscr{F}$, λ a σ-finite signed measure (that is, $|\lambda|$ is σ-finite). Then λ has a unique decomposition as $\lambda_1 + \lambda_2$, where λ_1 and λ_2 are signed measures such that $\lambda_1 \ll \mu$, $\lambda_2 \perp \mu$.

PROOF. First assume λ is a σ-finite measure. Let $m = \mu + \lambda$, also a σ-finite measure. Then μ and λ are each absolutely continuous with respect to m; hence by 2.2.1 and 2.2.2(c) there are nonnegative Borel measurable functions f and g such that $\mu(A) = \int_A f \, dm$, $\lambda(A) = \int_A g \, dm$, $A \in \mathscr{F}$.

Let $B = \{\omega : f(\omega) > 0\}$, $C = B^c = \{\omega : f(\omega) = 0\}$, and define, for each $A \in \mathscr{F}$,

$$\lambda_1(A) = \lambda(A \cap B), \qquad \lambda_2(A) = \lambda(A \cap C).$$

Thus $\lambda_1 + \lambda_2 = \lambda$. In fact, $\lambda_1 \ll \mu$ and $\lambda_2 \perp \mu$. To prove $\lambda_1 \ll \mu$, assume $\mu(A) = 0$. Then $\int_A f \, dm = 0$, hence $f = 0$ a.e. $[m]$ on A. But $f > 0$ on $A \cap B$; hence $m(A \cap B) = 0$, and consequently $\lambda(A \cap B) = 0$; in other words, $\lambda_1(A) = 0$. Thus $\lambda_1 \ll \mu$.

To prove $\lambda_2 \perp \mu$, observe that $\lambda_2(B) = 0$ and $\mu(B^c) = \mu(C) = \int_C 0 \, dm = 0$.

Now if λ is a σ-finite signed measure, the above argument applied to λ^+ and λ^- proves the existence of the desired decomposition.

To prove uniqueness, first assume λ finite. If $\lambda = \lambda_1 + \lambda_2 = \lambda_1' + \lambda_2'$, where $\lambda_1, \lambda_1' \ll \mu$, $\lambda_2, \lambda_2' \perp \mu$, then $\lambda_1 - \lambda_1' = \lambda_2' - \lambda_2$ is both absolutely continuous and singular with respect to μ; hence is identically 0 by 2.2.5(d). If λ is σ-finite and Ω is the disjoint union of sets A_n with $|\lambda|(A_n) < \infty$, apply the above argument to each A_n and put the results together to obtain uniqueness of λ_1 and λ_2. ∎

Problems

1. Give an example of a measure μ and a nonnegative finite-valued Borel measurable function g such that the measure λ defined by $\lambda(A) = \int_A g \, d\mu$ is not σ-finite.

2. If $\lambda(A) = \int_A g\, d\mu$, $A \in \mathscr{F}$, and g is μ-integrable, we know that λ is finite; in particular, $A = \{\omega : g(\omega) \neq 0\}$ has finite λ-measure. Show that A has σ-finite μ-measure, that is, it is a countable union of sets of finite μ-measure. Give an example to show that $\mu(A)$ need not be finite.
3. Give an example in which the conclusion of the Radon–Nikodym theorem fails; in other words, $\lambda \ll \mu$ but there is no Borel measurable g such that $\lambda(A) = \int_A g\, d\mu$ for all $A \in \mathscr{F}$. Of course μ cannot be σ-finite.
4. (*A chain rule*) Let $(\Omega, \mathscr{F}, \mu)$ be a measure space, and g a nonnegative Borel measurable function on Ω. Define a measure λ on $\mathscr{F}$ by

$$\lambda(A) = \int_A g\, d\mu, \qquad A \in \mathscr{F}.$$

Show that if f is a Borel measurable function on Ω,

$$\int_\Omega f\, d\lambda = \int_\Omega fg\, d\mu$$

in the sense that if one of the integrals exists, so does the other, and the two integrals are equal. (Intuitively, $d\lambda/d\mu = g$, so that $d\lambda = g\, d\mu$.)
5. Show that Theorem 2.2.5(e) fails if λ_1 is not finite.
6. (*Complex measures*) If $(\Omega, \mathscr{F})$ is a measurable space, a *complex measure* λ on $\mathscr{F}$ is a countably additive complex-valued set function; that is, $\lambda = \lambda_1 + i\lambda_2$, where λ_1 and λ_2 are finite signed measures.

(a) Define the *total variation* of λ as

$$|\lambda|(A) = \sup\left\{\sum_{i=1}^n |\lambda(E_i)| : E_1, \ldots, E_n \text{ disjoint measurable subsets of } A,\ n = 1, 2, \ldots\right\}.$$

Show that $|\lambda|$ is a measure on $\mathscr{F}$. (The definition is consistent with the earlier notion of total variation of a signed measure; see Section 2.1, Problem 4.)

In the discussion below, λ's, with various subscripts, denote arbitrary measures (real signed measures or complex measures), and μ denotes a nonnegative real measure. We define $\lambda \ll \mu$ in the usual way; if $A \in \mathscr{F}$ and $\mu(A) = 0$, then $\lambda(A) = 0$. Define $\lambda_1 \perp \lambda_2$ iff $|\lambda_1| \perp |\lambda_2|$. Establish the following results.

(b) $|\lambda_1 + \lambda_2| \leq |\lambda_1| + |\lambda_2|$; $|a\lambda| = |a|\,|\lambda|$ for any complex number a. In particular if $\lambda = \lambda_1 + i\lambda_2$ is a complex measure, then

$$|\lambda| \leq |\lambda_1| + |\lambda_2|; \qquad \text{hence} \quad |\lambda|(\Omega) < \infty \quad \text{by 2.1.3(b).}$$

(c) If $\lambda_1 \perp \mu$ and $\lambda_2 \perp \mu$, then $\lambda_1 + \lambda_2 \perp \mu$.
(d) If $\lambda \ll \mu$, then $|\lambda| \ll \mu$, and conversely.
(e) If $\lambda_1 \ll \mu$ and $\lambda_2 \perp \mu$, then $\lambda_1 \perp \lambda_2$.
(f) If $\lambda \ll \mu$ and $\lambda \perp \mu$, then $\lambda = 0$.
(g) If λ is finite, then $\lambda \ll \mu$ iff $\lim_{\mu(A)\to 0} \lambda(A) = 0$.

2.3 Applications to Real Analysis

We are going to apply the concepts of the previous section to some problems involving functions of a real variable. If $[a, b]$ is a closed bounded interval of reals and f: $[a, b] \to R$, f is said to be *absolutely continuous* iff for each $\varepsilon > 0$ there is a $\delta > 0$ such that for all positive integers n and all families $(a_1, b_1), \ldots, (a_n, b_n)$ of disjoint open subintervals of $[a, b]$ of total length at most δ, we have

$$\sum_{i=1}^{n} |f(b_i) - f(a_i)| \leq \varepsilon.$$

It is immediate that this property holds also for countably infinite families of disjoint open intervals of total length at most δ. It also follows from the definition that f is continuous.

We can connect absolute continuity of functions with the earlier notion of absolute continuity of measures, as follows:

2.3.1 ***Theorem.*** Suppose that F and G are distribution functions on $[a, b]$, with corresponding (finite) Lebesgue–Stieltjes measures μ_1 and μ_2. Let $f = F - G$, $\mu = \mu_1 - \mu_2$, so that μ is a finite signed measure on $\mathscr{B}[a, b]$, with $\mu(x, y] = f(y) - f(x)$, $x < y$. If m is Lebesgue measure on $\mathscr{B}[a, b]$, then $\mu \ll m$ iff f is absolutely continuous.

PROOF. Assume $\mu \ll m$. If $\varepsilon > 0$, by 2.2.5(b) and (e), there is a $\delta > 0$ such that $m(A) \leq \delta$ implies $|\mu|(A) \leq \varepsilon$. Thus if $(a_1, b_1), \ldots, (a_n, b_n)$ are disjoint open intervals of total length at most δ,

$$\sum_{i=1}^{n} |f(b_i) - f(a_i)| = \sum_{i=1}^{n} |\mu(a_i, b_i]| = \sum_{i=1}^{n} |\mu(a_i, b_i)| \leq \varepsilon.$$

(Note that $\mu\{b_i\} = 0$ since $\mu \ll m$.) Therefore f is absolutely continuous.

Now assume f absolutely continuous; if $\varepsilon > 0$, choose $\delta > 0$ as in the definition of absolute continuity. If $m(A) = 0$, we must show that $\mu(A) = 0$. We use Problem 12, Section 1.4:

$$m(A) = \inf\{m(V) : V \supset A, \quad V \text{ open}\},$$
$$\mu_i(A) = \inf\{\mu_i(V) : V \supset A, \quad V \text{ open}\}, \qquad i = 1, 2.$$

(This problem assumes that the measures are defined on $\mathscr{B}(R)$ rather than $\mathscr{B}[a, b]$. The easiest way out is to extend all measures to $\mathscr{B}(R)$ by assigning measure 0 to $R - [a, b]$.) Since a finite intersection of open sets is open, we can find a decreasing sequence $\{V_n\}$ of open sets such that $\mu(V_n) \to \mu(A)$ and $m(V_n) \to m(A) = 0$.

Choose n large enough so that $m(V_n) < \delta$; if V_n is the disjoint union of the open intervals (a_i, b_i), $i = 1, 2, \ldots$, then $|\mu(V_n)| \le \sum_i |\mu(a_i, b_i)|$. But f is continuous, hence

$$\mu\{b_i\} = \lim_{n\to\infty} \mu(b_i - 1/n, b_i] = \lim_{n\to\infty} [f(b_i) - f(b_i - 1/n)] = 0.$$

Therefore

$$|\mu(V_n)| \le \sum_i |\mu(a_i, b_i]| = \sum_i |f(b_i) - f(a_i)| \le \varepsilon.$$

Since ε is arbitrary and $\mu(V_n) \to \mu(A)$, we have $\mu(A) = 0$. ∎

If $f: R \to R$, absolute continuity of f is defined exactly as above. If F and G are bounded distribution functions on R with corresponding Lebesgue–Stieltjes measures μ_1 and μ_2, and $f = F - G$, $\mu = \mu_1 - \mu_2$ [a finite signed measure on $\mathscr{B}(R)$], then f is absolutely continuous iff μ is absolutely continuous with respect to Lebesgue measure; the proof is the same as in 2.3.1.

Any absolutely continuous function on $[a, b]$ can be represented as the difference of two absolutely continuous increasing functions. We prove this in a sequence of steps.

If $f: [a, b] \to R$ and $P: a = x_0 < x_1 < \cdots < x_n = b$ is a partition of $[a, b]$, define

$$V(P) = \sum_{i=1}^{n} |f(x_i) - f(x_{i-1})|.$$

The sup of $V(P)$ over all partitions of $[a, b]$ is called the *variation* of f on $[a, b]$, written $V_f(a, b)$, or simply $V(a, b)$ if f is understood. We say that f is of *bounded variation* on $[a, b]$ iff $V(a, b) < \infty$. If $a < c < b$, a brief argument shows that $V(a, b) = V(a, c) + V(c, b)$.

2.3.2 ***Lemma.*** If $f: [a, b] \to R$ and f is absolutely continuous on $[a, b]$, then f is of bounded variation on $[a, b]$.

Proof. Pick any $\varepsilon > 0$, and let $\delta > 0$ be chosen as in the definition of absolute continuity. If P is any partition of $[a, b]$, there is a refinement Q of P consisting of subintervals of length less than $\delta/2$. If $Q: a = x_0 < x_1 < \cdots < x_n = b$, let $i_0 = 0$, and let i_1 be the largest integer such that $x_{i_1} - x_{i_0} < \delta$; let i_2 be the largest integer greater than i_1 such that $x_{i_2} - x_{i_1} < \delta$, and continue in this fashion until the process terminates, say with $i_r = n$. Now $x_{i_k} - x_{i_{k-1}} \geq \delta/2$, $k = 1, 2, \ldots, r-1$, by construction of Q; hence

$$r \leq 1 + \frac{2(b-a)}{\delta} = M.$$

By absolute continuity, $V(Q) \leq M\varepsilon$. But $V(P) \leq V(Q)$ since the refining process can never decrease V; the result follows. ∎

It is immediate that a monotone function F on $[a, b]$ is of bounded variation: $V_F(a, b) = |F(b) - F(a)|$. Thus if $f = F - G$, where F and G are increasing, then f is of bounded variation. The converse is also true.

2.3.3 ***Lemma.*** If $f: [a, b] \to R$ and f is of bounded variation on $[a, b]$, then there are increasing functions F and G on $[a, b]$ such that $f = F - G$. If f is absolutely continuous, F and G may also be taken as absolutely continuous.

Proof. Let $F(x) = V_f(a, x)$, $a \leq x \leq b$; F is increasing, for if $h \geq 0$, $V(a, x+h) - V(a, x) = V(x, x+h) \geq 0$. If $G(x) = F(x) - f(x)$, then G is also increasing. For if $x_1 < x_2$, then

$$\begin{aligned} G(x_2) - G(x_1) &= F(x_2) - F(x_1) - (f(x_2) - f(x_1)) \\ &= V(x_1, x_2) - (f(x_2) - f(x_1)) \\ &\geq V(x_1, x_2) - |f(x_2) - f(x_1)| \\ &\geq 0 \qquad \text{by definition of} \quad V(x_1, x_2). \end{aligned}$$

Now assume f absolutely continuous. If $\varepsilon > 0$, choose $\delta > 0$ as in the definition of absolute continuity. Let $(a_1, b_1), \ldots, (a_n, b_n)$ be disjoint open intervals with total length at most δ. If P_i is a partition of $[a_i, b_i]$, $i = 1, 2, \ldots, n$, then

$$\sum_{i=1}^{n} V(P_i) \leq \varepsilon \qquad \text{by absolute continuity of } f.$$

Take the sup successively over $P_1, \ldots, P_n$ to obtain

$$\sum_{i=1}^{n} V(a_i, b_i) \leq \varepsilon;$$

in other words,

$$\sum_{i=1}^{n} [F(b_i) - F(a_i)] \leq \varepsilon.$$

Therefore F is absolutely continuous. Since sums and differences of absolutely continuous functions are absolutely continuous, G is also absolutely continuous. ▌

We have seen that there is a close connection between absolute continuity and indefinite integrals, via the Radon–Nikodym theorem. The connection carries over to real analysis, as follows:

2.3.4 ***Theorem.*** Let f: $[a, b] \to R$. Then f is absolutely continuous on $[a, b]$ iff f is an indefinite integral, that is, iff

$$f(x) - f(a) = \int_a^x g(t)\, dt, \qquad a \le x \le b,$$

where g: $[a, b] \to R$ is Borel measurable and integrable with respect to Lebesgue measure.

PROOF. First assume f absolutely continuous. By 2.3.3, it is sufficient to assume f increasing. If μ is the Lebesgue–Stieltjes measure corresponding to f, and m is Lebesgue measure, then $\mu \ll m$ by 2.3.1. By the Radon–Nikodym theorem, there is an m-integrable function g such that $\mu(A) = \int_A g\, dm$ for all Borel subsets A of $[a, b]$. Take $A = [a, x]$ to obtain $f(x) - f(a) = \int_a^x g(t)\, dt$.

Conversely, assume $f(x) - f(a) = \int_a^x g(t)\, dt$. It is sufficient to assume $g \ge 0$ (if not, consider g^+ and g^- separately). Define $\mu(A) = \int_A g\, dm$, $A \in \mathscr{B}[a, b]$; then $\mu \ll m$, and if F is a distribution function corresponding to μ, F is absolutely continuous by 2.3.1. But

$$F(x) - F(a) = \mu(a, x] = \int_a^x g(t)\, dt = f(x) - f(a).$$

Therefore f is absolutely continuous. ▌

If g is Lebesgue integrable on R, the "if" part of the proof of 2.3.4 shows that the function defined by $\int_{-\infty}^x g(t)\, dt$, $x \in R$, is absolutely continuous, hence continuous, on R. Another way of proving continuity is to observe that

$$\int_{-\infty}^{x+h} g(t)\, dt - \int_{-\infty}^{x} g(t)\, dt = \int_{-\infty}^{\infty} g(t) I_{(x,\, x+h)}(t)\, dt$$

if $h > 0$, and this approaches 0 as $h \to 0$, by the dominated convergence theorem.

If $f(x) - f(a) = \int_a^x g(t)\, dt$, $a \le x \le b$, and g is continuous at x, then f is differentiable at x and $f'(x) = g(x)$; the proof given in calculus carries over.

If the continuity hypothesis is dropped, we can prove that $f'(x) = g(x)$ for *almost* every $x \in [a, b]$. One approach to this result is via the theory of differentiation of measures, which we now describe.

2.3.5 ***Definition.*** For the remainder of this section, μ is a signed measure on the Borel sets of R^k, assumed finite on bounded sets; thus if μ is nonnegative, it is a Lebesgue–Stieltjes measure. If m is Lebesgue measure, we define, for each $x \in R^k$,

$$(\bar{D}\mu)(x) = \limsup_{r \to 0 \;\; C_r} \frac{\mu(C_r)}{m(C_r)}, \qquad (\underline{D}\mu)(x) = \liminf_{r \to 0 \;\; C_r} \frac{\mu(C_r)}{m(C_r)},$$

where the C_r range over all open cubes of diameter less than r that contain x. It will be convenient (although not essential) to assume that all cubes have edges parallel to the coordinate axes.

We say that μ is *differentiable* at x iff $\bar{D}\mu$ and $\underline{D}\mu$ are equal and finite at x; we write $(D\mu)(x)$ for the common value. Thus μ is differentiable at x iff for every sequence $\{C_n\}$ of open cubes containing x, with the diameter of C_n approaching 0, $\mu(C_n)/m(C_n)$ approaches a finite limit, independent of the particular sequence.

The following result will play an important role:

2.3.6 ***Lemma.*** If $\{C_1, \ldots, C_n\}$ is a family of open cubes in R^k, there is a disjoint subfamily $\{C_{i_1}, \ldots, C_{i_s}\}$ such that $m(\bigcup_{j=1}^n C_j) \le 3^k \sum_{p=1}^s m(C_{i_p})$.

PROOF. Assume that the diameter of C_i decreases with i. Set $i_1 = 1$, and take i_2 to be the smallest index greater than i_1 such that C_{i_2} is disjoint from C_{i_1}; let i_3 be the smallest index greater than i_2 such that C_{i_3} is disjoint from $C_{i_1} \cup C_{i_2}$. Continue in this fashion to obtain disjoint sets $C_{i_1}, \ldots, C_{i_s}$. Now for any $j = 1, \ldots, n$, we have $C_j \cap C_{i_p} \ne \varnothing$ for some $i_p \le j$, for if not, j is not one of the i_p, hence $i_p < j < i_{p+1}$ for some p (or $i_s < j$). But $C_j \cap (C_{i_1} \cup \cdots \cup C_{i_p})$ is assumed empty, contradicting the definition of i_{p+1}.

If B_p is the open cube with the same center as C_{i_p} and diameter three times as large, then since $C_j \cap C_{i_p} \ne \varnothing$ and diameter $C_j \le$ diameter C_{i_p}, we have $C_j \subset B_p$. Therefore,

$$m\left(\bigcup_{j=1}^{n} C_j\right) \le m\left(\bigcup_{p=1}^{s} B_p\right) \le \sum_{p=1}^{s} m(B_p) = 3^k \sum_{p=1}^{s} m(C_{i_p}). \quad \blacksquare$$

We now prove the first differentiation result.

2.3.7 ***Lemma.*** Let μ be a Lebesgue–Stieltjes measure on the Borel sets of R^k. If $\mu(A) = 0$, then $D\mu = 0$ a.e. $[m]$ on A.

PROOF. If $a > 0$, let $B = \{x \in A : (\bar{D}\mu)(x) > a\}$. (Note that $\{x : \sup_{C_r} \mu(C_r)/m(C_r) > a\}$ is open, and it follows that B is a Borel set.) Fix $r > 0$, and let K be a compact subset of B. If $x \in K$, there is an open cube C_r of diameter less than r with $x \in C_r$ and $\mu(C_r) > am(C_r)$. By compactness, K is covered by finitely many of the cubes, say $C_1, \ldots, C_n$. If $\{C_{i_1}, \ldots, C_{i_s}\}$ is the subcollection of 2.3.6, we have

$$m(K) \le m\left(\bigcup_{j=1}^{n} C_j\right) \le 3^k \sum_{p=1}^{s} m(C_{i_p}) \le \frac{3^k}{a} \sum_{p=1}^{s} \mu(C_{i_p}) = \frac{3^k}{a} \mu\left(\bigcup_{p=1}^{s} C_{i_p}\right) \le \frac{3^k}{a} \mu(K_r),$$

where $K_r = \{x \in R^k : \operatorname{dist}(x, K) < r\}$. Since r is arbitrary, we have $m(K) \le 3^k\mu(K)/a \le 3^k\mu(A)/a = 0$. Take the sup over K to obtain, by Problem 12, Section 1.4, $m(B) = 0$, and since a is arbitrary, it follows that $\bar{D}\mu \le 0$ a.e. [m] on A. But $\mu \ge 0$; hence $0 \le \underline{D}\mu \le \bar{D}\mu$, so that $D\mu = 0$ a.e. [m] on A. ∎

We are going to show that $D\mu$ exists a.e. [m], and to do this the Lebesgue decomposition theorem is helpful. We write $\mu = \mu_1 + \mu_2$, where $\mu_1 \ll m$, $\mu_2 \perp m$. If $|\mu_2|(A) = 0$ and $m(A^c) = 0$, then by 2.3.7, $D\mu_2^+ = D\mu_2^- = 0$ a.e. [m] on A; hence a.e. [m] on R^k. Thus $D\mu_2 = 0$ a.e. [m] on R^k.

By the Radon–Nikodym theorem, we have $\mu_1(E) = \int_E g\, dm$, $E \in \mathscr{B}(R^k)$, for some Borel measurable function g. As might be expected intuitively, g is (a.e.) the derivative of μ_1; hence $D\mu = g$ a.e. [m].

2.3.8 ***Theorem.*** Let μ be a signed measure on $\mathscr{B}(R^k)$ that is finite on bounded sets, and let $\mu = \mu_1 + \mu_2$, where $\mu_1 \ll m$ and $\mu_2 \perp m$. Then $D\mu$ exists a.e. [m] and coincides a.e. [m] with the Radon–Nikodym derivative $g = d\mu_1/dm$.

PROOF. If $a \in R$ and C is an open cube of diameter less than r,

$$\mu_1(C) - am(C) = \int_C (g - a)\, dm \le \int_{C \cap \{g \ge a\}} (g - a)\, dm.$$

If $\lambda(E) = \int_{E \cap \{g \ge a\}} (g - a)\, dm$, $E \in \mathscr{B}(R^k)$, and $A = \{g < a\}$, then $\lambda(A) = 0$; so by 2.3.7, $D\lambda = 0$ a.e. [m] on A. But

$$\frac{\mu_1(C)}{m(C)} \le a + \frac{\lambda(C)}{m(C)};$$

hence $\bar{D}\mu_1 \le a$ a.e. [m] on A. Therefore, if $E_a = \{x \in R^k : g(x) < a < (\bar{D}\mu_1)(x)\}$, then $m(E_a) = 0$. Since $\{\bar{D}\mu_1 > g\} \subset \bigcup \{E_a : a \text{ rational}\}$, we have $\bar{D}\mu_1 \le g$ a.e. [m]. Replace μ_1 by $-\mu_1$ and g by $-g$ to obtain $\underline{D}\mu_1 \ge g$ a.e. [m]. By 2.2.2(b), g is finite a.e. [m], and the result follows. ∎

We now return to functions on the real line.

2.3.9 ***Theorem.*** Let $f\colon [a, b] \to R$ be an increasing function. Then the derivative of f exists at almost every point of $[a, b]$ (with respect to Lebesgue measure). Thus by 2.3.3, a function of bounded variation is differentiable almost everywhere.

PROOF. Let μ by the Lebesgue–Stieltjes measure corresponding to f; by 2.3.8, $D\mu$ exists a.e. $[m]$; we show that $D\mu = f'$ a.e. $[m]$. If $a \le x \le b$ and μ is differentiable at x, then f is continuous at x by definition of μ. If $\lim_{h\to 0} [f(x+h) - f(x)]/h \ne (D\mu)(x) = c$, there is an $\varepsilon > 0$ and a sequence $h_n \to 0$ with all h_n of the same sign and $|[f(x+h_n) - f(x)]/h_n - c| \ge \varepsilon$ for all n. Assuming all $h_n > 0$, we can find numbers $k_n > 0$ such that

$$\left| \frac{f(x+h_n) - f(x-k_n)}{h_n + k_n} - c \right| \ge \frac{\varepsilon}{2}$$

for all n, and since f has only countably many discontinuities, it may be assumed that f is continuous at $x + h_n$ and $x - k_n$. Thus we conclude that $\mu(x - k_n, x + h_n)/(h_n + k_n) \nrightarrow c$, a contradiction. ∎

We now prove the main theorem on absolutely continuous functions.

2.3.10 ***Theorem.*** Let f be absolutely continuous on $[a, b]$, with $f(x) - f(a) = \int_a^x g(t)\, dt$, as in 2.3.4. Then $f' = g$ almost everywhere on $[a, b]$ (Lebesgue measure). Thus by 2.3.4, f is absolutely continuous iff f is the integral of its derivative, that is,

$$f(x) - f(a) = \int_a^x f'(t)\, dt, \qquad a \le x \le b.$$

PROOF. We may assume $g \ge 0$ (if not, consider g^+ and g^-). If $\mu_1(A) = \int_A g\, dm$, $A \in \mathscr{B}(R^k)$, then $D\mu_1 = g$ a.e. $[m]$ by 2.3.8. But if $a \le x \le y \le b$, then $\mu_1(x, y] = f(y) - f(x)$, so that μ_1 is the Lebesgue–Stieltjes measure corresponding to f. Thus by the proof of 2.3.9, $D\mu_1 = f'$ a.e. $[m]$. ∎

Problems

1. Let F be a bounded distribution function on R. Use the Lebesgue decomposition theorem to show that F may be represented uniquely (up to additive constants) as $F_1 + F_2 + F_3$, where the distribution functions F_j, $j = 1, 2, 3$ (and the corresponding Lebesgue–Stieltjes measures μ_j) have the following properties:

   (a) F_1 is discrete (that is, μ_1 is concentrated on a countable set of points).

(b) F_2 is absolutely continuous (μ_2 is absolutely continuous with respect to Lebesgue measure; see 2.3.1).

(c) F_3 is continuous and singular (that is, μ_3 is singular with respect to Lebesgue measure).

2. If f is an increasing function from $[a, b]$ to R, show that $\int_a^b f'(x)\,dx \le f(b) - f(a)$. The inequality may be strict, as Problem 3 shows. (Note that by 2.3.9, f' exists a.e.; for integration purposes, f' may be defined arbitrarily on the exceptional set of Lebesgue measure 0.)

3. (*The Cantor function*) Let $E_1, E_2, \ldots$ be the sets removed from $[0, 1]$ to form the Cantor ternary set (see Problem 7, Section 1.4). Define functions $F_n : [0, 1] \to [0, 1]$ as follows: Let $A_1, A_2, \ldots, A_{2^n-1}$ be the subintervals of $\bigcup_{i=1}^n E_i$, arranged in increasing order. For example, if $n = 3$,

$$\begin{aligned} E_1 \cup E_2 \cup E_3 &= (\tfrac{1}{27}, \tfrac{2}{27}) \cup (\tfrac{1}{9}, \tfrac{2}{9}) \cup (\tfrac{7}{27}, \tfrac{8}{27}) \cup (\tfrac{1}{3}, \tfrac{2}{3}) \\ &\quad \cup (\tfrac{19}{27}, \tfrac{20}{27}) \cup (\tfrac{7}{9}, \tfrac{8}{9}) \cup (\tfrac{25}{27}, \tfrac{26}{27}) \\ &= A_1 \cup A_2 \cup \cdots \cup A_7. \end{aligned}$$

Define

$$\begin{aligned} &F_n(0) = 0, \\ &F_n(x) = k/2^n \qquad \text{if} \quad x \in A_k, \quad k = 1, 2, \ldots, 2^n - 1, \\ &F_n(1) = 1. \end{aligned}$$

Complete the specification of F_n by interpolating linearly. For $n = 2$, see Fig. 2.1; in this case,

$$\begin{aligned} E_1 \cup E_2 &= (\tfrac{1}{9}, \tfrac{2}{9}) \cup (\tfrac{1}{3}, \tfrac{2}{3}) \cup (\tfrac{7}{9}, \tfrac{8}{9}) \\ &= A_1 \cup A_2 \cup A_3. \end{aligned}$$

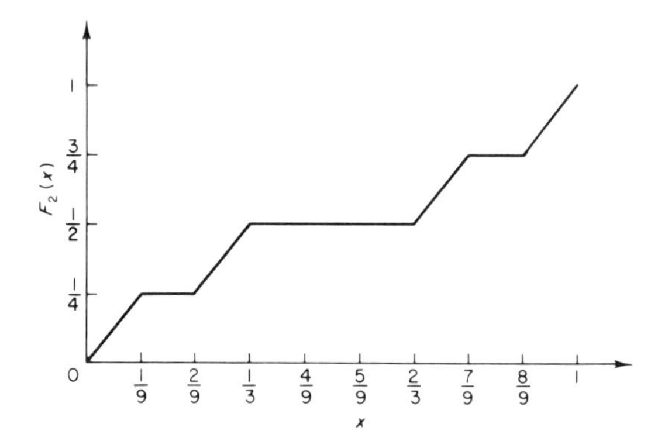

Figure 2.1. Approximation to the Cantor function.

Show that $F_n(x) \to F(x)$ for each x, where F, the Cantor function, has the following properties:

(a) F is continuous and increasing.
(b) $F' = 0$ almost everywhere (Lebesgue measure).
(c) F is not absolutely continuous.
In fact
(d) F is singular; that is, the corresponding Lebesgue–Stieltjes measure μ is singular with respect to Lebesgue measure.

4. Let f be a Lebesgue integrable real-valued function on R^k (or on an open subset of R^k). If $\mu(E) = \int_E f(x)\,dx$, $E \in \mathscr{B}(R^k)$, we know that $D\mu = f$ a.e. (Lebesgue measure). If $D\mu = f$ at x_0, then if C is an open cube containing x_0 and diam $C \to 0$, we have $\mu(C)/m(C) \to f(x_0)$; that is,

$$\frac{1}{m(C)} \int_C [f(x) - f(x_0)]\,dx \to 0 \quad \text{as} \quad \text{diam } C \to 0.$$

In fact, show that

$$\frac{1}{m(C)} \int_C |f(x) - f(x_0)|\,dx \to 0 \quad \text{as} \quad \text{diam } C \to 0$$

for almost every x_0. The set of favorable x_0 is called the *Lebesgue set* of f.

5. This problem relates various concepts discussed in Section 2.3. In all cases, f is a real-valued function defined on the closed bounded interval $[a, b]$. Establish the following:

(a) If f is continuous, f need not be of bounded variation.
(b) If f is continuous and increasing (hence of bounded variation), f need not be absolutely continuous.
(c) If f satisfies a Lipschitz condition, that is, $|f(x) - f(y)| \le L|x - y|$ for some fixed positive number L and all $x, y \in [a, b]$, then f is absolutely continuous.
(d) If f' exists everywhere and is bounded, f is absolutely continuous. [It can also be shown that if f' exists everywhere and is Lebesgue integrable on $[a, b]$, then f is absolutely continuous; see Titchmarsh (1939, p. 368).]
(e) If f is continuous and f' exists everywhere, f need not be absolutely continuous [consider $f(x) = x^2 \sin(1/x^2)$, $0 < x \le 1$, $f(0) = 0$].

6. The following problem considers the change of variable formula in a multiple integral. Throughout the problem, T will be a map from V onto W, where V and W are open subsets of R^k, T is assumed one-to-one, continuously differentiable, with a nonzero Jacobian. Thus T has a continuously differentiable inverse, by the inverse function theorem of

advanced calculus [see, for example, Apostol (1957, p. 144)]. It also follows from standard advanced calculus results that for all $x \in V$,

$$\frac{1}{|h|}\left[|T(x+h) - T(x) - A(x)h|\right] \to 0 \quad \text{as} \quad h \to 0, \tag{1}$$

where $A(x)$ is the linear transformation on R^k represented by the Jacobian matrix of T, evaluated at x. [See Apostol (1957, p. 118).]

(a) Let A be a nonsingular linear transformation on R^k, and define a measure λ on $\mathscr{B}(R^k)$ by $\lambda(E) = m(A(E))$ where m is Lebesgue measure. Show that $\lambda = c(A)m$ for some constant $c(A)$, and in fact $c(A)$ is the absolute value of the determinant of A. [Use translation-invariance of Lebesgue measure (Problem 5, Section 1.4) and the fact that any matrix can be represented as a product of matrices corresponding to elementary row operations.]

Now define a measure μ on $\mathscr{B}(V)$ by $\mu(E) = m(T(E))$. By continuity of T, if $\varepsilon > 0$, $x \in V$, and C is a sufficiently small open cube containing x, then $T(C)$ has diameter less than ε, in particular, $m(T(C)) < \infty$. It follows by a brief compactness argument that μ is a Lebesgue–Stieltjes measure on $\mathscr{B}(V)$.

Our objective is to show that μ is differentiable and $(D\mu)(x) = |J(x)|$ for every $x \in V$, where $J(x) = \det A(x)$, the Jacobian of the transformation T.

(b) Show that it suffices to prove that if $0 \in V$ and $T(0) = 0$, then $(D\mu)(0) = |\det A(0)|$.

(c) Show that it may be assumed without loss of generality that $A(0)$ is the identity transformation; hence $\det A(0) = 1$.

Now given $\varepsilon > 0$, choose $\alpha \in (0, \frac{1}{4})$ such that

$$1 - \varepsilon < (1 - 2\alpha)^k < (1 + 2\alpha)^k < 1 + \varepsilon.$$

Under the assumptions of (b) and (c), by Eq. (1), there is a $\delta > 0$ such that if $|x| < \delta$, then $|T(x) - x| \le \alpha|x|/\sqrt{k}$.

(d) If C is an open cube containing 0 with edge length β and diameter $\sqrt{k}\,\beta < \delta$, take C_1, C_2 as open cubes concentric with C, with edge lengths $\beta_1 = (1 - 2\alpha)\beta$ and $\beta_2 = (1 + 2\alpha)\beta$. Establish the following:

(i) If $x \in \overline{C}$, then $T(x) \in C_2$.
(ii) If x belongs to the boundary of C, then $T(x) \notin C_1$.
(iii) If x is the center of C, then $T(x) \in C_1$.
(iv) $C_1 - T(C) = C_1 - T(\overline{C})$.

Use a connectedness argument to conclude that $C_1 \subset T(C) \subset C_2$, and complete the proof that $(D\mu)(0) = 1$.

(e) If λ is any measure on $\mathscr{B}(V)$ and $\bar{D}\lambda < \infty$ on V, show that λ is absolutely continuous with respect to Lebesgue measure. It therefore follows from Theorem 2.3.8 that

$$m(T(E)) = \int_E |J(x)|\, dx, \qquad E \in \mathscr{B}(V).$$

[If this is false, find a compact set K and positive integers n and j such that $m(K) = 0$, $\lambda(K) > 0$, and $\lambda(C) < nm(C)$ for all open cubes C containing a point of K and having diameter less than $1/j$. Essentially, the idea is to cover K by such cubes and conclude that $\lambda(K) = 0$, a contradiction.]

(f) If f is a real-valued Borel measurable function on W, show that

$$\int_W f(y)\, dy = \int_V f(T(x))|J(x)|\, dx$$

in the sense that if one of the two integrals exists, so does the other, and the two integrals are equal.

7. (*Fubini's differentiation theorem*) Let $f_1, f_2, \ldots$ be increasing functions from R to R, and assume that for each x, $\sum_{n=1}^{\infty} f_n(x)$ converges to a finite number $f(x)$. Show that $\sum_{n=1}^{\infty} f_n'(x) = f'(x)$ almost everywhere (Lebesgue measure).

   Outline:

   (a) It suffices to restrict the domain of all functions to $[0,1]$ and to assume all functions nonnegative. Use Fatou's lemma to show that $\sum_{n=1}^{\infty} f_n'(x) \le f'(x)$ a.e.; hence $f_n'(x) \to 0$ a.e.

   (b) Choose $n_1, n_2, \ldots$ such that $\sum_{j>n_k} f_j(1) \le 2^{-k}$, $k = 1, 2, \ldots$, and apply part (a) to the functions $g_k(x) = f(x) - \sum_{j=1}^{n_k} f_j(x) = \sum_{j>n_k} f_j(x)$.

2.4 L^p Spaces

If $(\Omega, \mathscr{F}, \mu)$ is a measure space and p is a real number with $p \ge 1$, the set of all Borel measurable functions f such that $|f|^p$ is μ-integrable has many important properties. In order to fully develop these properties, it will be convenient to work with complex-valued functions.

2.4.1 ***Definitions.*** If $(\Omega, \mathscr{F})$ is a measurable space, a *complex-valued Borel measurable function* on $(\Omega, \mathscr{F})$ is a mapping $f\colon (\Omega, \mathscr{F}) \to (R^2, \mathscr{B}(R^2))$. If $p_1(x, y) = x$ and $p_2(x, y) = y$, $x, y \in R$, we may identify $p_1 \circ f$ and $p_2 \circ f$ with the real and imaginary parts of f. If μ is a measure on $\mathscr{F}$, we define

$$\int_\Omega f\, d\mu = \int_\Omega \operatorname{Re} f\, d\mu + i \int_\Omega \operatorname{Im} f\, d\mu,$$

provided $\int_\Omega \operatorname{Re} f\,d\mu$ and $\int_\Omega \operatorname{Im} f\,d\mu$ are *both* finite. In this case we say that f is *μ-integrable*. Thus in working with complex-valued functions, we do not consider any cases in which integrals exist but are not finite.

The following result was established earlier for real-valued f [see 1.5.9(c)]; it is still valid in the complex case, but the proof must be modified.

2.4.2 ***Lemma.*** If f is μ-integrable,

$$\left|\int_\Omega f\,d\mu\right| \le \int_\Omega |f|\,d\mu.$$

PROOF. If $\int_\Omega f\,d\mu = re^{i\theta}$, $r \ge 0$, then $\int_\Omega e^{-i\theta} f\,d\mu = r = |\int_\Omega f\,d\mu|$. But if $f(\omega) = \rho(\omega)e^{i\varphi(\omega)}$ (taking $\rho \ge 0$), then

$$\begin{aligned}\int_\Omega e^{-i\theta} f\,d\mu &= \int_\Omega \rho e^{i(\varphi-\theta)}\,d\mu \\ &= \int_\Omega \rho\cos(\varphi-\theta)\,d\mu \qquad \text{since } r \text{ is real} \\ &\le \int_\Omega \rho\,d\mu = \int_\Omega |f|\,d\mu. \quad \blacksquare\end{aligned}$$

Many other standard properties of the integral carry over to the complex case, in particular 1.5.5(b), 1.5.9(a) and (e), 1.6.1, 1.6.3, 1.6.4(b) and (c), 1.6.5, 1.6.9, 1.6.10, and 1.7.1. In almost all cases, the result is an immediate consequence of the fact that integrating a complex-valued function is equivalent to integrating the real and imaginary parts separately. Only two theorems require additional comment. To prove that h is integrable iff $|h|$ is integrable [1.6.4(b)], use the fact that $|\operatorname{Re} h|$, $|\operatorname{Im} h| \le |h| \le |\operatorname{Re} h| + |\operatorname{Im} h|$. Finally, to prove the dominated convergence theorem (1.6.9), apply the real version of the theorem to $|f_n - f|$, and note that $|f_n - f| \le |f_n| + |f| \le 2g$.

If $p > 0$, we define the space $L^p = L^p(\Omega, \mathscr{F}, \mu)$ as the collection of all complex-valued Borel measurable functions f such that $\int_\Omega |f|^p\,d\mu < \infty$. We set

$$\|f\|_p = \left(\int_\Omega |f|^p\,d\mu\right)^{1/p}, \qquad f \in L^p.$$

It follows that for any complex number a, $\|af\|_p = |a|\,\|f\|_p$, $f \in L^p$.

We are going to show that L^p forms a linear space over the complex field. The key steps in the proof are the Hölder and Minkowski inequalities, which we now develop.

2.4.3 ***Lemma.*** If $a, b, \alpha, \beta > 0$, $\alpha + \beta = 1$, then $a^\alpha b^\beta \le \alpha a + \beta b$.

PROOF. The statement to be proved is equivalent to $-\log(\alpha a + \beta b) \le \alpha(-\log a) + \beta(-\log b)$, which holds since $-\log$ is convex. [If g has a nonnegative second derivative on the interval $I \subset R$, then g is convex on I, that is $g(\alpha x + \beta y) \le \alpha g(x) + \beta g(y)$, $x, y \ni I$, $\alpha, \beta > 0$, $\alpha + \beta = 1$. To see this, assume $x < y$ and write

$$\begin{aligned} g(\alpha x + \beta y) - \alpha g(x) - \beta g(y) &= \alpha[g(\alpha x + \beta y) - g(x)] + \beta[g(\alpha x + \beta y) - g(y)] \\ &= \alpha\beta(y - x)[g'(u) - g'(v)] \qquad \text{for some} \quad u, v \end{aligned}$$

with $x \le u \le \alpha x + \beta y$, $\alpha x + \beta y \le v \le y$. But $g'(u) - g'(v) \le 0$ since g' is increasing on I.] ∎

2.4.4 ***Corollary.*** If $c, d > 0$, $p, q > 1$, $(1/p) + (1/q) = 1$, then $cd \le (c^p/p) + (d^q/q)$.

PROOF. In 2.4.3, let $\alpha = 1/p$, $\beta = 1/q$, $a = c^p$, $b = d^q$. ∎

2.4.5 ***Hölder Inequality.*** Let $1 < p < \infty$, $1 < q < \infty$, $(1/p) + (1/q) = 1$. If $f \in L^p$ and $g \in L_q$, then $fg \in L^1$ and $\|fg\|_1 \le \|f\|_p \|g\|_q$.

PROOF. In 2.4.4, take $c = |f(\omega)|/\|f\|_p$, $d = |g(\omega)|/\|g\|_q$ (the inequality is immediate if $\|f\|_p$ or $\|g\|_q = 0$). Then

$$\frac{|f(\omega)g(\omega)|}{\|f\|_p\|g\|_q} \le \frac{|f(\omega)|^p}{p\|f\|_p^p} + \frac{|g(\omega)|^q}{q\|g\|_q^q};$$

integrate to obtain

$$\frac{\int_\Omega |fg|\, d\mu}{\|f\|_p\|g\|_q} \le \frac{1}{p} + \frac{1}{q} = 1. \ ∎$$

When $p = q = 2$, we obtain

$$\int_\Omega |fg|\, d\mu \le \left[\int_\Omega |f|^2\, d\mu \int_\Omega |g|^2\, d\mu\right]^{1/2}$$

and thus, using 2.4.2, we have the *Cauchy–Schwarz inequality*: If f and $g \in L^2$, then $fg \in L^1$ and

$$\left|\int_\Omega f\bar{g}\, d\mu\right| \le \left[\int_\Omega |f|^2\, d\mu \int_\Omega |g|^2\, d\mu\right]^{1/2},$$

where $\bar{g}$ is the complex conjugate of g. (The reason for replacing g by $\bar{g}$ is to make the inequality agree with the Hilbert space result to be discussed in Chapter 3.)

2.4.6 ***Lemma.*** If $a, b \geq 0$, $p \geq 1$, then $(a+b)^p \leq 2^{p-1}(a^p + b^p)$.

PROOF. Let $h(x) = d[(a+x)^p - 2^{p-1}(a^p + x^p)]/dx = p(a+x)^{p-1} - 2^{p-1}px^{p-1}$; since $p \geq 1$,

$$\begin{aligned} h(x) &> 0 \quad \text{for} \quad a + x > 2x, \quad \text{that is,} \quad x < a,\\ h(x) &= 0 \quad \text{at} \quad x = a,\\ h(x) &< 0 \quad \text{for} \quad x > a. \end{aligned}$$

The maximum therefore occurs at $x = a$; hence

$$(a+b)^p - 2^{p-1}(a^p + b^p) \leq (a+a)^p - 2^{p-1}(a^p + a^p) = 0. \blacksquare$$

2.4.7 ***Minkowski Inequality.*** If $f, g \in L^p$ $(1 \leq p < \infty)$, then $f + g \in L^p$ and $\|f+g\|_p \leq \|f\|_p + \|g\|_p$.

PROOF. By 2.4.6, $|f+g|^p \leq (|f| + |g|)^p \leq 2^{p-1}(|f|^p + |g|^p)$, hence f, $g \in L^p$ implies $f + g \in L^p$. Now the inequality is clear when $p = 1$, so assume $p > 1$ and choose q such that $(1/p) + (1/q) = 1$. Then

$$|f+g|^p = |f+g|\ |f+g|^{p-1} \leq |f|\ |f+g|^{p-1} + |g|\ |f+g|^{p-1}. \tag{1}$$

Now $|f+g|^{p-1} \in L^q$; for

$$(p-1)q = \frac{p-1}{1/q} = \frac{p-1}{1-1/p} = p;$$

hence

$$\int_\Omega [|f+g|^{p-1}]^q \, d\mu = \int_\Omega |f+g|^p \, d\mu < \infty.$$

Since f and g belong to L^p and $|f+g|^{p-1} \in L^q$, Hölder's inequality implies that $|f|\ |f+g|^{p-1}$ and $|g|\ |f+g|^{p-1} \in L^1$, and

$$\int_\Omega |f|\ |f+g|^{p-1} \, d\mu \leq \|f\|_p \left[\int_\Omega (|f+g|^{p-1})^q \, d\mu\right]^{1/q} = \|f\|_p \|f+g\|_p^{p/q}, \tag{2}$$

$$\int_\Omega |g|\ |f+g|^{p-1} \, d\mu \leq \|g\|_p \|f+g\|_p^{p/q}. \tag{3}$$

By Eq. (1), $\|f+g\|_p^p \leq (\|f\|_p + \|g\|_p)(\|f+g\|_p^{p/q})$. Since $p - (p/q) = 1$, the result follows. $\blacksquare$

By Minkowski's inequality and the fact that $\|af\|_p = |a|\ \|f\|_p$ for $f \in L^p$, L^p $(1 \leq p < \infty)$ is a vector space over the complex field. Furthermore, there is a natural notion of distance in L^p, by virtue of the fact that $\|\ \|_p$ is a seminorm.

2.4.8 ***Definitions and Comments.*** A *seminorm* on a vector space L (over the real or complex field) is a real-valued function $\| \ \|$ on L, with the following properties.

$$\|f\| \geq 0,$$

$$\|af\| = |a| \ \|f\| \qquad \text{for each scalar } a;$$

consequently, if $f = 0$, then $\|f\| = 0$.

$$\|f + g\| \leq \|f\| + \|g\|$$

(f and g are arbitrary elements of L). If $\| \ \|$ is a seminorm with the additional property that $\|f\| = 0$ implies $f = 0$, $\| \ \|$ is said to be a *norm*.

Now $\| \ \|_p$ is a seminorm on L^p; the first two properties follow from the definition of $\| \ \|_p$, and the last property is a consequence of Minkowski's inequality.

We can, in effect, change $\| \ \|_p$ into a norm by passing to equivalence classes as follows.

If $f, g \in L^p(\Omega, \mathscr{F}, \mu)$, define $f \sim g$ iff $f = g$ a.e. $[\mu]$. Then $\|f\|_p$ is the same for all f in a given equivalence class, by 1.6.5(b). Thus if $\mathbf{L}^p$ is the collection of equivalence classes, $\mathbf{L}^p$ becomes a linear space, and $\| \ \|_p$ is a seminorm on $\mathbf{L}^p$. In fact $\| \ \|_p$ is a norm, since $\|f\|_p = 0$ implies $f = 0$ a.e. $[\mu]$, by 1.6.6(b).

If $\| \ \|$ is a seminorm on a vector space, we have a natural notion of distance: $d(f, g) = \|f - g\|$. By definition of seminorm we have

$$d(f, g) \geq 0,$$

$$d(f, g) = 0 \qquad \text{if} \quad f = g,$$

$$d(f, g) = d(g, f),$$

$$d(f, h) \leq d(f, g) + d(g, h).$$

Thus d has all the properties of a metric, except that $d(f, g) = 0$ does not necessarily imply $f = g$; we call d a *pseudometric*. If $\| \ \|$ is a norm, d is actually a metric. [There is an asymmetry of terminology between seminorm and pseudometric, but these terms seem to be most popular, although "pseudonorm" is sometimes used, as is "semimetric."]

One of the first questions that arises in any metric space is the problem of completeness; we ask whether or not Cauchy sequences converge. We are going to show that the L^p spaces are complete. The following result will be needed; students of probability are likely to recognize it immediately, but it appears in other parts of analysis as well.

2.4.9 ***Chebyshev's Inequality.*** Let f be a nonnegative, extended real-valued, Borel measurable function on $(\Omega, \mathscr{F}, \mu)$. If $0 < p < \infty$ and $0 < \varepsilon < \infty$,

$$\mu\{\omega : f(\omega) \geq \varepsilon\} \leq \frac{1}{\varepsilon^p} \int_\Omega f^p \, d\mu.$$

The following version is often applied in probability. If g is an extended real-valued Borel measurable function on $(\Omega, \mathscr{F})$ and P is a probability measure on $\mathscr{F}$, define

$$m = \int_\Omega g\, dP \qquad \text{(assumed finite, so that } g \text{ is finite a.e. } [P]),$$

$$\sigma^2 = \int_\Omega (g - m)^2\, dP.$$

If $0 < k < \infty$,

$$P\{\omega: |g(\omega) - m| \geq k\sigma\} \leq \frac{1}{k^2}.$$

This follows from the first version with $f = |g - m|$, $\varepsilon = k\sigma$, $p = 2$.

PROOF.

$$\int_\Omega f^p\, d\mu \geq \int_{\{\omega:\, f(\omega) \geq \varepsilon\}} f^p\, d\mu \geq \varepsilon^p \mu\{\omega: f(\omega) \geq \varepsilon\}. \quad ∎$$

One more auxiliary result will be needed.

2.4.10 ***Lemma.*** If $g_1, g_2, \ldots \in L^p$ $(p > 0)$ and $\|g_k - g_{k+1}\|_p < (\frac{1}{4})^k$, $k = 1, 2, \ldots$, then $\{g_k\}$ converges a.e.

PROOF. Let $A_k = \{\omega: |g_k(\omega) - g_{k+1}(\omega)| \geq 2^{-k}\}$. Then by 2.4.9,

$$\mu(A_k) \leq 2^{kp} \|g_k - g_{k+1}\|_p^p < 2^{-kp}.$$

By 2.2.4, $\mu(\limsup_n A_n) = 0$. But if $\omega \notin \limsup_n A_n$, then $|g_k(\omega) - g_{k+1}(\omega)| < 2^{-k}$ for large k, so $\{g_k(\omega)\}$ is a Cauchy sequence of complex numbers, and therefore converges. ∎

Now, the main result:

2.4.11 ***Completeness of L^p, $1 \leq p < \infty$.*** If $f_1, f_2, \ldots$ form a Cauchy sequence in L^p, that is, $\|f_n - f_m\|_p \to 0$ as $n, m \to \infty$, there is an $f \in L^p$ such that $\|f_n - f\|_p \to 0$.

PROOF. Let n_1 be such that $\|f_n - f_m\|_p < \frac{1}{4}$ for $n, m \geq n_1$, and let $g_1 = f_{n_1}$. In general, having chosen $g_1, \ldots, g_k$ and $n_1, \ldots, n_k$, let $n_{k+1} > n_k$ be such that $\|f_n - f_m\|_p < (\frac{1}{4})^{k+1}$ for $n, m \geq n_{k+1}$, and let $g_{k+1} = f_{n_{k+1}}$. By 2.4.10, g_k converges a.e. to a limit function f.

Given $\varepsilon > 0$, choose N such that $\|f_n - f_m\|_p^p < \varepsilon$ for $n, m \geq N$. Fix $n \geq N$ and let $m \to \infty$ through values in the subsequence, that is, let $m = n_k$, $k \to \infty$. Then

$$\begin{aligned}
\varepsilon \geq \liminf_{k\to\infty} \|f_n - f_{n_k}\|_p^p &= \liminf_{k\to\infty} \int_\Omega |f_n - f_{n_k}|^p \, d\mu \\
&\geq \int_\Omega \liminf_{k\to\infty} |f_n - g_k|^p \, d\mu \qquad \text{by Fatou's lemma} \\
&= \|f_n - f\|_p^p .
\end{aligned}$$

Thus $\|f_n - f\|_p \to 0$. Since $f = f - f_n + f_n$, we have $f \in L^p$. ∎

2.4.12 ***Examples and Comments.*** Let Ω be the positive integers; take $\mathscr{F}$ as all subsets of Ω, and let μ be counting measure. A real-valued function on Ω may be represented as a sequence of real numbers; we write $f = \{a_n, n = 1, 2, \ldots\}$. An integral on this space is really a sum [see Problem 1(a)]:

$$\int_\Omega f \, d\mu = \sum_{n=1}^\infty a_n,$$

where the series is interpreted as $\sum_{n=1}^\infty a_n^+ - \sum_{n=1}^\infty a_n^-$ if this is not of the form $+\infty - \infty$ (if it is, the integral does not exist). Thus the following cases occur:

(1) $\sum_{n=1}^\infty a_n^+ = \infty$, $\sum_{n=1}^\infty a_n^- < \infty$. The series diverges to ∞ and the integral is ∞.

(2) $\sum_{n=1}^\infty a_n^+ < \infty$, $\sum_{n=1}^\infty a_n^- = \infty$. The series diverges to $-\infty$ and the integral is $-\infty$.

(3) $\sum_{n=1}^\infty a_n^+ < \infty$, $\sum_{n=1}^\infty a_n^- < \infty$. The series is absolutely convergent and the integral equals the sum of the series.

(4) $\sum_{n=1}^\infty a_n^+ = \infty$, $\sum_{n=1}^\infty a_n^- = \infty$. The series is not absolutely convergent; it may or may not converge conditionally. Whether it does or not, the integral does not exist. Thus when summation is considered from the point of view of Lebesgue integration theory, series that converge conditionally but not absolutely are ignored.

If μ is changed so that $\mu\{n\}$ is a nonnegative number p_n, not necessarily 1 as in the case of counting measure, the same analysis shows that

$$\int_\Omega f \, d\mu = \sum_{n=1}^\infty p_n a_n,$$

where the series is interpreted as $\sum_{n=1}^\infty p_n a_n^+ - \sum_{n=1}^\infty p_n a_n^-$.

If $f = \{a_n, n = 1, 2, \ldots\}$ is a sequence of complex numbers and μ is counting measure,

$$\int_\Omega f \, d\mu = \sum_{n=1}^\infty a_n = \sum_{n=1}^\infty \operatorname{Re} a_n + i \sum_{n=1}^\infty \operatorname{Im} a_n;$$

the integral is defined provided $\sum_{n=1}^\infty |a_n| < \infty$.

Now let Ω be an *arbitrary* set, and take $\mathscr{F}$ as all subsets of Ω and μ as counting measure. If $f = (f(\alpha), \alpha \in \Omega)$ is a nonnegative real-valued function on Ω, then [Problem 1(b)]

$$\int_\Omega f\, d\mu = \sum_\alpha f(\alpha), \tag{1}$$

where the series is defined as $\sup\{\sum_{\alpha \in F} f(\alpha)\colon F \subset \Omega,\ F \text{ finite}\}$. If $f(\alpha) > 0$ for uncountably many α, then for some $\delta > 0$ we have $f(\alpha) \geq \delta$ for infinitely many α, so that $\sum_\alpha f(\alpha) = \infty$.

If the nonnegativity hypothesis is dropped, we apply the above results to f^+ and f^- to again obtain Eq. (1), where the series is interpreted as $\sum_\alpha f^+(\alpha) - \sum_\alpha f^-(\alpha)$. If f is complex-valued, Eq. (1) still applies, with the series interpreted as $\sum_\alpha \operatorname{Re} f(\alpha) + i \sum_\alpha \operatorname{Im} f(\alpha)$. The integral is defined provided $\sum_\alpha |f(\alpha)| < \infty$.

The space $L^p(\Omega, \mathscr{F}, \mu)$ will be denoted by $l^p(\Omega)$; it consists of all complex-valued functions $(f(\alpha), \alpha \in \Omega)$ such that $f(\alpha) = 0$ for all but countably many α, and

$$\|f\|_p^p = \sum_\alpha |f(\alpha)|^p < \infty.$$

If Ω is the set of positive integers, the space $l^p(\Omega)$ will be denoted simply by l^p; it consists of all sequences $f = \{a_n\}$ of complex numbers such that

$$\|f\|_p^p = \sum_{n=1}^{\infty} |a_n|^p < \infty.$$

It will be useful to state the Hölder and Minkowski inequalities for sums. If $f \in l^p(\Omega)$ and $g \in l^q(\Omega)$, where $1 < p < \infty$, $1 < q < \infty$, $(1/p) + (1/q) = 1$, then $fg \in l^1(\Omega)$ and

$$\sum_\alpha |f(\alpha)g(\alpha)| \leq \left(\sum_\alpha |f(\alpha)|^p\right)^{1/p} \left(\sum_\alpha |g(\alpha)|^q\right)^{1/q}.$$

If $f, g \in l^p(\Omega)$, $1 \leq p < \infty$, then $f + g \in l^p(\Omega)$ and

$$\left(\sum_\alpha |f(\alpha) + g(\alpha)|^p\right)^{1/p} \leq \left(\sum_\alpha |f(\alpha)|^p\right)^{1/p} + \left(\sum_\alpha |g(\alpha)|^p\right)^{1/p}.$$

As in 2.4.5, we obtain the Cauchy–Schwarz inequality for sums from the Hölder inequality. If $f, g \in l^2(\Omega)$, then $fg \in l^1(\Omega)$ and

$$\left|\sum_\alpha f(\alpha)\overline{g(\alpha)}\right| \leq \left(\sum_\alpha |f(\alpha)|^2\right)^{1/2} \left(\sum_\alpha |g(\alpha)|^2\right)^{1/2}.$$

If in the above discussion we replace Ω by $\{1, 2, \ldots, n\}$, all convergence difficulties are eliminated, and all the spaces $l^p(\Omega)$ coincide with C^n.

If $0 < p < 1$, $\|\ \|_p$ is not a seminorm on $L^p(\Omega, \mathscr{F}, \mu)$. For let A and B be disjoint sets with $a = \mu(A)$ and $b = \mu(B)$ assumed finite and positive. If $f = I_A$, $g = I_B$, then

$$\|f+g\|_p = \left(\int_\Omega |f+g|^p \, d\mu\right)^{1/p} = \left(\int_\Omega (I_A + I_B) \, d\mu\right)^{1/p} = (a+b)^{1/p},$$

$$\|f\|_p = a^{1/p}, \qquad \|g\|_p = b^{1/p}.$$

But $(a+b)^{1/p} > a^{1/p} + b^{1/p}$ if $a, b > 0$, $0 < p < 1$, since $(a+x)^r - a^r - x^r$ is strictly increasing for $r > 1$, and has the value 0 when $x = 0$. Thus the triangle inequality fails. We can, however, describe convergence in L^p, $0 < p < 1$, in the following way. We use the inequality

$$(a+b)^p \le a^p + b^p, \qquad a, b \ge 0, \quad 0 < p < 1,$$

which is proved by considering $(a+x)^p - a^p - x^p$. It follows that

$$\int_\Omega |f+g|^p \, d\mu \le \int_\Omega |f|^p \, d\mu + \int_\Omega |g|^p \, d\mu, \qquad f, g \in L^p, \tag{2}$$

and therefore $d(f, g) = \int_\Omega |f-g|^p \, d\mu$ defines a pseudometric on L^p. In fact the pseudometric is complete (every Cauchy sequence converges); for Eq. (2) implies that if $f, g \in L^p$, then $f + g \in L^p$, so that the proof of 2.4.11 goes through.

If Ω is an interval of reals, $\mathscr{F}$ is the class of Borel sets of Ω, and μ is Lebesgue measure, the space $L^p(\Omega, \mathscr{F}, \mu)$ will be denoted by $L^p(\Omega)$. Thus, for example, $L^p[a, b]$ is the set of all complex-valued Borel measurable functions f on $[a, b]$ such that

$$\|f\|_p^p = \int_a^b |f(x)|^p \, dx < \infty.$$

If f is a complex valued Borel measurable function on $(\Omega, \mathscr{F}, \mu)$ and $f_1, f_2, \ldots \in L^p(\Omega, \mathscr{F}, \mu)$, we say that the sequence $\{f_n\}$ *converges to f in L^p* iff $\|f_n - f\|_p \to 0$, that is, iff $\int_\Omega |f_n - f|^p \, d\mu \to 0$ as $n \to \infty$. We use the notation $f_n \xrightarrow{L^p} f$. In Section 2.5, we shall compare various types of convergence of sequences of measurable functions. We show now that any $f \in L^p$ is an L^p-limit of simple functions.

2.4.13 ***Theorem.*** Let $f \in L^p$, $0 < p < \infty$. If $\varepsilon > 0$, there is a simple function $g \in L^p$ such that $\|f - g\|_p < \varepsilon$; g can be chosen to be finite-valued and to satisfy $|g| \le |f|$. Thus the finite-valued simple functions are dense in L^p.

PROOF. This follows from 1.5.5(b) and 1.6.10. ∎

If we specialize to functions on R^n and Lebesgue–Stieltjes measures, we may obtain another basic approximation theorem.

2.4.14 ***Theorem.*** Let $f \in L^p(\Omega, \mathscr{F}, \mu)$, $0 < p < \infty$, where $\Omega = R^n$, $\mathscr{F} = \mathscr{B}(R^n)$, and μ is a Lebesgue–Stieltjes measure. If $\varepsilon > 0$, there is a continuous function $g \in L^p(\Omega, \mathscr{F}, \mu)$ such that $\|f - g\|_p < \varepsilon$; furthermore, g can be chosen so that $\sup |g| \le \sup |f|$. Thus the continuous functions are dense in L^p.

PROOF. By 2.4.13, it suffices to show that an indicator I_A in L^p can be approximated in the L^p sense by a continuous function with absolute value at most 1. Now $I_A \in L^p$ means that $\mu(A) < \infty$; hence by Problem 12, Section 1.4, there is a closed set $C \subset A$ and an open set $V \supset A$ such that $\mu(V - C) < \varepsilon^p 2^{-p}$. Let g be a continuous map of Ω into $[0, 1]$ with $g = 1$ on C and $g = 0$ on V^c (g exists by Urysohn's lemma). Then

$$\int_\Omega |I_A - g|^p \, d\mu = \int_{\{I_A \neq g\}} |I_A - g|^p \, d\mu.$$

But $\{I_A \neq g\} \subset V - C$ and $|I_A - g| \leq 2$; hence

$$\|I_A - g\|_p^p \leq 2^p \mu(V - C) < \varepsilon^p.$$

Since $g = g - I_A + I_A$, we have $g \in L^p$. ∎

Theorem 2.4.14 shows that the continuous functions in L^p do not form a closed subset, for if they did, every function in L^p would be continuous. Equivalently, the continuous functions in L^p are not complete, in other words, there are Cauchy sequences of continuous functions in L^p that do not converge in L^p to a continuous limit. Explicit examples may be given without making use of 2.4.14; see Problem 2.

2.4.15 ***The Space L^∞.*** If we wish to define L^p spaces for $p = \infty$, we must proceed differently. We define the *essential supremum* of the real-valued Borel measurable function g on $(\Omega, \mathscr{F}, \mu)$ as

$$\text{ess sup } g = \inf\{c \in \overline{R} : \mu\{\omega : g(\omega) > c\} = 0\}$$

that is, the smallest number c such that $g \leq c$ a.e. $[\mu]$.

If f is a complex-valued Borel measurable function on $(\Omega, \mathscr{F}, \mu)$, we define

$$\|f\|_\infty = \text{ess sup } |f|.$$

The space $L^\infty(\Omega, \mathscr{F}, \mu)$ is the collection of all f such that $\|f\|_\infty < \infty$. Thus $f \in L^\infty$ iff f is essentially bounded, that is, bounded outside a set of measure 0.

Now $|f + g| \leq |f| + |g| \leq \|f\|_\infty + \|g\|_\infty$ a.e.; hence

$$\|f + g\|_\infty \leq \|f\|_\infty + \|g\|_\infty.$$

In particular, $f, g \in L^\infty$ implies $f + g \in L^\infty$. The other properties of a seminorm are easily checked. Thus L^∞ is a vector space over the complex field, $\| \ \|_\infty$ is a seminorm on L^∞, and becomes a norm if we pass to equivalence classes as before.

If $f, f_1, f_2, \ldots \in L^\infty$ and $\|f_n - f\|_\infty \to 0$, we write $f_n \xrightarrow{L^\infty} f$; we claim that:

$\|f_n - f\|_\infty \to 0$ iff there is a set $A \in \mathscr{F}$ with $\mu(A) = 0$ such that $f_n \to f$ uniformly on A^c.

For, assume $\|f_n - f\|_\infty \to 0$. Given a positive integer m, $\|f_n - f\|_\infty \le 1/m$ for sufficiently large n; hence $|f_n(\omega) - f(\omega)| \le 1/m$ for almost every ω, say for $\omega \notin A_m$, where $\mu(A_m) = 0$. If $A = \bigcup_{m=1}^\infty A_m$, then $\mu(A) = 0$ and $f_n \to f$ uniformly on A^c. Conversely, assume $\mu(A) = 0$ and $f_n \to f$ uniformly on A^c. Given $\varepsilon > 0$, $|f_n - f| \le \varepsilon$ on A^c for sufficiently large n, so that $|f_n - f| \le \varepsilon$ a.e. Thus $\|f_n - f\|_\infty \le \varepsilon$ for large enough n, and the result follows.

An identical argument shows that $\{f_n\}$ is a Cauchy sequence in L^∞ ($\|f_n - f_m\|_\infty \to 0$ as $n, m \to \infty$) iff there is a set $A \in \mathscr{F}$ with $\mu(A) = 0$ and $f_n - f_m \to 0$ uniformly on A^c.

It is immediate that the Hölder inequality still holds when $p = 1$, $q = \infty$, and we have shown above that the Minkowski inequality holds when $p = \infty$.

To show that L^∞ is complete, let $\{f_n\}$ be a Cauchy sequence in L^∞, and let A be a set of measure 0 such that $f_n(\omega) - f_m(\omega) \to 0$ uniformly for $\omega \in A^c$. But then $f_n(\omega)$ converges to a limit $f(\omega)$ for each $\omega \in A^c$, and the convergence is uniform on A^c. If we define $f(\omega) = 0$ for $\omega \in A$, we have $f \in L^\infty$ and $f_n \xrightarrow{L^\infty} f$.

Theorem 2.4.13 holds also when $p = \infty$. For if f is a function in L^∞, the standard approximating sequence $\{f_n\}$ of simple functions (see 1.5.5) converges to f uniformly, outside a set of measure 0. However, Theorem 2.4.14 fails when $p = \infty$ (see Problem 12).

If Ω is an arbitrary set, $\mathscr{F}$ consists of all subsets of Ω, and μ is counting measure, then $L^\infty(\Omega, \mathscr{F}, \mu)$ is the set of all bounded complex-valued functions $f = (f(\alpha), \alpha \in \Omega)$, denoted by $l^\infty(\Omega)$. The essential supremum is simply the supremum; in other words, $\|f\|_\infty = \sup\{|f(\alpha)| : \alpha \in \Omega\}$. If Ω is the set of positive integers, $l^\infty(\Omega)$ is the space of bounded sequences of complex numbers, denoted simply by l^∞.

Problems

1. (a) If $f = \{a_n, n = 1, 2, \ldots\}$, the a_n are real or complex numbers, and μ is counting measure on subsets of the positive integers, show that $\int_\Omega f\, d\mu = \sum_{n=1}^\infty a_n$, where the sum is interpreted as in 2.4.12.
   (b) If $f = (f(\alpha), \alpha \in \Omega)$ is a real- or complex-valued function on the arbitrary set Ω, and μ is counting measure on subsets of Ω, show that $\int_\Omega f\, d\mu = \sum_\alpha f(\alpha)$, where the sum is interpreted as in 2.4.12.
2. Give an example of functions $f, f_1, f_2, \ldots$ from R to $[0, 1]$ such that
   (a) each f_n is continuous on R,
   (b) $f_n(x)$ converges to $f(x)$ for all x, $\int_{-\infty}^\infty |f_n(x) - f(x)|^p\, dx \to 0$ for every $p \in (0, \infty)$, and
   (c) f is discontinuous at some point of R.
3. For each $n = 1, 2, \ldots$, let $f_n = \{a_1^{(n)}, a_2^{(n)}, \ldots\}$ be a sequence of complex numbers.

(a) If the $a_k^{(n)}$ are real and $0 \le a_k^{(n)} \le a_k^{(n+1)}$ for all k and n, show that

$$\lim_{n\to\infty} \sum_{k=1}^{\infty} a_k^{(n)} = \sum_{k=1}^{\infty} \lim_{n\to\infty} a_k^{(n)}.$$

Show that the same conclusion holds if the $a_k^{(n)}$ are complex and $|a_k^{(n)}| \le b_k$ for all k and n, where $\sum_{k=1}^{\infty} b_k < \infty$.

(b) If the $a_k^{(n)}$ are real and nonnegative, show that

$$\sum_{k=1}^{\infty} \sum_{n=1}^{\infty} a_k^{(n)} = \sum_{n=1}^{\infty} \sum_{k=1}^{\infty} a_k^{(n)}.$$

(c) If the $a_k^{(n)}$ are complex and $\sum_{n=1}^{\infty} \sum_{k=1}^{\infty} |a_k^{(n)}| < \infty$, show that $\sum_{n=1}^{\infty} \sum_{k=1}^{\infty} a_k^{(n)}$ and $\sum_{k=1}^{\infty} \sum_{n=1}^{\infty} a_k^{(n)}$ both converge to the same finite number.

4. Show that there is equality in the Hölder inequality iff $|f|^p$ and $|g|^q$ are linearly dependent, that is, iff $A|f|^p = B|g|^q$ a.e. for some constants A and B, not both 0.
5. If f is a complex-valued μ-integrable function, show that $|\int_\Omega f\,d\mu| = \int_\Omega |f|\,d\mu$ iff arg f is a.e. constant on $\{\omega : f(\omega) \ne 0\}$.
6. Show that equality holds in the Cauchy–Schwarz inequality iff f and g are linearly dependent.
7. (a) If $1 < p < \infty$, show that equality holds in the Minkowski inequality iff $Af = Bg$ a.e. for some nonnegative constants A and B, not both 0.
   (b) What are the conditions for equality if $p = 1$?
8. If $1 \le r < s < \infty$, and $f \in L^s(\Omega, \mathscr{F}, \mu)$, μ finite, show that $\|f\|_r \le k\|f\|_s$ for some finite positive constant k. Thus $L^s \subset L^r$ and L^s convergence implies L^r convergence. (We may take $k = 1$ if μ is a probability measure.) Note that finiteness of μ is essential here; if μ is Lebesgue measure on $\mathscr{B}(R)$ and $f(x) = 1/x$ for $x \ge 1$, $f(x) = 0$ for $x < 1$, then $f \in L^2$ but $f \notin L^1$.
9. If μ is finite, show that $\|f\|_p \to \|f\|_\infty$ as $p \to \infty$. Give an example to show that this fails if $\mu(\Omega) = \infty$.
10. (*Radon–Nikodym theorem, complex case*) If μ is a σ-finite (nonnegative, real) measure, λ a complex measure on $(\Omega, \mathscr{F})$, and $\lambda \ll \mu$, show that there is a complex-valued μ-integrable function g such that $\lambda(A) = \int_A g\,d\mu$ for all $A \in \mathscr{F}$. If h is another such function, $g = h$ a.e.

    Show also that the Lebesgue decomposition theorem holds if λ is a complex measure and μ is a σ-finite measure. (See Problem 6, Section 2.2, for properties of complex measures.)
11. (a) Let f be a complex-valued μ-integrable function, where μ is a nonnegative real measure. If S is a closed set of complex numbers

and $[1/\mu(E)] \int_E f \, d\mu \in S$ for all measurable sets E such that $\mu(E) > 0$, show that $f(\omega) \in S$ for almost every ω. [If D is a closed disk with center at z and radius r, and $D \subset S^c$, take $E = f^{-1}(D)$. Show that $|\int_E (f - z) \, d\mu| \leq r\mu(E)$, and conclude that $\mu(E) = 0$.]

(b) If λ is a complex measure, then $\lambda \ll |\lambda|$ by definition of $|\lambda|$; hence by the Radon–Nikodym theorem, there is a $|\lambda|$-integrable complex-valued function h such that $\lambda(E) = \int_E h \, d|\lambda|$ for all $E \in \mathscr{F}$. Show that $|h| = 1$ a.e. $[|\lambda|]$. [Let $A_r = \{\omega: |h(\omega)| < r\}$, $0 < r < 1$, and use the definition of $|\lambda|$ to show that $|h| \geq 1$ a.e. Use part (a) to show $|h| \leq 1$ a.e.]

(c) Let μ be a nonnegative real measure, g a complex-valued μ-integrable function, and $\lambda(E) = \int_E g \, d\mu$, $E \in \mathscr{F}$. If $h = d\lambda/d|\lambda|$ as in part (b), show that $|\lambda|(E) = \int_E \bar{h} g \, d\mu$. (Intuitively, $\bar{h} g \, d\mu = \bar{h} \, d\lambda = \bar{h} h \, d|\lambda| = |h|^2 \, d|\lambda| = d|\lambda|$. Formally, show that $\int_\Omega fh \, d|\lambda| = \int_\Omega fg \, d\mu$ if f is a bounded, complex-valued, Borel measurable function, and set $f = \bar{h} I_E$.)

(d) Under the hypothesis of (c), show that

$$|\lambda|(E) = \int_E |g| \, d\mu \qquad \text{for all} \quad E \in \mathscr{F}.$$

12. Give an example of a bounded real-valued function f on R such that there is no sequence of continuous functions f_n such that $\|f - f_n\|_\infty \to 0$. Thus the continuous functions are not dense in $L^\infty(R)$.

2.5 Convergence of Sequences of Measurable Functions

In the previous section we introduced the notion of L^p convergence; we are also familiar with convergence almost everywhere. We now consider other types of convergence and make comparisons.

Let $f, f_1, f_2, \ldots$ be complex-valued Borel measurable functions on $(\Omega, \mathscr{F}, \mu)$. We say that $f_n \to f$ *in measure* (or in μ-measure if we wish to emphasize the dependence on μ) iff for every $\varepsilon > 0$, $\mu\{\omega: |f_n(\omega) - f(\omega)| \geq \varepsilon\} \to 0$ as $n \to \infty$. (Notation: $f_n \xrightarrow{\mu} f$.) When μ is a probability measure, the convergence is called *convergence in probability*.

The first result shows that L^p convergence is stronger than convergence in measure.

2.5.1 ***Theorem.*** If $f, f_1, f_2, \ldots \in L^p$ $(0 < p < \infty)$, then $f_n \xrightarrow{L^p} f$ implies $f_n \xrightarrow{\mu} f$.

PROOF. Apply Chebyshev's inequality (2.4.9) to $|f_n - f|$. ∎

The same argument shows that if $\{f_n\}$ is a Cauchy sequence in L^p, then $\{f_n\}$ is *Cauchy in measure*, that is, given $\varepsilon > 0$, $\mu\{\omega: |f_n(\omega) - f_m(\omega)| \geq \varepsilon\} \to 0$ as $n, m \to \infty$.

If $f, f_1, f_2, \ldots$ are complex-valued Borel measurable functions on $(\Omega, \mathscr{F}, \mu)$, we say that $f_n \to f$ *almost uniformly* iff, given $\varepsilon > 0$, there is a set $A \in \mathscr{F}$ such that $\mu(A) < \varepsilon$ and $f_n \to f$ uniformly on A^c.

Almost uniform convergence is stronger than both a.e. convergence and convergence in measure, as we now prove.

2.5.2 ***Theorem.*** If $f_n \to f$ almost uniformly, then $f_n \to f$ in measure and almost everywhere.

PROOF. If $\varepsilon > 0$, let $f_n \to f$ uniformly on A^c, with $\mu(A) < \varepsilon$. If $\delta > 0$, then eventually $|f_n - f| < \delta$ on A^c, so $\{|f_n - f| \geq \delta\} \subset A$. Thus $\mu\{|f_n - f \geq \delta\} \leq \mu(A) < \varepsilon$, proving convergence in measure.

To prove almost everywhere convergence, choose, for each positive integer k, a set A_k with $\mu(A_k) < 1/k$ and $f_n \to f$ uniformly on $A_k{}^c$. If $B = \bigcup_{k=1}^{\infty} A_k{}^c$, then $f_n \to f$ on B and $\mu(B^c) = \mu(\bigcap_{k=1}^{\infty} A_k) \leq \mu(A_k) \to 0$ as $k \to \infty$. Thus $\mu(B^c) = 0$ and the result follows. ∎

The converse to 2.5.2 does not hold in general, as we shall see in 2.5.6(c), but we do have the following result.

2.5.3 ***Theorem.*** If $\{f_n\}$ is convergent in measure, there is a subsequence converging almost uniformly (in particular, a.e. and in measure) to the same limit function.

PROOF. First note that $\{f_n\}$ is Cauchy in measure, because if $|f_n - f_m| \geq \varepsilon$, then either $|f_n - f| \geq \varepsilon/2$ or $|f - f_m| \geq \varepsilon/2$. Thus

$$\mu\{|f_n - f_m| \geq \varepsilon\} \leq \mu\left\{|f_n - f| \geq \frac{\varepsilon}{2}\right\} + \mu\left\{|f - f_m| \geq \frac{\varepsilon}{2}\right\} \to 0 \quad \text{as} \quad n, m \to \infty.$$

Now for each positive integer k, choose a positive integer N_k such that $N_{k+1} > N_k$ for all k and

$$\mu\{\omega: |f_n(\omega) - f_m(\omega)| \geq 2^{-k}\} \leq 2^{-k} \quad \text{for} \quad n, m \geq N_k.$$

Pick integers $n_k \geq N_k$, $k = 1, 2, \ldots$; then if $g_k = f_{n_k}$,

$$\mu\{\omega: |g_k(\omega) - g_{k+1}(\omega)| \geq 2^{-k}\} \leq 2^{-k}.$$

Let $A_k = \{|g_k - g_{k+1}| \geq 2^{-k}\}$, $A = \limsup_k A_k$. Then $\mu(A) = 0$ by 2.2.4; but if $\omega \notin A$, then $\omega \in A_k$ for only finitely many k; hence $|g_k(\omega) - g_{k+1}(\omega)| < 2^{-k}$

for large k, and it follows that $g_k(\omega)$ converges to a limit $g(\omega)$. Since $\mu(A) = 0$ we have $g_k \to g$ a.e.

If $B_r = \bigcup_{k=r}^{\infty} A_k$, then $\mu(B_r) \le \sum_{k=r}^{\infty} \mu(A_k) < \varepsilon$ for large r. If $\omega \notin B_r$, then $|g_k(\omega) - g_{k+1}(\omega)| < 2^{-k}$, $k = r, r+1, r+2, \ldots$. By the Weierstrass M-test, $g_k \to g$ uniformly on B_r, which proves almost uniform convergence.

Now by hypothesis, we have $f_n \xrightarrow{\mu} f$ for some f, hence $f_{n_k} \xrightarrow{\mu} f$. But by 2.5.2, $f_{n_k} \xrightarrow{\mu} g$ as well, hence $f = g$ a.e. (see Problem 1). Thus f_{n_k} converges almost uniformly to f, completing the proof. ▮

There is a partial converse to 2.5.2, but before discussing this it will be convenient to look at a condition equivalent to a.e. convergence:

2.5.4 ***Lemma.*** If μ is finite, then $f_n \to f$ a.e. iff for every $\delta > 0$,

$$\mu\left(\bigcup_{k=n}^{\infty} \{\omega\colon |f_k(\omega) - f(\omega)| \ge \delta\}\right) \to 0 \qquad \text{as} \quad n \to \infty.$$

PROOF. Let $B_{n\delta} = \{\omega\colon |f_n(\omega) - f(\omega)| \ge \delta\}$, $B_\delta = \limsup_n B_{n\delta} = \bigcap_{n=1}^{\infty} \bigcup_{k=n}^{\infty} B_{k\delta}$. Now $\bigcup_{k=n}^{\infty} B_{k\delta} \downarrow B_\delta$; hence $\mu(\bigcup_{k=n}^{\infty} B_{k\delta}) \to \mu(B_\delta)$ as $n \to \infty$ by 1.2.7(b). Now

$$\{\omega\colon f_n(\omega) \not\to f(\omega)\} = \bigcup_{\delta > 0} B_\delta$$

$$= \bigcup_{m=1}^{\infty} B_{1/m} \qquad \text{since} \quad B_{\delta_1} \subset B_{\delta_2} \quad \text{for} \quad \delta_1 > \delta_2.$$

Therefore,

$$f_n \to f \quad \text{a.e.} \quad \text{iff} \quad \mu(B_\delta) = 0 \qquad \text{for all} \quad \delta > 0$$

$$\text{iff} \quad \mu\left(\bigcup_{k=n}^{\infty} B_{k\delta}\right) \to 0 \qquad \text{for all} \quad \delta > 0. \quad ▮$$

2.5.5 ***Egoroff's Theorem.*** If μ is finite and $f_n \to f$ a.e., then $f_n \to f$ almost uniformly. Hence by 2.5.2, if μ is finite, then almost everywhere convergence implies convergence in measure.

PROOF. It follows from 2.5.4 that given $\varepsilon > 0$ and a positive integer j, for sufficiently large $n = n(j)$, the set $A_j = \bigcup_{k=n(j)}^{\infty} \{|f_k - f| \ge 1/j\}$ has measure less than $\varepsilon/2^j$. If $A = \bigcup_{j=1}^{\infty} A_j$, then $\mu(A) \le \sum_{j=1}^{\infty} \mu(A_j) < \varepsilon$. Also, if $\delta > 0$ and j is chosen so that $1/j < \delta$, we have, for any $k \ge n(j)$ and $\omega \in A^c$ (hence $\omega \notin A_j$), $|f_k(\omega) - f(\omega)| < 1/j < \delta$. Thus $f_n \to f$ uniformly on A^c. ▮

We now give some examples to illustrate the relations between the various types of convergence. In all cases, we assume that $\mathscr{F}$ is the class of Borel sets and μ is Lebesgue measure.

2.5.6 ***Examples.*** (a) Let $\Omega = [0, 1]$ and define

$$f_n(x) = \begin{cases} e^n & \text{if } 0 \le x \le \dfrac{1}{n}, \\ 0 & \text{elsewhere.} \end{cases}$$

Then $f_n \to 0$ a.e., hence in measure by 2.5.5. But for each $p \in (0, \infty]$, f_n fails to converge in L^p. For if $p < \infty$,

$$\|f_n\|_p^p = \int_0^1 |f_n(x)|^p \, dx = \frac{1}{n} e^{np} \to \infty,$$

and

$$\|f_n\|_\infty = e^n \to \infty.$$

(b) Let $\Omega = R$, and define

$$f_n(x) = \begin{cases} \dfrac{1}{n} & \text{if } 0 \le x \le e^n, \\ 0 & \text{elsewhere.} \end{cases}$$

Then $f_n \to 0$ uniformly on R, so that $f_n \xrightarrow{L^\infty} 0$. It follows quickly that $f_n \to 0$ a.e. and in measure. But for each $p \in (0, \infty)$, f_n fails to converge in L^p, since $\|f_n\|_p^p = n^{-p} e^n \to \infty$.

(c) Let $\Omega = [0, \infty)$ and define

$$f_n(x) = \begin{cases} 1 & \text{if } n \le x \le n + \dfrac{1}{n}, \\ 0 & \text{elsewhere.} \end{cases}$$

Then $f_n \to 0$ a.e. and in measure (as well as in L^p, $0 < p < \infty$), but does not converge almost uniformly. For, if $f_n \to 0$ uniformly on A and $\mu(A^c) < \varepsilon$, then eventually $f_n < 1$ on A; hence if $A_n = [n, n + (1/n)]$ we have $A \cap \bigcup_{k \ge n} A_k = \varnothing$ for sufficiently large n. Therefore, $A^c \supset \bigcup_{k \ge n} A_k$, and consequently $\mu(A^c) \ge \sum_{k=n}^\infty \mu(A_k) = \infty$, a contradiction.

(d) Let $\Omega = [0, 1]$, and define

$$f_{nm}(x) = \begin{cases} 1 & \text{if } \dfrac{m-1}{n} < x \le \dfrac{m}{n}, \quad m = 1, \dots, n, \quad n = 1, 2, \dots, \\ 0 & \text{elsewhere.} \end{cases}$$

Then $\|f_{nm}\|_p^p = 1/n \to 0$, so for each $p \in (0, \infty)$, the sequence $f_{11}, f_{21}, f_{22}, f_{31}, f_{32}, f_{33}, \ldots$ converges to 0 in L^p (hence converges in measure by 2.5.1). But the sequence does not converge a.e., hence by 2.5.2, does not converge almost uniformly. To see this, observe that for any $x \neq 0$, the sequence $\{f_{nm}(x)\}$ has infinitely many zeros and infinitely many ones. Thus the set on which f_{nm} converges has measure 0. Also, f_{nm} does not converge in L^∞, for if $f_{nm} \xrightarrow{L^\infty} f$, then $f_{nm} \xrightarrow{\mu} f$, hence $f = 0$ a.e. (see Problem 1). But $\|f_{nm}\|_\infty \equiv 1$, a contradiction.

Problems

1. If f_n converges to both f and g in measure, show that $f = g$ a.e.
2. Show that a sequence is Cauchy in measure iff it is convergent in measure.
3. (a) If μ is finite, show that L^∞ convergence implies L^p convergence for all $p \in (0, \infty)$.
   (b) Show that any real-valued function in $L^p[a, b]$, $-\infty < a < b < \infty$, $0 < p < \infty$, can be approximated in L^p by a polynomial, in fact by a polynomial with rational coefficients.
4. If μ is finite, show that $\{f_n\}$ is Cauchy a.e. (for almost every ω, $\{f_n(\omega)\}$ is a Cauchy sequence) iff for every $\delta > 0$,

$$\mu\left(\bigcup_{j,k=n}^{\infty} \{\omega : |f_j(\omega) - f_k(\omega)| \geq \delta\}\right) \to 0 \quad \text{as} \quad n \to \infty.$$

5. (*Extension of the dominated convergence theorem*) If $|f_n| \leq g$ for all $n = 1, 2, \ldots$, where g is μ-integrable, and $f_n \xrightarrow{\mu} f$, show that f is μ-integrable and $\int_\Omega f_n \, d\mu \to \int_\Omega f \, d\mu$.

2.6 Product Measures and Fubini's Theorem

Lebesgue measure on R^n is in a sense the product of n copies of one-dimensional Lebesgue measure, since the volume of an n-dimensional rectangular box is the product of the lengths of the sides. In this section we develop this idea in a general setting. We shall be interested in two constructions. First, suppose that $(\Omega_j, \mathscr{F}_j, \mu_j)$ is a measure space for $j = 1, 2, \ldots, n$. We wish to construct a measure on subsets of $\Omega_1 \times \Omega_2 \times \cdots \times \Omega_n$ such that the measure of the "rectangle" $A_1 \times A_2 \times \cdots \times A_n$ [with each $A_j \in \mathscr{F}_j$] is $\mu_1(A_1)\mu_2(A_2) \cdots \mu_n(A_n)$. The second construction involves compound experiments in probability. Suppose that two observations are made, with the first observation resulting in a point $\omega_1 \in \Omega_1$, the second in a point $\omega_2 \in \Omega_2$. The probability that the first observation falls into the set A is, say, $\mu_1(A)$. Furthermore, if the first observation is ω_1, the probability that the second observation falls into B is, say, $\mu(\omega_1, B)$, where $\mu(\omega_1, \cdot)$ is a probability measure defined on $\mathscr{F}_2$ for each $\omega_1 \in \Omega_1$. The probability that the first observation will belong to A and the second will belong to B should be given by

$$\mu(A \times B) = \int_A \mu(\omega_1, B)\mu_1(d\omega_1),$$

and we would like to construct a probability measure on subsets of $\Omega_1 \times \Omega_2$ such that $\mu(A \times B)$ is given by this formula for each $A \in \mathscr{F}_1$ and $B \in \mathscr{F}_2$. [Intuitively, the probability that the first observation will fall near ω_1 is $\mu_1(d\omega_1)$; given that the first observation is ω_1, the second observation will fall in B with probability $\mu(\omega_1, B)$. Thus $\mu(\omega_1, B)\mu_1(d\omega_1)$ represents the probability of one possible favorable outcome of the experiment. The total probability is found by adding the probabilities of favorable outcomes, in other words, by integrating over A. Reasoning of this type may not appear natural at this point, since we have not yet talked in detail about probability theory. However, it may serve to indicate the motivation behind the theorems of this section.]

2.6.1 ***Definitions.*** Let $\mathscr{F}_j$ be a σ-field of subsets of Ω_j, $j = 1, 2, \ldots, n$, and let $\Omega = \Omega_1 \times \Omega_2 \times \cdots \times \Omega_n$. A *measurable rectangle* in Ω is a set $A = A_1 \times A_2 \times \cdots \times A_n$, where $A_j \in \mathscr{F}_j$ for each $j = 1, 2, \ldots, n$. The smallest σ-field containing the measurable rectangles is called the *product* σ-field, written $\mathscr{F}_1 \times \mathscr{F}_2 \times \cdots \times \mathscr{F}_n$. If all $\mathscr{F}_j$ coincide with a fixed σ-field $\mathscr{F}$, the product σ-field is denoted by $\mathscr{F}^n$. Note that in spite of the notation, $\mathscr{F}_1 \times \mathscr{F}_2 \times \cdots \times \mathscr{F}_n$ is not the Cartesian product of the $\mathscr{F}_j$; the Cartesian product is the set of measurable rectangles, while the product σ-field is the minimal σ-field over the measurable rectangles. Note also that the collection of finite disjoint unions of measurable rectangles forms a field (see Problem 1).

The next theorem is stated in such a way that both constructions described above become special cases.

2.6.2 ***Product Measure Theorem.*** Let $(\Omega_1, \mathscr{F}_1, \mu_1)$ be a measure space, with μ_1 σ-finite on $\mathscr{F}_1$, and let Ω_2 be a set with σ-field $\mathscr{F}_2$. Assume that for each $\omega_1 \in \Omega_1$ we are given a measure $\mu(\omega_1, \cdot)$ on $\mathscr{F}_2$. Assume that $\mu(\omega_1, B)$, besides being a measure in B for each fixed $\omega_1 \in \Omega_1$, is Borel measurable in ω_1 for each fixed $B \in \mathscr{F}_2$. Assume that the $\mu(\omega_1 \cdot)$ are *uniformly σ-finite*; that is, Ω_2 can be written as $\bigcup_{n=1}^{\infty} B_n$, where for some positive (finite) constants k_n we have $\mu(\omega_1, B_n) \leq k_n$ for all $\omega_1 \in \Omega_1$. [The case in which the $\mu(\omega_1, \cdot)$ are uniformly bounded, that is, $\mu(\omega_1, \Omega_2) \leq k < \infty$ for all ω_1, is of course included.]

Then there is a unique measure μ on $\mathscr{F} = \mathscr{F}_1 \times \mathscr{F}_2$ such that

$$\mu(A \times B) = \int_A \mu(\omega_1, B)\mu_1(d\omega_1) \qquad \text{for all} \quad A \in \mathscr{F}_1, \quad B \in \mathscr{F}_2,$$

namely,

$$\mu(F) = \int_{\Omega_1} \mu(\omega_1, F(\omega_1))\mu_1(d\omega_1), \qquad F \in \mathscr{F},$$

where $F(\omega_1)$ denotes the *section* of F at ω_1:

$$F(\omega_1) = \{\omega_2 \in \Omega_2 : (\omega_1, \omega_2) \in F\}.$$

Furthermore, μ is σ-finite on $\mathscr{F}$; if μ_1 and all the $\mu(\omega_1, \cdot)$ are probability measures, so is μ.

PROOF. First assume that the $\mu(\omega_1, \cdot)$ are finite.

(1) If $C \in \mathscr{F}$, then $C(\omega_1) \in \mathscr{F}_2$ for each $\omega_1 \in \Omega_1$.

To prove this, let $\mathscr{C} = \{C \in \mathscr{F} : C(\omega_1) \in \mathscr{F}_2\}$. Then $\mathscr{C}$ is a σ-field since

$$\left(\bigcup_{n=1}^{\infty} C_n\right)(\omega_1) = \bigcup_{n=1}^{\infty} C_n(\omega_1), \qquad C^c(\omega_1) = (C(\omega_1))^c.$$

If $A \in \mathscr{F}_1$, $B \in \mathscr{F}_2$, then $(A \times B)(\omega_1) = B$ if $\omega_1 \in A$ and $\varnothing$ if $\omega_1 \notin A$. Thus $\mathscr{C}$ contains all measurable rectangles; hence $\mathscr{C} = \mathscr{F}$.

(2) If $C \in \mathscr{F}$, then $\mu(\omega_1, C(\omega_1))$ is Borel measurable in ω_1.

To prove this, let $\mathscr{C}$ be the class of sets in $\mathscr{F}$ for which the conclusion of (2) holds. If $C = A \times B$, $A \in \mathscr{F}_1$, $B \in \mathscr{F}_2$, then

$$\mu(\omega_1, C(\omega_1)) = \begin{cases} \mu(\omega_1, B) & \text{if} \quad \omega_1 \in A, \\ \mu(\omega_1, \varnothing) = 0 & \text{if} \quad \omega_1 \notin A. \end{cases}$$

Thus $\mu(\omega_1, C(\omega_1)) = \mu(\omega_1, B)I_A(\omega_1)$, and is Borel measurable by hypothesis. Therefore measurable rectangles belong to $\mathscr{C}$. If $C_1, \ldots, C_n$ are disjoint measurable rectangles,

$$\mu\left(\omega_1, \left(\bigcup_{i=1}^{n} C_i\right)(\omega_1)\right) = \mu\left(\omega_1, \bigcup_{i=1}^{n} C_i(\omega_1)\right)$$
$$= \sum_{i=1}^{n} \mu(\omega_1, C_i(\omega_1))$$

is a finite sum of Borel measurable functions, and hence is Borel measurable in ω_1. Thus $\mathscr{C}$ contains the field of finite disjoint unions of measurable rectangles. But $\mathscr{C}$ is a monotone class, for if $C_n \in \mathscr{C}$, $n = 1, 2, \ldots$, and $C_n \uparrow C$, then $C_n(\omega_1) \uparrow C(\omega_1)$; hence $\mu(\omega_1, C_n(\omega_1)) \to \mu(\omega_1, C(\omega_1))$. Thus $\mu(\omega_1, C(\omega_1))$, a limit of measurable functions, is measurable in ω_1. If $C_1 \downarrow C$, the same conclusion holds since the $\mu(\omega_1, \cdot)$ are finite. Thus $\mathscr{C} = \mathscr{F}$.

(3) Define

$$\mu(F) = \int_{\Omega_1} \mu(\omega_1, F(\omega_1))\mu_1(d\omega_1), \qquad F \in \mathscr{F}$$

[the integral exists by (2)]. Then μ is a measure on $\mathscr{F}$, and

$$\mu(A \times B) = \int_A \mu(\omega_1, B)\mu_1\,(d\omega_1) \qquad \text{for all} \quad A \in \mathscr{F}_1, \quad B \in \mathscr{F}_2.$$

To prove this, let $F_1, F_2, \ldots$ be disjoint sets in $\mathscr{F}$. Then

$$\begin{aligned} \mu\left(\bigcup_{n=1}^{\infty} F_n\right) &= \int_{\Omega_1} \mu\left(\omega_1, \bigcup_{n=1}^{\infty} F_n(\omega_1)\right)\mu_1(d\omega_1) \\ &= \int_{\Omega_1} \sum_{n=1}^{\infty} \mu(\omega_1, F_n(\omega_1))\mu_1(d\omega_1) \\ &= \sum_{n=1}^{\infty} \int_{\Omega_1} \mu(\omega_1, F_n(\omega_1))\mu_1(d\omega_1) \\ &= \sum_{n=1}^{\infty} \mu(F_n), \end{aligned}$$

proving that μ is a measure. Now

$$\begin{aligned} \mu(A \times B) &= \int_{\Omega_1} \mu(\omega_1, (A \times B)(\omega_1))\mu_1(d\omega_1) \\ &= \int_{\Omega_1} \mu(\omega_1, B) I_A(\omega_1)\mu_1(d\omega_1) \qquad \text{[see (2)]} \\ &= \int_A \mu(\omega_1, B)\mu_1(d\omega_1) \end{aligned}$$

as desired.

Now assume the $\mu(\omega_1, \cdot)$ uniformly σ-finite. Let $\Omega_2 = \bigcup_{n=1}^{\infty} B_n$, where the B_n are disjoint sets in $\mathscr{F}_2$ and $\mu(\omega_1, B_n) \le k_n < \infty$ for all $\omega_1 \in \Omega_1$. If we set

$$\mu_n'(\omega_1, B) = \mu(\omega_1, B \cap B_n), \qquad B \in \mathscr{F}_2,$$

the $\mu_n'(\omega_1, \cdot)$ are finite, and the above construction gives a measure μ_n' on $\mathscr{F}$ such that

$$\mu_n'(A \times B) = \int_A \mu_n'(\omega_1, B)\mu_1(d\omega_1), \qquad A \in \mathscr{F}_1, \quad B \in \mathscr{F}_2$$

$$= \int_A \mu(\omega_1, B \cap B_n)\mu_1(d\omega_1),$$

namely,

$$\mu_n'(F) = \int_{\Omega_1} \mu_n'(\omega_1, F(\omega_1))\mu_1(d\omega_1) = \int_{\Omega_1} \mu(\omega_1, F(\omega_1) \cap B_n)\mu_1(d\omega_1).$$

Let $\mu = \sum_{n=1}^{\infty} \mu_n'$; μ has the desired properties.

For the uniqueness proof, assume the $\mu(\omega_1, \cdot)$ to be uniformly σ-finite. If λ is a measure on $\mathscr{F}$ such that $\lambda(A \times B) = \int_A \mu(\omega_1, B)\mu_1(d\omega_1)$ for all $A \in \mathscr{F}_1$, $B \in \mathscr{F}_2$, then $\lambda = \mu$ on the field $\mathscr{F}_0$ of finite disjoint unions of measurable rectangles. Now μ is σ-finite on $\mathscr{F}_0$, for if $\Omega_2 = \bigcup_{n=1}^{\infty} B_n$ with $B_n \in \mathscr{F}_2$ and $\mu(\omega_1, B_n) \le k_n < \infty$ for all ω_1, and $\Omega_1 = \bigcup_{m=1}^{\infty} A_m$, where the A_m belong to $\mathscr{F}_1$ and $\mu_1(A_m) < \infty$, then $\Omega_1 \times \Omega_2 = \bigcup_{m,n=1}^{\infty} (A_m \times B_n)$ and

$$\mu(A_m \times B_n) = \int_{A_m} \mu(\omega_1, B_n)\mu_1(d\omega_1) \le k_n \mu_1(A_m) < \infty.$$

Thus $\lambda = \mu$ on $\mathscr{F}$ by the Carathéodory extension theorem.

We have just seen that μ is σ-finite on $\mathscr{F}_0$, hence on $\mathscr{F}$. If μ_1 and all the $\mu(\omega_1, \cdot)$ are probability measures, it is immediate that μ is also. ▮

2.6.3 ***Corollary: Classical Product Measure Theorem.*** Let $(\Omega_j, \mathscr{F}_j, \mu_j)$ be a measure space for $j = 1, 2$, with μ_j σ-finite on $\mathscr{F}_j$. If $\Omega = \Omega_1 \times \Omega_2$, $\mathscr{F} = \mathscr{F}_1 \times \mathscr{F}_2$, the set function given by

$$\mu(F) = \int_{\Omega_1} \mu_2(F(\omega_1))\, d\mu_1(\omega_1) = \int_{\Omega_2} \mu_1(F(\omega_2))\, d\mu_2(\omega_2)$$

is the unique measure on $\mathscr{F}$ such that $\mu(A \times B) = \mu_1(A)\mu_2(B)$ for all $A \in \mathscr{F}_1$, $B \in \mathscr{F}_2$. Furthermore, μ is σ-finite on $\mathscr{F}$, and is a probability measure if μ_1 and μ_2 are. The measure μ is called the *product* of μ_1 and μ_2, written $\mu = \mu_1 \times \mu_2$.

PROOF. In 2.6.2, take $\mu(\omega_1, \cdot) = \mu_2$ for all ω_1. The second formula for $\mu(F)$ is obtained by interchanging μ_1 and μ_2. ▮

As a special case, Let $\Omega_1 = \Omega_2 = R$, $\mathscr{F}_1 = \mathscr{F}_2 = \mathscr{B}(R)$, $\mu_1 = \mu_2 =$ Lebesgue measure. Then $\mathscr{F}_1 \times \mathscr{F}_2 = \mathscr{B}(R^2)$ (Problem 2), and $\mu = \mu_1 \times \mu_2$ agrees with Lebesgue measure on intervals $(a, b] = (a_1, b_1] \times (a_2, b_2]$. By the Carathéodory extension theorem, μ is Lebesgue measure on $\mathscr{B}(R^2)$, so we have another method of constructing two-dimensional Lebesgue measure. We shall generalize to n dimensions later in the section.

The integration theory we have developed thus far includes the notion of a multiple integral on R^n; this is simply an integral with respect to n-dimensional Lebesgue measure. However, in calculus, integrals of this type are evaluated by computing iterated integrals. The general theorem which justifies this process is Fubini's theorem, which is a direct consequence of the product measure theorem.

2.6.4 ***Fubini's Theorem.*** Assume the hypothesis of the product measure theorem 2.6.2. Let $f: (\Omega, \mathscr{F}) \to (\bar{R}, \mathscr{B}(\bar{R}))$.

(a) If f is nonnegative, then $\int_{\Omega_2} f(\omega_1, \omega_2)\mu(\omega_1, d\omega_2)$ exists and defines a Borel measurable function of ω_1. Also

$$\int_{\Omega} f\, d\mu = \int_{\Omega_1}\left(\int_{\Omega_2} f(\omega_1, \omega_2)\mu(\omega_1, d\omega_2)\right)\mu_1(d\omega_1).$$

(b) If $\int_{\Omega} f\, d\mu$ exists (respectively, is finite), then $\int_{\Omega_2} f(\omega_1, \omega_2)\mu(\omega_1, d\omega_2)$ exists (respectively, is finite) for μ_1-almost every ω_1, and defines a Borel measurable function of ω_1 if it is taken as 0 (or as any Borel measurable function of ω_1) on the exceptional set. Also,

$$\int_{\Omega} f\, d\mu = \int_{\Omega_1}\left(\int_{\Omega_2} f(\omega_1, \omega_2)\mu(\omega_1, d\omega_2)\right)\mu_1(d\omega_1).$$

[The notation $\int_{\Omega_2} f(\omega_1, \omega_2)\mu(\omega_1, d\omega_2)$ indicates that for a fixed ω_1, the function given by $g(\omega_2) = f(\omega_1, \omega_2)$ is to be integrated with respect to the measure $\mu(\omega_1, \cdot)$.]

PROOF. (a) First note that:

(1) For each fixed ω_1 we have $f(\omega_1, \cdot): (\Omega_2, \mathscr{F}_2) \to (\bar{R}, \mathscr{B}(\bar{R}))$. In other words if f is jointly measurable, that is, measurable relative to the product σ-field $\mathscr{F}_1 \times \mathscr{F}_2$, it is measurable in each variable separately. For if $B \in \mathscr{B}(\bar{R})$, $\{\omega_2 : f(\omega_1, \omega_2) \in B\} = \{\omega_2 : (\omega_1, \omega_2) \in f^{-1}(B)\} = f^{-1}(B)(\omega_1) \in \mathscr{F}_2$ by part (1) of the proof of 2.6.2. Thus $\int_{\Omega_2} f(\omega_1, \omega_2)\mu(\omega_1, d\omega_2)$ exists.

Now let I_F, $F \in \mathscr{F}$, be an indicator. Then

$$\int_{\Omega_2} I_F(\omega_1, \omega_2)\mu(\omega_1, d\omega_2) = \int_{\Omega_2} I_{F(\omega_1)}(\omega_2)\mu(\omega_1, d\omega_2) = \mu(\omega_1, F(\omega_1)),$$

and this is Borel measurable in ω_1 by part (2) of the proof of 2.6.2. Also

$$\int_{\Omega} I_F\, d\mu = \mu(F) = \int_{\Omega_1} \mu(\omega_1, F(\omega_1))\mu_1(d\omega_1) \qquad \text{by 2.6.2}$$

$$= \int_{\Omega_1}\int_{\Omega_2} I_F(\omega_1, \omega_2)\mu(\omega_1, d\omega_2)\mu_1(d\omega_1).$$

Now if $f = \sum_{j=1}^{n} x_j I_{F_j}$, the F_j disjoint sets in $\mathscr{F}$, is a nonnegative simple function, then

$$\int_{\Omega_2} f(\omega_1, \omega_2)\mu(\omega_1, d\omega_2) = \sum_{j=1}^{n} x_j \mu(\omega_1, F_j(\omega_1)),$$

Borel measurable in ω_1,

and

$$\int_{\Omega} f\, d\mu = \sum_{j=1}^{n} x_j \int_{\Omega} I_{Fj}\, d\mu = \sum_{j=1}^{n} x_j \int_{\Omega_1} \int_{\Omega_2} I_{F_j}(\omega_1, \omega_2)\mu(\omega_1, d\omega_2)\mu_1(d\omega_1)$$

by what we have proved for indicators

$$= \int_{\Omega_1} \int_{\Omega_2} f(\omega_1, \omega_2)\mu(\omega_1, d\omega_2)\mu_1(d\omega_1).$$

Finally, if $f: (\Omega, \mathscr{F}) \to (\bar{R}, \mathscr{B}(\bar{R}))$, $f \geq 0$, let $0 \leq f_n \uparrow f$, f_n simple. Then

$$\int_{\Omega_2} f(\omega_1, \omega_2)\mu(\omega_1, d\omega_2) = \lim_{n\to\infty} \int_{\Omega_2} f_n(\omega_1, \omega_2)\mu(\omega_1, d\omega_2),$$

which is Borel measurable in ω_1, and

$$\int_{\Omega} f\, d\mu = \lim_{n\to\infty} \int_{\Omega} f_n\, d\mu = \lim_{n\to\infty} \int_{\Omega_1} \int_{\Omega_2} f_n(\omega_1, \omega_2)\mu(\omega_1, d\omega_2)\mu_1(d\omega_1)$$

by what we have proved for simple functions

$$= \int_{\Omega_1} \int_{\Omega_2} f(\omega_1, \omega_2)\mu(\omega_1, d\omega_2)\mu_1(d\omega_1)$$

using the monotone convergence theorem twice.

This proves (a).

(b) Suppose that $\int_{\Omega} f^-\, d\mu < \infty$. By (a),

$$\int_{\Omega_1} \int_{\Omega_2} f^-(\omega_1, \omega_2)\mu(\omega_1, d\omega_2)\mu_1(d\omega_1) = \int_{\Omega} f^-\, d\mu < \infty$$

so that $\int_{\Omega_2} f^-(\omega_1, \omega_2)\mu(\omega_1, d\omega_2)$ is μ_1-integrable; hence finite a.e. $[\mu_1]$; thus:

(2) For μ_1-almost every ω_1 we may write:

$$\int_{\Omega_2} f(\omega_1, \omega_2)\mu(\omega_1, d\omega_2) = \int_{\Omega_2} f^+(\omega_1, \omega_2)\mu(\omega_1, d\omega_2) - \int_{\Omega_2} f^-(\omega_1, \omega_2)\mu(\omega_1, d\omega_2).$$

If $\int_{\Omega} f\, d\mu$ is finite, both integrals on the right side of (2) are finite a.e. $[\mu_1]$. In any event, we may define all integrals in (2) to be 0 (or any other Borel measurable function of ω_1) on the exceptional set, and (2) will then be valid for all ω_1, and will define a Borel measurable function of ω_1. If we integrate (2) with respect to μ_1, we obtain, by (a) and the additivity theorem for integrals,

$$\int_{\Omega_1}\int_{\Omega_2} f(\omega_1, \omega_2)\mu(\omega_1, d\omega_2)\mu_1(d\omega_1) = \int_\Omega f^+ \, d\mu - \int_\Omega f^- \, d\mu = \int_\Omega f \, d\mu. \quad ▮$$

2.6.5 ***Corollary.*** If $f: (\Omega, \mathscr{F}) \to (\bar{R}, \mathscr{B}(\bar{R}))$ and the iterated integral $\int_{\Omega_1}\int_{\Omega_2} |f(\omega_1, \omega_2)| \mu(\omega_1, d\omega_2) \, \mu_1(d\omega_1) < \infty$, then $\int_\Omega f \, d\mu$ is finite, and thus Fubini's theorem applies.

PROOF. By 2.6.4(a), $\int_\Omega |f| \, d\mu < \infty$, and thus the hypothesis of 2.6.4(b) is satisfied. ▮

As a special case, we obtain the following classical result.

2.6.6 ***Classical Fubini Theorem.*** Let $\Omega = \Omega_1 \times \Omega_2$, $\mathscr{F} = \mathscr{F}_1 \times \mathscr{F}_2$, $\mu = \mu_1 \times \mu_2$, where μ_j is a σ-finite measure on $\mathscr{F}_j$, $j = 1, 2$. If f is a Borel measurable function on $(\Omega, \mathscr{F})$ such that $\int_\Omega f \, d\mu$ exists, then

$$\begin{aligned} \int_\Omega f \, d\mu &= \int_{\Omega_1}\int_{\Omega_2} f \, d\mu_2 \, d\mu_1 \\ &= \int_{\Omega_2}\int_{\Omega_1} f \, d\mu_1 \, d\mu_2 \qquad \text{by symmetry.} \end{aligned}$$

PROOF. Apply 2.6.4 with $\mu(\omega_1, \cdot) = \mu_2$ for all ω_1. ▮

Note that by 2.6.5, if $\int_{\Omega_1}\int_{\Omega_2} |f| \, d\mu_2 \, d\mu_1 < \infty$ (or $\int_{\Omega_2}\int_{\Omega_1} |f| \, d\mu_1 \, d\mu_2 < \infty$), the iterated integration formula 2.6.6 holds.

In 2.6.4(b), if we wish to define $\int_{\Omega_2} f(\omega_1, \omega_2)\mu(\omega_1, d\omega_2)$ in a completely arbitrary fashion on the exceptional set where the integral does not exist, and still produce a Borel measurable function of ω_1, we should assume that $(\Omega_1, \mathscr{F}_1, \mu_1)$ is a complete measure space. The situation is as follows. We have $h: (\Omega_1, \mathscr{F}_1) \to (\bar{R}, \mathscr{B}(\bar{R}))$, where h is the above integral, taken as 0 on the exceptional set A. We set $g(\omega_1) = h(\omega_1)$, $\omega_1 \notin A$; $g(\omega_1) = q(\omega_1)$ arbitrary, $\omega_1 \in A$ (q not necessarily Borel measurable). If B is a Borel subset of $\bar{R}$, then

$$\{\omega_1 : g(\omega_1) \in B\} = [A^c \cap \{\omega_1 : h(\omega_1) \in B\}] \cup [A \cap \{\omega_1 : q(\omega_1) \in B\}].$$

The first set of the union belongs to $\mathscr{F}_1$, and the second is a subset of A, with $\mu_1(A) = 0$, and hence belongs to $\mathscr{F}_1$ by completeness. Thus g is Borel measurable.

In the classical Fubini theorem, if we want to define $\int_{\Omega_2} f(\omega_1, \omega_2)\, d\mu_2(\omega_2)$ and $\int_{\Omega_1} f(\omega_1, \omega_2)\, d\mu_1(\omega_1)$ in a completely arbitrary fashion on the exceptional sets, we should assume completeness of both spaces $(\Omega_1, \mathscr{F}_1, \mu_1)$ and $(\Omega_2, \mathscr{F}_2, \mu_2)$.

The product measure theorem and Fubini's theorem may be extended to n factors, as follows:

2.6.7 ***Theorem.*** Let $\mathscr{F}_j$ be a σ-field of subsets of Ω_j, $j = 1, \ldots, n$. Let μ_1 be a σ-finite measure on $\mathscr{F}_1$, and, for each $(\omega_1, \ldots, \omega_j) \in \Omega_1 \times \cdots \times \Omega_j$, let $\mu(\omega_1, \ldots, \omega_j, B)$, $B \in \mathscr{F}_{j+1}$, be a measure on $\mathscr{F}_{j+1}$ $(j = 1, 2, \ldots, n-1)$. Assume the $\mu(\omega_1, \ldots, \omega_j, \cdot)$ to be uniformly σ-finite, and assume that $\mu(\omega_1, \ldots, \omega_j, C)$ is measurable: $(\Omega_1 \times \cdots \times \Omega_j, \mathscr{F}_1 \times \cdots \times \mathscr{F}_j) \to (\overline{R}, \mathscr{B}(\overline{R}))$ for each fixed $C \in \mathscr{F}_{j+1}$.

Let $\Omega = \Omega_1 \times \cdots \times \Omega_n$, $\mathscr{F} = \mathscr{F}_1 \times \cdots \times \mathscr{F}_n$.

(a) There is a unique measure μ on $\mathscr{F}$ such that for each measurable rectangle $A_1 \times \cdots \times A_n \in \mathscr{F}$,

$$
\begin{aligned}
\mu(A_1 \times \cdots \times A_n) &= \int_{A_1} \mu_1(d\omega_1) \int_{A_2} \mu(\omega_1, d\omega_2) \\
&\quad \cdots \int_{A_{n-1}} \mu(\omega_1, \ldots, \omega_{n-2}, d\omega_{n-1}) \int_{A_n} \mu(\omega_1, \ldots, \omega_{n-1}, d\omega_n).
\end{aligned}
$$

[Note that the last factor on the right is $\mu(\omega_1, \ldots, \omega_{n-1}, A_n)$.] The measure μ is σ-finite on $\mathscr{F}$, and is a probability measure if μ_1 and all the $\mu(\omega_1, \ldots, \omega_j, \cdot)$ are probability measures.

(b) Let $f: (\Omega, \mathscr{F}) \to (\overline{R}, \mathscr{B}(\overline{R}))$. If $f \geq 0$, then

$$
\begin{aligned}
\int_\Omega f\, d\mu = \int_{\Omega_1} \mu_1(d\omega_1) \int_{\Omega_2} \mu(\omega_1, d\omega_2) \cdots \int_{\Omega_{n-1}} \mu(\omega_1, \ldots, \omega_{n-2}, d\omega_{n-1}) \\
\int_{\Omega_n} f(\omega_1, \ldots, \omega_n) \mu(\omega_1, \ldots, \omega_{n-1}, d\omega_n), \quad (1)
\end{aligned}
$$

where, after the integration with respect to $\mu(\omega_1, \ldots, \omega_j, \cdot)$ is performed $(j = n-1, n-2, \ldots, 1)$, the result is a Borel measurable function of $(\omega_1, \ldots, \omega_j)$.

If $\int_\Omega f\, d\mu$ exists (respectively, is finite), then Eq. (1) holds in the sense that for each $j = n-1, n-2, \ldots, 1$, the integral with respect to $\mu(\omega_1, \ldots, \omega_j, \cdot)$ exists (respectively, is finite) except for $(\omega_1, \ldots, \omega_j)$ in a set of λ_j-measure 0, where λ_j is the measure determined [see (a)] by μ_1 and the measures $\mu(\omega_1, \cdot), \ldots, \mu(\omega_1, \ldots, \omega_{j-1}, \cdot)$. If the integral is defined on the exceptional set as 0 [or any Borel measurable function on the space $(\Omega_1 \times \cdots \times \Omega_j, \mathscr{F}_1 \times \cdots \times \mathscr{F}_j)$], it becomes Borel measurable in $(\omega_1, \ldots \omega_j)$.

PROOF. By 2.6.2 and 2.6.4, the result holds for $n = 2$. Assuming that (a) and (b) hold up to $n - 1$ factors, we consider the n-dimensional case. By the induction hypothesis, there is a unique measure λ_{n-1} on $\mathscr{F}_1 \times \mathscr{F}_2 \times \cdots \times \mathscr{F}_{n-1}$ such that for all $A_1 \in \mathscr{F}_1, \ldots, A_{n-1} \in \mathscr{F}_{n-1}$,

$$\lambda_{n-1}(A_1 \times \cdots \times A_{n-1}) = \int_{A_1} \mu_1(d\omega_1) \int_{A_2} \mu(\omega_1, d\omega_2) \cdots \int_{A_{n-1}} \mu(\omega_1, \ldots, \omega_{n-2}, d\omega_{n-1})$$

and λ_{n-1} is σ-finite. By the $n = 2$ case, there is a unique measure μ on $(\mathscr{F}_1 \times \cdots \times \mathscr{F}_{n-1}) \times \mathscr{F}_n$ (which equals $\mathscr{F}_1 \times \cdots \times \mathscr{F}_n$; see Problem 3) such that for each $A \in \mathscr{F}_1 \times \cdots \times \mathscr{F}_{n-1}$, $A_n \in \mathscr{F}_n$,

$$\begin{aligned}\mu(A \times A_n) &= \int_A \mu(\omega_1, \ldots, \omega_{n-1}, A_n)\, d\lambda_{n-1}(\omega_1, \ldots, \omega_{n-1}) \\ &= \int_{\Omega_1 \times \cdots \times \Omega_{n-1}} I_A(\omega_1, \ldots \omega_{n-1})\mu(\omega_1, \ldots, \omega_{n-1}, A_n) \\ &\qquad d\lambda_{n-1}(\omega_1, \ldots, \omega_{n-1}). \qquad (2)\end{aligned}$$

If A is a measurable rectangle $A_1 \times \cdots \times A_{n-1}$, then $I_A(\omega_1, \ldots, \omega_{n-1}) = I_{A_1}(\omega_1) \cdots I_{A_{n-1}}(\omega_{n-1})$; thus (2) becomes, with the aid of the induction hypothesis on (b),

$$\mu(A_1 \times \cdots \times A_n) = \int_{A_1} \mu_1(d\omega_1) \cdots \int_{A_{n-1}} \mu(\omega_1, \ldots, \omega_{n-1}, A_n)\mu(\omega_1, \ldots, \omega_{n-2}, d\omega_{n-1})$$

which proves the existence of the desired measure μ on $\mathscr{F}$. To show that μ is σ-finite on $\mathscr{F}_0$, the field of finite disjoint unions of measurable rectangles, and consequently μ is unique, let $\Omega_j = \bigcup_{r=1}^{\infty} A_{jr}, j = 1, \ldots, n,$ where $\mu(\omega_1, \ldots, \omega_{j-1}, A_{jr}) \le k_{jr} < \infty$ for all $\omega_1, \ldots, \omega_{j-1}, j = 2, \ldots, n,$ and $\mu_1(A_{1r}) = k_{1r} < \infty$. Then

$$\Omega = \bigcup_{i_1, \ldots, i_n = 1}^{\infty} (A_{1i_1} \times A_{2i_2} \times \cdots \times A_{ni_n}),$$

with

$$\mu(A_{1i_1} \times A_{2i_2} \times \cdots \times A_{ni_n}) \le k_{1i_1} k_{2i_2} \cdots k_{ni_n} < \infty.$$

This proves (a).

To prove (b), note that the measure μ constructed in (a) is determined by λ_{n-1} and the measures $\mu(\omega_1, \ldots, \omega_{n-1}, \cdot)$. Thus by the $n = 2$ case.

$$\int_\Omega f\,d\mu = \int_{\Omega_1 \times \cdots \times \Omega_{n-1}} \int_{\Omega_n} f(\omega_1, \ldots, \omega_n)\mu(\omega_1, \ldots, \omega_{n-1}, d\omega_n)$$

$$d\lambda_{n-1}(\omega_1, \ldots, \omega_{n-1})$$

where the inner integral is Borel measurable in $(\omega_1, \ldots, \omega_{n-1})$, or becomes so after adjustment on a set of λ_{n-1}-measure 0. The desired result now follows by the induction hypothesis. ∎

2.6.8 ***Comments.*** (a) If we take $f = I_F$ in formula (1) of 2.6.7(b), we obtain an explicit formula for $\mu(F)$, $F \in \mathscr{F}$, namely,

$$\mu(F) = \int_{\Omega_1} \mu_1(d\omega_1) \int_{\Omega_2} \mu(\omega_1, d\omega_2) \cdots \int_{\Omega_{n-1}} \mu(\omega_1, \ldots, \omega_{n-2}, d\omega_{n-1})$$

$$\int_{\Omega_n} I_F(\omega_1, \ldots, \omega_n)\mu(\omega_1, \ldots, \omega_{n-1}, d\omega_n)$$

(b) We obtain the classical product measure and Fubini theorems by taking $\mu(\omega_1, \ldots, \omega_j, \cdot) \equiv \mu_{j+1}$, $j = 1, 2, \ldots, n-1$ (with μ_{j+1} σ-finite). We obtain a unique measure μ on $\mathscr{F}$ such that on measurable rectangles,

$$\mu(A_1 \times \cdots \times A_n) = \mu_1(A_1)\mu_2(A_2) \cdots \mu_n(A_n).$$

If $f\colon (\Omega, \mathscr{F}) \to (\overline{R}, \mathscr{B}(\overline{R}))$ and $f \geq 0$ or $\int_\Omega f\,d\mu$ exists, then

$$\int_\Omega f\,d\mu = \int_{\Omega_1} d\mu_1 \int_{\Omega_2} d\mu_2 \cdots \int_{\Omega_n} f\,d\mu_n$$

and by symmetry, the integration may be performed in any order. The measure μ is called the *product* of $\mu_1, \ldots, \mu_n$, written $\mu = \mu_1 \times \cdots \times \mu_n$. In particular, if each μ_j is Lebesgue measure on $\mathscr{B}(R)$, then $\mu_1 \times \cdots \times \mu_n$ is Lebesgue measure on $\mathscr{B}(R^n)$, just as in the discussion after 2.6.3.

Problems

1. Show that the collection of finite disjoint unions of measurable rectangles in $\Omega_1 \times \cdots \times \Omega_n$ forms a field.
2. Show that $\mathscr{B}(R^n) = \mathscr{B}(R) \times \cdots \times \mathscr{B}(R)$ (n times).
3. If $\mathscr{F}_1, \ldots, \mathscr{F}_n$ are arbitrary σ-fields, show that

$$(\mathscr{F}_1 \times \cdots \times \mathscr{F}_{n-1}) \times \mathscr{F}_n = \mathscr{F}_1 \times \mathscr{F}_2 \times \cdots \times \mathscr{F}_n.$$

4. Let μ be the product of the σ-finite measures μ_1 and μ_2. If $C \in \mathscr{F}_1 \times \mathscr{F}_2$, show that the following are equivalent:
   (a) $\mu(C) = 0$,
   (b) $\mu_2(C(\omega_1)) = 0$ for μ_1-almost all $\omega_1 \in \Omega_1$,
   (c) $\mu_1(C(\omega_2)) = 0$ for μ_2-almost all $\omega_2 \in \Omega_2$.
5. In Problem 4, let $(\Omega', \mathscr{F}', \mu')$ be the completion of $(\Omega, \mathscr{F}, \mu)$, and assume μ_1, μ_2 complete. If $B \in \mathscr{F}'$, show that $B(\omega_1) \in \mathscr{F}_2$ for μ_1-*almost* all $\omega_1 \in \Omega_1$ [and $B(\omega_2) \in \mathscr{F}_1$ for μ_2-almost all $\omega_2 \in \Omega_2$]. Give an example in which $B(\omega_1) \notin \mathscr{F}_2$ for some $\omega_1 \in \Omega_1$.
6. (a) Let $\Omega_1 = \Omega_2 =$ the set of positive integers, $\mathscr{F}_1 = \mathscr{F}_2 =$ all subsets, $\mu_1 = \mu_2 =$ counting measure, $f(n, n) = n$, $f(n, n+1) = -n$, $n = 1, 2, \ldots$, $f(i, j) = 0$ if $j \neq i$ or $i + 1$. Show that $\int_{\Omega_1} \int_{\Omega_2} f \, d\mu_2 \, d\mu_1 = 0$, $\int_{\Omega_2} \int_{\Omega_1} f \, d\mu_1 \, d\mu_2 = \infty$. (Fubini's theorem fails since the integral of f with respect to $\mu_1 \times \mu_2$ does not exist.)
   (b) Let $\Omega_1 = \Omega_2 = R$, $\mathscr{F}_1 = \mathscr{F}_2 = \mathscr{B}(R)$, $\mu_1 =$ Lebesgue measure, $\mu_2 =$ counting measure. Let $A = \{(\omega_1, \omega_2) \colon \omega_1 = \omega_2\} \in \mathscr{F}_1 \times \mathscr{F}_2$. Show that

$$\int_{\Omega_1} \int_{\Omega_2} I_A \, d\mu_2 \, d\mu_1 = \int_{\Omega_1} \mu_2(A(\omega_1)) \, d\mu_1(\omega_1) = \infty,$$

   but

$$\int_{\Omega_2} \int_{\Omega_1} I_A \, d\mu_1 \, d\mu_2 = \int_{\Omega_2} \mu_1(A(\omega_2)) \, d\mu_2(\omega_2) = 0.$$

   [Fubini's theorem fails since μ_2 is not σ-finite; the product measure theorem fails also since $\int_{\Omega_1} \mu_2(F(\omega_1)) \, d\mu_1(\omega_1)$ and

   $\int_{\Omega_2} \mu_1(F(\omega_2)) \, d\mu_2(\omega_2)$ do not agree on $\mathscr{F}_1 \times \mathscr{F}_2$.]
7.† Let $\Omega_1 = \Omega_2 =$ the first uncountable ordinal, $\mathscr{F}_1 = \mathscr{F}_2 =$ all subsets, $\Omega = \Omega_1 \times \Omega_2$, $\mathscr{F} = \mathscr{F}_1 \times \mathscr{F}_2$. Assume the continuum hypothesis, which identifies Ω_1 and Ω_2 with $[0, 1]$.
   (a) If f is any function from Ω_1 (or from a subset of Ω_1) to $[0, 1]$ and $G = \{(x, y) \colon x \in \Omega_1, y = f(x)\}$ is the graph of f, show that $G \in \mathscr{F}$.
   (b) Let $C_1 = \{(x, y) \in \Omega \colon y \leq x\}$, $C_2 = \{(x, y) \in \Omega \colon y > x\}$. If $B \subset C_1$ or $B \subset C_2$, show that $B \in \mathscr{F}$. (The relation $y \leq x$ refers to the ordering of y and x as *ordinals*, not as real numbers.)
   (c) Show that $\mathscr{F}$ consists of all subsets of Ω.
8. Show that a measurable function of one variable is jointly measurable. Specifically, if $g \colon (\Omega_1, \mathscr{F}_1) \to (\Omega', \mathscr{F}')$ and we define $f \colon \Omega_1 \times \Omega_2 \to \Omega'$ by $f(\omega_1, \omega_2) = g(\omega_1)$, then f is measurable relative to $\mathscr{F}_1 \times \mathscr{F}_2$ and $\mathscr{F}'$, regardless of the nature of $\mathscr{F}_2$.

† Rao, B. V., *Bull. Amer. Math. Soc.* **75**, 614 (1969).

9. Give an example of a function f: $[0. 1] \times [0, 1] \to [0, 1]$ such that
   (a) $f(x, y)$ is Borel measurable in y for each fixed x and Borel measurable in x for each fixed y,
   (b) f is not jointly measurable, that is, f is not measurable relative to the product σ-field $\mathscr{B}[0, 1] \times \mathscr{B}[0, 1]$, and
   (c) $\int_0^1 \left(\int_0^1 f(x, y)\, dy\right) dx$ and $\int_0^1 \left(\int_0^1 f(x, y)\, dx\right) dy$ exist but are unequal. (One example is suggested by Problem 7.)

2.7 Measures on Infinite Product Spaces

The n-dimensional product measure theorem formalizes the notion of an n-stage random experiment, where the probability of an event associated with the nth stage depends on the result of the first $n - 1$ trials. It will be convenient later to have a single probability space which is adequate to handle n-stage experiments for n arbitrarily large (not fixed in advance). Such a space can be constructed if the product measure theorem can be extended to infinitely many dimensions. Our first task is to construct the product of infinitely many σ-fields.

2.7.1 ***Definitions.*** For each $j = 1, 2, \ldots$, let $(\Omega_j, \mathscr{F}_j)$ be a measurable space. Let $\Omega = \prod_{j=1}^{\infty} \Omega_j$, the set of all sequences $(\omega_1, \omega_2, \ldots)$ such that $\omega_j \in \Omega_j$, $j = 1, 2, \ldots$. If $B^n \subset \prod_{j=1}^{n} \Omega_j$, we define

$$B_n = \{\omega \in \Omega: (\omega_1, \ldots, \omega_n) \in B^n\}.$$

The set B_n is called the *cylinder* with *base* B^n; the cylinder is said to be *measurable* if $B^n \in \prod_{j=1}^{n} \mathscr{F}_j$. If $B^n = A_1 \times \cdots \times A_n$, where $A_i \subset \Omega_i$ for each i, B_n is called a *rectangle*, a *measurable rectangle* if $A_i \in \mathscr{F}_i$ for each i.

A cylinder with an n-dimensional base may always be regarded as having a higher dimensional base. For example, if

$$B = \{\omega \in \Omega: (\omega_1, \omega_2, \omega_3) \in B^3\},$$

then

$$\begin{aligned} B &= \{\omega \in \Omega: (\omega_1, \omega_2, \omega_3) \in B^3, \quad \omega_4 \in \Omega_4\} \\ &= \{\omega \in \Omega: (\omega_1, \omega_2, \omega_3, \omega_4) \in B^3 \times \Omega_4\}. \end{aligned}$$

It follows that the measurable cylinders form a field. It is also true that finite disjoint unions of measurable rectangles form a field; the argument is the same as in Problem 1 of Section 2.6.

The minimal σ-field over the measurable cylinders is called the *product* of the σ-fields $\mathscr{F}_j$, written $\prod_{j=1}^{\infty} \mathscr{F}_j$; $\prod_{j=1}^{\infty} \mathscr{F}_j$ is also the minimal σ-field over the measurable rectangles (see Problem 1). If all $\mathscr{F}_j$ coincide with a fixed σ-field $\mathscr{F}$, then $\prod_{j=1}^{\infty} \mathscr{F}_j$ is denoted by $\mathscr{F}^{\infty}$, and if all Ω_j coincide with a fixed set S, $\prod_{j=1}^{\infty} \Omega_j$ is denoted by S^{∞}.

The infinite-dimensional version of the product measure theorem will be used only for probability measures, and is therefore stated in that context. (In fact the construction to be described below runs into trouble for non-probability measures.)

2.7.2 ***Theorem.*** Let $(\Omega_j, \mathscr{F}_j)$, $j = 1, 2, \ldots$, be arbitrary measurable spaces; let $\Omega = \prod_{j=1}^{\infty} \Omega_j$, $\mathscr{F} = \prod_{j=1}^{\infty} \mathscr{F}_j$. Suppose that we are given an arbitrary probability measure P_1 on $\mathscr{F}_1$, and for each $j = 1, 2, \ldots$ and each $(\omega_1, \ldots, \omega_j) \in \Omega_1 \times \cdots \times \Omega_j$ we are given a probability measure $P(\omega_1, \ldots, \omega_j, \cdot)$ on $\mathscr{F}_{j+1}$. Assume that $P(\omega_1, \ldots, \omega_j, C)$ is measurable: $(\prod_{i=1}^{j} \Omega_i, \prod_{i=1}^{j} \mathscr{F}_i) \to (R, \mathscr{B}(R))$ for each fixed $C \in \mathscr{F}_{j+1}$.

If $B^n \in \prod_{j=1}^{n} \mathscr{F}_j$, define

$$P_n(B^n) = \int_{\Omega_1} P_1(d\omega_1) \int_{\Omega_2} P(\omega_1, d\omega_2) \cdots \int_{\Omega_{n-1}} P(\omega_1, \ldots, \omega_{n-2}, d\omega_{n-1})$$

$$\int_{\Omega_n} I_{B^n}(\omega_1, \ldots, \omega_n) P(\omega_1, \ldots, \omega_{n-1}, d\omega_n).$$

Note that P_n is a probability measure on $\prod_{j=1}^{n} \mathscr{F}_j$ by 2.6.7 and 2.6.8(a).

There is a unique probability measure P on $\mathscr{F}$ such that for all n, P agrees with P_n on n-dimensional cylinders, that is, $P\{\omega \in \Omega: (\omega_1, \ldots, \omega_n) \in B^n\} = P_n(B^n)$ for all $n = 1, 2, \ldots$ and all $B^n \in \prod_{j=1}^{n} \mathscr{F}_j$.

PROOF. Any measurable cylinder can be represented in the form $B_n = \{\omega \in \Omega: \{\omega_1, \ldots, \omega_n) \in B^n\}$ for some n and some $B^n \in \prod_{j=1}^{n} \mathscr{F}_j$; define $P(B_n) = P_n(B^n)$. We must show that P is well-defined on measurable cylinders. For suppose that B_n can also be expressed as $\{\omega \in \Omega: (\omega_1, \ldots, \omega_m) \in C^m\}$ where $C^m \in \prod_{j=1}^{m} \mathscr{F}_j$; we must show that $P_n(B^n) = P_m(C^m)$. Say $m < n$; then $(\omega_1, \ldots, \omega_m) \in C^m$ iff $(\omega_1, \ldots, \omega_n) \in B^n$, hence $B^n = C^m \times \Omega_{m+1} \times \cdots \times \Omega_n$. It follows from the definition of P_n that $P_n(B^n) = P_m(C^m)$. (The fact that the $P(\omega_1, \ldots, \omega_j, \cdot)$ are probability measures is used here.)

Since P_n is a measure on $\prod_{j=1}^{n} \mathscr{F}_j$, it is immediate that P is finitely additive on the field $\mathscr{F}_0$ of measurable cylinders. If we can show that P is continuous from above at the empty set, 1.2.8(b) implies that P is countably additive on $\mathscr{F}_0$, and the Carathéodory extension theorem extends P to a probability measure on $\prod_{j=1}^{\infty} \mathscr{F}_j$; by construction, P agrees with P_n on n-dimensional cylinders.

Let $\{B_n, n = n_1, n_2, \ldots\}$ be a sequence of measurable cylinders decreasing to $\varnothing$ (we may assume $n_1 < n_2 < \cdots$, and in fact nothing is lost if we take $n_i = i$ for all i). Assume $\lim_{n\to\infty} P(B_n) > 0$. Then for each $n > 1$,

$$P(B_n) = \int_{\Omega_1} g_n^{(1)}(\omega_1) P_1(d\omega_1),$$

where

$$g_n^{(1)}(\omega_1) = \int_{\Omega_2} P(\omega_1, d\omega_2) \cdots \int_{\Omega_n} I_{B^n}(\omega_1, \ldots, \omega_n) P(\omega_1, \ldots, \omega_{n-1}, d\omega_n).$$

Since $B_{n+1} \subset B_n$, it follows that $B^{n+1} \subset B^n \times \Omega_{n+1}$; hence

$$I_{B^{n+1}}(\omega_1, \ldots, \omega_{n+1}) \le I_{B^n}(\omega_1, \ldots, \omega_n).$$

Therefore $g_n^{(1)}(\omega_1)$ decreases as n increases (ω_1 fixed); say $g_n^{(1)}(\omega_1) \to h_1(\omega_1)$. By the extended monotone convergence theorem (or the dominated convergence theorem), $P(B_n) \to \int_{\Omega_1} h_1(\omega_1) P_1(d\omega_1)$. If $\lim_{n\to\infty} P(B_n) > 0$, then $h_1(\omega_1{}') > 0$ for some $\omega_1{}' \in \Omega_1$. In fact $\omega_1{}' \in B^1$, for if not, $I_{B^n}(\omega_1{}', \omega_2, \ldots, \omega_n) = 0$ for all n; hence $g_n^{(1)}(\omega_1{}') = 0$ for all n, and $h_1(\omega_1{}') = 0$, a contradiction.

Now for each $n > 2$,

$$g_n^{(1)}(\omega_1{}') = \int_{\Omega_2} g_n^{(2)}(\omega_2) P(\omega_1{}', d\omega_2),$$

where

$$g_n^{(2)}(\omega_2) = \int_{\Omega_3} P(\omega_1{}', \omega_2, d\omega_3)$$

$$\cdots \int_{\Omega_n} I_{B^n}(\omega_1{}', \omega_2, \ldots, \omega_n) P(\omega_1{}', \ldots, \omega_{n-1}, d\omega_n).$$

As above, $g_n^{(2)}(\omega_2) \downarrow h_2(\omega_2)$; hence

$$g_n^{(1)}(\omega_1{}') \to \int_{\Omega_2} h_2(\omega_2) P(\omega_1{}', d\omega_2).$$

Since $g_n^{(1)}(\omega_1{}') \to h_1(\omega_1{}') > 0$, we have $h_2(\omega_2{}') > 0$ for some $\omega_2{}' \in \Omega_2$, and as above we have $(\omega_1{}', \omega_2{}') \in B^2$.

The process may be repeated inductively to obtain points $\omega_1{}', \omega_2{}', \ldots$ such that for each n, $(\omega_1{}', \ldots, \omega_n{}') \in B^n$. But then $(\omega_1{}', \omega_2{}', \ldots) \in \bigcap_{n=1}^{\infty} B_n = \varnothing$, a contradiction. This proves the existence of the desired probability measure P. If Q is another such probability measure, then $P = Q$ on measurable cylinders, hence $P = Q$ on $\mathscr{F}$ by the uniqueness part of the Carathéodory extension theorem. ∎

The classical product measure theorem extends as follows:

2.7.3 ***Corollary.*** For each $j = 1, 2, \ldots$, let $(\Omega_j, \mathscr{F}_j, P_j)$ be an arbitrary probability space. Let $\Omega = \prod_{j=1}^{\infty} \Omega_j$, $\mathscr{F} = \prod_{j=1}^{\infty} \mathscr{F}_j$. There is a unique probability measure P on $\mathscr{F}$ such that

$$P\{\omega \in \Omega\colon \omega_1 \in A_1, \ldots, \omega_n \in A_n\} = \prod_{j=1}^{n} P_j(A_j)$$

for all $n = 1, 2, \ldots$ and all $A_j \in \mathscr{F}_j$, $j = 1, 2, \ldots$. We call P the *product* of the P_j, and write $P = \prod_{j=1}^{\infty} P_j$.

PROOF. In 2.7.2, take $P(\omega_1, \ldots, \omega_j, B) = P_{j+1}(B)$, $B \in \mathscr{F}_{j+1}$. Then $P_n(A_1 \times \cdots \times A_n) = \prod_{j=1}^{n} P_j(A_j)$, and thus the probability measure P of 2.7.2 has the desired properties. If Q is another such probability measure, then $P = Q$ on the field of finite disjoint unions of measurable rectangles; hence $P = Q$ on $\mathscr{F}$ by the Carathéodory extension theorem. ∎

Problems

1. Show that $\prod_{j=1}^{\infty} \mathscr{F}_j$ is the minimal σ-field over the measurable rectangles.
2. Let $\mathscr{F} = \mathscr{B}(R)$; show that the following sets belong to $\mathscr{F}^{\infty}$:
   (a) $\{x \in R^{\infty}\colon \sup_n x_n < a\}$,
   (b) $\{x \in R^{\infty}\colon \sum_{n=1}^{\infty} |x_n| < a\}$,
   (c) $\{x \in R^{\infty}\colon \lim_{n\to\infty} x_n$ exists and is finite$\}$,
   (d) $\{x \in R^{\infty}\colon \limsup_{n\to\infty} x_n \le a\}$,
   (e) $\{x \in R^{\infty}\colon \sum_{k=1}^{n} x_k = 0$ for at least one $n > 0\}$.
3. Let $\mathscr{F}$ be a σ-field of subsets of a set S, and assume $\mathscr{F}$ is countably generated, that is, there is a sequence of sets $A_1, A_2, \ldots$ in $\mathscr{F}$ such that the smallest σ-field containing the A_j is $\mathscr{F}$. Show that $\mathscr{F}^{\infty}$ is also countably generated. In particular, $\mathscr{B}(R)^{\infty}$ is countably generated; take the A_j as intervals with rational endpoints.
4. How many sets are there in $\mathscr{B}(R)^{\infty}$?
5. Define $f\colon R^{\infty} \to \overline{R}$ as follows:

$$f(x_1, x_2, \ldots) = \begin{cases} \text{the smallest positive integer } n \\ \text{such that } x_1 + \cdots + x_n \ge 1, \\ \text{if such an } n \text{ exists,} \\ \infty \text{ if } x_1 + \cdots + x_n < 1 \text{ for all } n. \end{cases}$$

Show that $f\colon (R^{\infty}, \mathscr{B}(R)^{\infty}) \to (\overline{R}, \mathscr{B}(\overline{R}))$.

2.8 References

The presentation in Chapters 1 and 2 has been strongly influenced by several sources. The first systematic presentation of measure theory appeared in Halmos (1950). Halmos achieves slightly greater generality at the expense of technical complications by replacing σ-fields by σ-rings. (A σ-ring is a class of sets closed under differences and countable unions.) However, σ-fields will be completely adequate for our purposes. The first account of measure theory specifically oriented toward probability was given by Loève (1955). Several useful refinements were made by Royden (1963), Neveu (1965), and Rudin (1966). Neveu's book emphasizes probability while Rudin's book is particularly helpful as a preparation for work in harmonic analysis.

For further properties of finitely additive set functions, and a development of integration theory for functions with values in a Banach space, see Dunford and Schwartz (1958).

3

Introduction to Functional Analysis

3.1 Introduction

An important part of analysis consists of the study of vector spaces endowed with an additional structure of some kind. In Chapter 2, for example, we studied the vector space $L^p(\Omega, \mathscr{F}, \mu)$. If $1 \le p \le \infty$, the seminorm $\| \ \|_p$ allowed us to talk about such notions as distance, convergence, and completeness.

In this chapter, we look at various structures that can be defined on vector spaces. The most general concept studied is that of a topological vector space, which is a vector space endowed with a topology compatible with the algebraic operations, that is, the topology makes vector addition and scalar multiplication continuous. Special cases are Banach and Hilbert spaces. In a Banach space there is a notion of length of a vector, and in a Hilbert space, length is in turn determined by a "dot product" of vectors. Hilbert spaces are a natural generalization of finite-dimensional Euclidean spaces.

We now list the spaces we are going to study. The term "vector space" will always mean vector space over the complex field C; "real vector space" indicates that the scalar field is R; no other fields will be considered.

3.1.1 ***Definitions.*** Let L be a vector space. A *seminorm* on L is a function $\| \ \|$ from L to the nonnegative reals satisfying

$$\|ax\| = |a| \, \|x\| \qquad \text{for all} \quad a \in C, \quad x \in L,$$

$$\|x + y\| \le \|x\| + \|y\| \qquad \text{for all} \quad x, y \in L.$$

The first property is called *absolute homogeneity*, the second *subadditivity*. Note that absolute homogeneity implies that $\|0\| = 0$. (We use the same symbol for the zero vector and the zero scalar.) If, in addition, $\|x\| = 0$ implies that $x = 0$, the seminorm is called a *norm* on L and L is said to be a *normed linear space*.

If $\| \ \|$ is a seminorm on L, and $d(x, y) = \|x - y\|$, $x, y \in L$, d is a pseudometric on L; a metric if $\| \ \|$ is a norm. A *Banach space* is a complete normed linear space, that is, relative to the metric d induced by the norm, every Cauchy sequence converges.

An *inner product* on L is a function from $L \times L$ to C, denoted by $(x, y) \to \langle x, y \rangle$, satisfying

$$\langle ax + by, z \rangle = a\langle x, z \rangle + b\langle y, z \rangle \qquad \text{for all} \quad a, b \in C, \quad x, y, z \in L,$$

$$\langle x, y \rangle = \overline{\langle y, x \rangle} \qquad \text{for all} \quad x, y \in L$$

$$\langle x, x \rangle \geq 0 \qquad \text{for all} \quad x \in L,$$

$$\langle x, x \rangle = 0 \qquad \text{if and only if} \quad x = 0$$

(the over-bar indicates complex conjugation). A vector space endowed with an inner product is called an *inner product space* or *pre-Hilbert space*. If L is an inner product space, $\|x\| = (\langle x, x \rangle)^{1/2}$ defines a norm on L; this is a consequence of the Cauchy–Schwarz inequality, to be proved in Section 3.2. If, with this norm, L is complete, L is said to be a *Hilbert space*. Thus a Hilbert space is a Banach space whose norm is determined by an inner product.

Finally, a *topological vector space* is a vector space L with a topology such that addition and scalar multiplication are continuous, in other words, the mappings

$$(x, y) \to x + y \qquad \text{of} \quad L \times L \quad \text{into} \quad L$$

and

$$(a, x) \to ax \qquad \text{of} \quad C \times L \quad \text{into} \quad L$$

are continuous, with the product topology on $L \times L$ and $C \times L$. In many books, the topology is required to be Hausdorff, but we find it more convenient not to make this assumption.

A Banach space is a topological vector space with the topology induced by the metric $d(x, y) = \|x - y\|$. For if $x_n \to x$ and $y_n \to y$, then

$$\|x_n + y_n - (x + y)\| \leq \|x_n - x\| + \|y_n - y\| \to 0;$$

if $a_n \to a$ and $x_n \to x$, then

$$\begin{aligned} \|a_n x_n - ax\| &\leq \|a_n x_n - a_n x\| + \|a_n x - ax\| \\ &= |a_n| \, \|x_n - x\| + |a_n - a| \, \|x\| \to 0. \end{aligned}$$

The above definitions remain unchanged if L is a real vector space, except of course that C is replaced by R. Also, we may drop the complex conjugate in the symmetry requirement for inner product and simply write $\langle x, y\rangle = \langle y, x\rangle$ for all $x, y \in L$.

3.1.2 ***Examples.*** (a) If $(\Omega, \mathscr{F}, \mu)$ is a measure space and $1 \le p \le \infty$, $\| \ \|_p$ is a seminorm on the vector space $L^p(\Omega, \mathscr{F}, \mu)$. If we pass to equivalence classes by identifying functions that agree a.e. $[\mu]$, we obtain $\mathbf{L}^p(\Omega, \mathscr{F}, \mu)$, a Banach space (see 2.4). When $p = 2$, the norm $\| \ \|_p$ is determined by an inner product

$$\langle f, g\rangle = \int_\Omega f\bar{g}\, d\mu.$$

Hence $\mathbf{L}^2(\Omega, \mathscr{F}, \mu)$ is a Hilbert space.

If $\mathscr{F}$ consists of all subsets of Ω and μ is counting measure, then $f = g$ a.e. $[\mu]$ implies $f \equiv g$. Thus it is not necessary to pass to equivalence classes; $L^p(\Omega, \mathscr{F}, \mu)$ is a Banach space, denoted for simplicity by $l^p(\Omega)$.

By 2.4.12, if $1 \le p < \infty$, then $l^p(\Omega)$ consists of all functions $f = (f(\alpha), \alpha \in \Omega)$ from Ω to C such that $f(\alpha) = 0$ for all but countably many α, and $\|f\|_p^p = \sum_\alpha |f(\alpha)|^p < \infty$. When $p = 2$, the norm on $l^2(\Omega)$ is induced by the inner product

$$\langle f, g\rangle = \sum_\alpha f(\alpha)\overline{g(\alpha)}.$$

When $p = \infty$, the situation is slightly different. The space $l^\infty(\Omega)$ is the collection of all bounded complex-valued functions on Ω, with the *sup norm*

$$\|f\| = \sup\{|f(x)| : x \in \Omega\}.$$

Similarly, if Ω is a topological space and L is the class of all bounded continuous complex-valued functions on Ω, then L is a Banach space under the sup norm, for we may verify directly that the sup norm is actually a norm, or equally well we may use the fact that $L \subset l^\infty(\Omega)$. Thus we need only check completeness, and this follows because a uniform limit of continuous functions is continuous.

(b) Let c be the set of all convergent sequences of complex numbers, and put the sup norm on c; if $f = \{a_n, n \ge 1\} \in c$, then

$$\|f\| = \sup\{|a_n| : n = 1, 2, \ldots\}.$$

Again, to show that c is a Banach space we need only establish completeness. Let $\{f_n\}$ be a Cauchy sequence in c; if $f_n = \{a_{nk}, k \ge 1\}$, then

$\lim_{k\to\infty} a_{nk}$ exists from each n since $f_n \in c$, and $b_k = \lim_{n\to\infty} a_{nk}$ exists, uniformly in k, since $|a_{nk} - a_{mk}| \le \|f_n - f_m\| \to 0$ as $n, m \to \infty$. By the standard double limit theorem,

$$\lim_{n\to\infty} \lim_{k\to\infty} a_{nk} = \lim_{k\to\infty} \lim_{n\to\infty} a_{nk}.$$

In particular, $\lim_{k\to\infty} b_k$ exists, so if $f = \{b_k, k \ge 1\}$, then $f \in c$. But $\|f_n - f\| = \sup_k |a_{nk} - b_k| \to 0$ as $n \to \infty$ since $a_{nk} \to b_k$ uniformly in k. This proves completeness.

(c) Let L be the collection of all complex valued functions on S, where S is an arbitrary set. Put the topology of pointwise convergence on L, so that a sequence or net $\{f_n\}$ of functions in L converges to the function $f \in L$ if and only if $f_n(x) \to f(x)$ for each $x \in S$. (See the appendix on general topology, Section A1, for properties of nets.) With this topology, L is a topological vector space. To show that addition is continuous, observe that if $f_n \to f$ and $g_n \to g$ pointwise, then $f_n + g_n \to f + g$ pointwise. Similarly if $a_n \in C$, $n = 1, 2, \ldots$, $a_n \to a$, and $f_n \to f$ pointwise, then $a_n f_n \to af$ pointwise.

3.2 Basic Properties of Hilbert Spaces

Hilbert spaces are a natural generalization of finite-dimensional Euclidean spaces in the sense that many of the familiar geometric results in R^n carry over. First recall the definition of the inner product (or "dot product") on R^n: If $x = (x_1, \ldots, x_n)$ and $y = (y_1, \ldots, y_n)$, then $\langle x, y\rangle = \sum_{j=1}^n x_j y_j$. (This becomes $\sum_{j=1}^n x_j \bar{y}_j$ in the space C^n of all n-tuples of complex numbers.) The length of a vector in R^n is given by $\|x\| = (\langle x, x\rangle)^{1/2} = (\sum_{j=1}^n x_j^2)^{1/2}$, and the distance between two points of R^n is $d(x, y) = \|x - y\|$. In order to show that d is a metric, the triangle inequality must be established; this in turn follows from the Cauchy–Schwarz inequality $|\langle x, y\rangle| \le \|x\|\, \|y\|$. In fact the Cauchy–Schwarz inequality holds in any inner product space, as we now prove:

3.2.1 ***Cauchy–Schwarz Inequality.*** If L is an inner product space, and $\|x\| = (\langle x, x\rangle)^{1/2}$, $x \in L$, then

$$|\langle x, y\rangle| \le \|x\|\, \|y\| \quad \text{for all} \quad x, y \in L.$$

Equality holds iff x and y are linearly dependent.

PROOF. For any $a \in C$,

$$\begin{aligned} 0 \le \langle x + ay, x + ay\rangle &= \langle x + ay, x\rangle + \langle x + ay, ay\rangle \\ &= \langle x, x\rangle + a\langle y, x\rangle + \bar{a}\langle x, y\rangle + |a|^2\langle y, y\rangle. \end{aligned}$$

Set $a = -\langle x, y\rangle/\langle y, y\rangle$ (if $\langle y, y\rangle = 0$, then $y = 0$ and the result is trivial). Since $\langle y, x\rangle = \overline{\langle x, y\rangle}$, we have

$$0 \leq \langle x, x\rangle - 2\frac{|\langle x, y\rangle|^2}{\langle y, y\rangle} + \frac{|\langle x, y\rangle|^2}{\langle y, y\rangle}$$

proving the inequality.

Since $\langle x + ay, x + ay\rangle = 0$ iff $x + ay = 0$, equality holds iff x and y are linearly dependent. ∎

3.2.2 ***Corollary.*** If L is an inner product space and $\|x\| = (\langle x, x\rangle)^{1/2}$, $x \in L$, then $\| \ \|$ is a norm on L.

PROOF. It is immediate that $\|x\| \geq 0$, $\|ax\| = |a| \ \|x\|$, and $\|x\| = 0$ iff $x = 0$. Now

$$\begin{aligned} \|x + y\|^2 = \langle x + y, x + y\rangle &= \|x\|^2 + \|y\|^2 + \langle x, y\rangle + \langle y, x\rangle \\ &= \|x\|^2 + \|y\|^2 + 2 \operatorname{Re} \langle x, y\rangle \\ &\leq \|x\|^2 + \|y\|^2 + 2\|x\| \ \|y\| \qquad \text{by 3.2.1.} \end{aligned}$$

Therefore $\|x + y\|^2 \leq (\|x\| + \|y\|)^2$. ∎

3.2.3 ***Corollary.*** An inner product is (jointly) continuous in both variables, that is, $x_n \to x$, $y_n \to y$ implies $\langle x_n, y_n\rangle \to \langle x, y\rangle$ ($\{(x_n, y_n)\}$ can be a net as well as a sequence).

PROOF.

$$\begin{aligned} |\langle x_n, y_n\rangle - \langle x, y\rangle| &= |\langle x_n, y_n - y\rangle + \langle x_n - x, y\rangle| \\ &\leq \|x_n\| \ \|y_n - y\| + \|x_n - x\| \ \|y\| \qquad \text{by 3.2.1.} \end{aligned}$$

But by subadditivity of the norm,

$$| \, \|x_n\| - \|x\| \, | \leq \|x_n - x\| \to 0;$$

hence $\|x_n\| \to \|x\|$. It follows that $\langle x_n, y_n\rangle \to \langle x, y\rangle$. ∎

The computation of 3.2.2 establishes the following result, which says geometrically that the sum of the squares of the lengths of the diagonals of a parallelogram is twice the sum of the squares of the lengths of the sides:

3.2.4 ***Parallelogram Law.*** In an inner product space,

$$\|x + y\|^2 + \|x - y\|^2 = 2(\|x\|^2 + \|y\|^2).$$

PROOF.

$$\|x + y\|^2 = \|x\|^2 + \|y\|^2 + 2 \operatorname{Re}\langle x, y\rangle,$$

and

$$\|x - y\|^2 = \|x\|^2 + \|y\|^2 - 2 \operatorname{Re}\langle x, y\rangle. \blacksquare$$

Now suppose that $x_1, \ldots, x_n$ are mutually perpendicular unit vectors in R^k, $k \geq n$. If x is an arbitrary vector in R^k, we try to approximate x by a linear combination $\sum_{j=1}^n a_j x_j$. The reader may recall that $\sum_{j=1}^n a_j x_j$ will be closest to x in the sense of Euclidean distance when $a_j = \langle x, x_j\rangle$. This result holds in an arbitrary inner product space.

3.2.5 ***Definition.*** Two elements x and y in an inner product space L are said to be *orthogonal* or *perpendicular* iff $\langle x, y\rangle = 0$. If $B \subset L$, B is said to be *orthogonal* iff $\langle x, y\rangle = 0$ for all $x, y \in B$ such that $x \neq y$; B is *orthonormal* iff it is orthogonal and $\|x\| = 1$ for all $x \in B$.

The computation of 3.2.2 shows that if $x_1, x_2, \ldots, x_n$ are orthogonal, the *Pythagorean relation* holds: $\|\sum_{i=1}^n x_i\|^2 = \sum_{i=1}^n \|x_i\|^2$.

3.2.6 ***Theorem.*** If $\{x_1, \ldots, x_n\}$ is an orthonormal set in the inner product space L, and $x \in L$,

$$\left\|x - \sum_{j=1}^n a_j x_j\right\| \quad \text{is minimized when} \quad a_j = \langle x, x_j\rangle, \quad j = 1, \ldots, n.$$

PROOF.

$$\begin{aligned}\left\|x - \sum_{j=1}^n a_j x_j\right\|^2 &= \left\langle x - \sum_{j=1}^n a_j x_j, x - \sum_{k=1}^n a_k x_k\right\rangle \\ &= \|x\|^2 - \sum_{k=1}^n \bar{a}_k\langle x, x_k\rangle - \sum_{j=1}^n a_j\langle x_j, x\rangle \\ &\quad + \left\langle \sum_{j=1}^n a_j x_j, \sum_{k=1}^n a_k x_k\right\rangle.\end{aligned}$$

The last term on the right is $\sum_{j=1}^n |a_j|^2$ since the x_j are orthonormal. Furthermore, $-\bar{a}_j\langle x, x_j\rangle - a_j\langle x_j, x\rangle + |a_j|^2 = -|\langle x, x_j\rangle|^2 + |a_j - \langle x, x_j\rangle|^2$. Thus

$$0 \le \left\| x - \sum_{j=1}^n a_j x_j \right\|^2 = \|x\|^2 - \sum_{j=1}^n |\langle x, x_j\rangle|^2 + \sum_{j=1}^n |a_j - \langle x, x_j\rangle|^2, \quad (1)$$

so that we can do no better than to take $a_j = \langle x, x_j\rangle$. ∎

The above computation establishes the following important inequality.

3.2.7 ***Bessel's Inequality.*** If B is an arbitrary orthonormal subset of the inner product space L and x is an element of L, then

$$\|x\|^2 \ge \sum_{y \in B} |\langle x, y\rangle|^2.$$

In other words, $\langle x, y\rangle = 0$ for all but countably many $y \in B$, say $y = x_1, x_2, \ldots$, and

$$\|x\|^2 \ge \sum_j |\langle x, x_j\rangle|^2.$$

Equality holds iff $\sum_{j=1}^n \langle x, x_j\rangle x_j \to x$ as $n \to \infty$.

PROOF. If $x_1, \ldots, x_n \in B$, set $a_j = \langle x, x_j\rangle$ in Eq. (1) of 3.2.6 to obtain $\|x - \sum_{j=1}^n \langle x, x_j\rangle x_j\|^2 = \|x\|^2 - \sum_{j=1}^n |\langle x, x_j\rangle|^2 \ge 0$. ∎

We now consider another basic geometric idea, that of *projection*. If M is a subspace of R^n and x is any vector in R^n, x can be resolved into a component in M and a component perpendicular to M. In other words, $x = y + z$ where $y \in M$ and z is orthogonal to every vector in M. Before generalizing to an arbitrary space, we indicate some terminology.

3.2.8 ***Definitions.*** A *subspace* or *linear manifold* of a vector space L is a subset M of L that is also a vector space; that is, M is closed under addition and scalar multiplication. If L is a topological vector space, M is said to be a *closed subspace* of L if M is a subspace and is also a closed set in the topology of L.

A subset M of the vector space L is said to be *convex* iff for all $x, y \in M$, we have $ax + (1 - a)y \in M$ for all real $a \in [0, 1]$.

The key fact that we need is that if M is a closed convex subset of the Hilbert space H and x is an arbitrary point of H, there is a unique point of M closest to x.

3.2.9 ***Theorem.*** Let M be a nonempty closed convex subset of the Hilbert space H. If $x \in H$, there is a unique element $y_0 \in M$ such that

$$\|x - y_0\| = \inf\{\|x - y\| : y \in M\}.$$

PROOF. Let $d = \inf\{\|x - y\| : y \in M\}$, and pick points $y_1, y_2, \ldots \in M$ with $\|x - y_n\| \to d$ as $n \to \infty$; we show that $\{y_n\}$ is a Cauchy sequence.

Since $\|u + v\|^2 + \|u - v\|^2 = 2\|u\|^2 + 2\|v\|^2$ for all $u, v \in H$ by the parallelogram law 3.2.4, we may set $u = y_n - x$, $v = y_m - x$ to obtain

$$\|y_n + y_m - 2x\|^2 + \|y_n - y_m\|^2 = 2\|y_n - x\|^2 + 2\|y_m - x\|^2$$

or

$$\|y_n - y_m\|^2 = 2\|y_n - x\|^2 + 2\|y_m - x\|^2 - 4\|\tfrac{1}{2}(y_n + y_m) - x\|^2.$$

Since $\frac{1}{2}(y_n + y_m) \in M$ by convexity, $\|\frac{1}{2}(y_n + y_m) - x\|^2 \geq d^2$, and it follows that $\|y_n - y_m\| \to 0$ as $n, m \to \infty$.

By completeness of H, y_n approaches a limit y_0, hence $\|x - y_n\| \to \|x - y_0\|$. But then $\|x - y_0\| = d$, and $y_0 \in M$ since M is closed; this finishes the existence part of the proof.

To prove uniqueness, let $y_0, z_0 \in M$, with $\|x - y_0\| = \|x - z_0\| = d$. In the parallelogram law, take $u = y_0 - x$, $v = z_0 - x$, to obtain

$$\|y_0 + z_0 - 2x\|^2 + \|y_0 - z_0\|^2 = 2\|y_0 - x\|^2 + 2\|z_0 - x\|^2 = 4d^2.$$

But $\|y_0 + z_0 - 2x\|^2 = 4\|\frac{1}{2}(y_0 + z_0) - x\|^2 \geq 4d^2$; hence $\|y_0 - z_0\| = 0$, so $y_0 = z_0$. ∎

If M is a closed subspace of H, the element y_0 found in 3.2.9 is called the *projection* of x on M. The following result helps to justify this terminology.

3.2.10 ***Theorem.*** Let M be a closed subspace of the Hilbert space H, and y_0 an element of M. Then

$$\|x - y_0\| = \inf\{\|x - y\| : y \in M\} \quad \text{iff} \quad x - y_0 \perp M,$$

that is, $\langle x - y_0, y\rangle = 0$ for all $y \in M$.

PROOF. Assume $x - y_0 \perp M$. If $y \in M$, then

$$\begin{aligned}\|x - y\|^2 &= \|x - y_0 - (y - y_0)\|^2 \\ &= \|x - y_0\|^2 + \|y - y_0\|^2 - 2\,\mathrm{Re}\langle x - y_0, y - y_0\rangle \\ &= \|x - y_0\|^2 + \|y - y_0\|^2 \qquad \text{since} \quad y - y_0 \in M \\ &\geq \|x - y_0\|^2.\end{aligned}$$

Therefore, $\|x - y_0\| = \inf\{\|x - y\| : y \in M\}$.

Conversely, assume $\|x - y_0\| = \inf\{\|x - y\|: y \in M\}$. Let $y \in M$ and let c be an arbitrary complex number. Then $y_0 + cy \in M$ since M is a subspace, hence $\|x - y_0 - cy\| \geq \|x - y_0\|$. But

$$\|x - y_0 - cy\|^2 = \|x - y_0\|^2 + |c|^2\|y\|^2 - 2\,\mathrm{Re}\langle x - y_0, cy\rangle;$$

hence

$$|c|^2\|y\|^2 - 2\,\mathrm{Re}\langle x - y_0, cy\rangle \geq 0.$$

Take $c = b\langle x - y_0, y\rangle$, b real. Then $\langle x - y_0, cy\rangle = b|\langle x - y_0, y\rangle|^2$. Thus $|\langle x - y_0, y\rangle|^2[b^2\|y\|^2 - 2b] \geq 0$. But the expression in square brackets is negative if b is positive and sufficiently close to 0; hence $\langle x - y_0, y\rangle = 0$. ∎

We may give still another way of characterizing the projection of x on M.

3.2.11 ***Projection Theorem.*** Let M be a closed subspace of the Hilbert space H. If $x \in H$, then x has a unique representation $x = y + z$ where $y \in M$ and $z \perp M$. Furthermore, y is the projection of x on M.

PROOF. Let y_0 be the projection of x on M, and take $y = y_0$, $z = x - y_0$. By 3.2.10, $z \perp M$, proving the existence of the desired representation. To prove uniqueness, let $x = y + z = y' + z'$ where $y, y' \in M$, $z, z' \perp M$. Then $y - y' \in M$ since M is a subspace, and $y - y' \perp M$ since $y - y' = z' - z$. Thus $y - y'$ is orthogonal to itself, hence $y = y'$. But then $z = z'$, proving uniqueness. ∎

If M is any subset of H, the set $M^\perp = \{x \in H: x \perp M\}$ is a closed subspace by definition of the inner product and 3.2.3. If M is a closed subspace, $M^\perp$ is called the *orthogonal complement* of H, and the projection theorem is expressed by saying that H is the *orthogonal direct sum* of M and $M^\perp$, written $H = M \oplus M^\perp$.

In R^n, it is possible to construct an orthonormal basis, that is, a set $\{x_1, \ldots, x_n\}$ of n mutually perpendicular unit vectors. Any vector x in R^n may then be represented as $x = \sum_{i=1}^n \langle x, x_i\rangle x_i$, so that $\langle x, x_i\rangle$ is the component of x in the direction of x_i. We are now able to generalize this idea to an arbitrary Hilbert space. The following terminology will be used.

3.2.12 ***Definitions.*** If B is a subset of the topological vector space L, the *space spanned by* B, denoted by $S(B)$, is the smallest closed subspace of L containing all elements of B. If $L(B)$ is the linear manifold generated by B, that is, $L(B)$ consists of all elements $\sum_{i=1}^{n} a_i x_i$, $a_i \in C$, $x_i \in B$, $i = 1, \ldots, n$, $n = 1, 2, \ldots$, then $S(B) = \overline{L(B)}$.

If B is a subset of the Hilbert space H, B is said to be an *orthonormal basis* for H iff B is a maximal orthonormal subset of H, in other words, B is not a proper subset of any other orthonormal subset of H. An orthonormal set $B \subset H$ is maximal iff $S(B) = H$, and there are several other conditions equivalent to this, as we now prove.

3.2.13 ***Theorem.*** Let $B = \{x_\alpha, \alpha \in I\}$ be an orthonormal subset of the Hilbert space H. The following conditions are equivalent:

(a) B is an orthonormal basis.

(b) B is a "complete orthonormal set," that is, the only $x \in H$ such that $x \perp B$ is $x = 0$.

(c) B spans H, that is, $S(B) = H$.

(d) For all $x \in H$, $x = \sum_\alpha \langle x, x_\alpha \rangle x_\alpha$. (Let us explain this notation. By 3.2.7, $\langle x, x_\alpha \rangle = 0$ for all but countably many x_α, say for $x_1, x_2, \ldots$; the assertion is that $\sum_{j=1}^{n} \langle x, x_j \rangle x_j \to x$, and this holds regardless of the order in which the x_j are listed.)

(e) For all $x, y \in H$, $\langle x, y \rangle = \sum_\alpha \langle x, x_\alpha \rangle \langle x_\alpha, y \rangle$.

(f) For all $x \in H$, $\|x\|^2 = \sum_\alpha |\langle x, x_\alpha \rangle|^2$.

Condition (f) [and sometimes (e) as well] is referred to as the *Parseval relation.*

PROOF. (a) *implies* (b): If $x \perp B$, $x \neq 0$, let $y = x/\|x\|$. Then $B \cup \{y\}$ is an orthonormal set, contradicting the maximality of B.

(b) *implies* (c): If $x \in H$, write $x = y + z$ where $y \in S(B)$ and $z \perp S(B)$ (see 3.2.11). By (b), $z = 0$; hence $x \in S(B)$.

(c) *implies* (d): Since $S(B) = \overline{L(B)}$, given $x \in H$ and $\varepsilon > 0$ there is a finite set $F \subset I$ and complex numbers a_α, $\alpha \in F$, such that

$$\left\| x - \sum_{\alpha \in F} a_\alpha x_\alpha \right\| \leq \varepsilon.$$

By 3.2.6, if G is any finite subset of I such that $F \subset G$,

$$\left\| x - \sum_{\alpha \in G} \langle x, x_\alpha \rangle x_\alpha \right\| \leq \left\| x - \sum_{\alpha \in G} a_\alpha x_\alpha \right\| \qquad \text{where} \quad a_\alpha = 0 \quad \text{for} \quad \alpha \notin F$$

$$= \left\| x - \sum_{\alpha \in F} a_\alpha x_\alpha \right\| \leq \varepsilon.$$

Thus if $x_1, x_2, \ldots$ is any ordering of the points $x_\alpha \in B$ for which $\langle x, x_\alpha \rangle \neq 0$, $\|x - \sum_{j=1}^{n} \langle x, x_j \rangle x_j\| \leq \varepsilon$ for sufficiently large n, as desired.

(d) *implies* (e): This is immediate from 3.2.3.

(e) *implies* (f): Set $x = y$ in (e).

(f) *implies* (a): Let C be an orthonormal set with $B \subset C$, $B \neq C$. If $x \in C$, $x \notin B$, we have $\|x\|^2 = \sum_\alpha |\langle x, x_\alpha \rangle|^2 = 0$ since by orthonormality of C, x is orthogonal to everything in B. This is a contradiction because $\|x\| = 1$ for all $x \in C$. ∎

3.2.14 ***Corollary.*** Let $B = \{x_\alpha, \alpha \in I\}$ be an orthonormal subset of H, not necessarily a basis.

(a) B is an orthonormal basis for $S(B)$. [Note that $S(B)$ is a closed subspace of H, hence is itself a Hilbert space with the same inner product.]

(b) If $x \in H$ and y is the projection of x on $S(B)$, then

$$y = \sum_\alpha \langle x, x_\alpha \rangle x_\alpha$$

[see 3.2.13(d) for the interpretation of the series].

PROOF. (a) Let $x \in S(B)$, $x \perp B$; then $x \perp L(B)$. If $y \in S(B)$, let $y_1, y_2, \ldots \in L(B)$ with $y_n \to y$. Since $\langle x, y_n \rangle = 0$ for all n, we have $\langle x, y \rangle = 0$ by 3.2.3. Thus $x \perp S(B)$, so that $\langle x, x \rangle = 0$, hence $x = 0$. The result follows from 3.2.13(b).

(b) By part (a) and 3.2.13(d), $y = \sum_\alpha \langle y, x_\alpha \rangle x_\alpha$. But $x - y \perp S(B)$ by 3.2.11, hence $\langle x, x_\alpha \rangle = \langle y, x_\alpha \rangle$ for all α. ∎

A standard application of Zorn's lemma shows that every Hilbert space has an orthonormal basis; an additional argument shows that any two orthonormal bases have the same cardinality (see Problem 5). This fact may be used to classify all possible Hilbert spaces, as follows:

3.2.15 ***Theorem.*** Let S be an arbitrary set, and let H be a Hilbert space with an orthonormal basis B having the same cardinality as S. Then there is an isometric isomorphism (a one-to-one-onto, linear, norm-preserving map) between H and $l^2(S)$.

PROOF. We may write $B = \{x_\alpha, \alpha \in S\}$. If $x \in H$, 3.2.13(d) then gives $x = \sum_\alpha \langle x, x_\alpha \rangle x_\alpha$, where $\sum_\alpha |\langle x, x_\alpha \rangle|^2 = \|x\|^2 < \infty$ by 3.2.13(f). The map $x \to (\langle x, x_\alpha \rangle, \alpha \in S)$ of H into $l^2(S)$ is therefore norm-preserving; since it is also linear, it must be one-to-one. To show that the map is onto, consider any collection of complex numbers a_α, $\alpha \in S$, with $\sum_\alpha |a_\alpha|^2 < \infty$. Say $a_\alpha = 0$ except for $\alpha = \alpha_1, \alpha_2, \ldots$, and let $x = \sum_\alpha a_{\alpha_j} x_{\alpha_j}$. [The series converges to an element of H because of the following fact, which occurs often enough to be stated separately: If $\{y_1, y_2, \ldots\}$ is an orthonormal subset of H, the series $\sum_j c_j y_j$ converges to some element of H iff $\sum_j |c_j|^2 < \infty$. To see this, observe that $\|\sum_{j=n}^m c_j y_j\|^2 = \sum_{j=n}^m |c_j|^2 \|y_j\|^2 = \sum_{j=n}^m |c_j|^2$; thus the partial sums form a Cauchy sequence iff $\sum_j |c_j|^2 < \infty$.]

Since the x_α are orthonormal, it follows that $\langle x, x_\alpha \rangle = a_\alpha$ for all α, so that x maps onto $(a_\alpha, \alpha \in S)$. ∎

We may also characterize Hilbert spaces that are separable, that is, have a countable dense set.

3.2.16 ***Theorem.*** A Hilbert space H is separable iff it has a countable orthonormal basis. If the orthonormal basis has n elements, H is isometrically isomorphic to C^n; if the orthonormal basis is infinite, H is isometrically isomorphic to l^2, that is, $l^2(S)$ with $S = \{1, 2, \ldots\}$.

PROOF. Let B be an orthonormal basis for H. Now $\|x - y\|^2 = \|x\|^2 + \|y\|^2 = 2$ for all $x, y \in B$, $x \neq y$, hence the balls $A_x = \{y\colon \|y - x\| < \frac{1}{2}\}$, $x \in B$, are disjoint. If D is dense in H, D must contain a point in each A_x, so that if B is uncountable, D must be also, and therefore H cannot be separable.

Now assume B is a countable set $\{x_1, x_2, \ldots\}$. If U is a nonempty open subset of H $[= S(B) = \overline{L(B)}]$, U contains an element of the form $\sum_{j=1}^n a_j x_j$ with the $a_j \in C$; in fact the a_j may be assumed to be rational, in other words, to have rational numbers as real and imaginary parts. Thus

$$D = \left\{ \sum_{j=1}^n a_j x_j : n = 1, 2, \ldots, \quad \text{the } a_j \text{ rational} \right\}$$

is a countable dense set, so that H is separable. The remaining statements of the theorem follow from 3.2.15. ∎

A linear norm-preserving map from one Hilbert space to another automatically preserves inner products; this is a consequence of the following proposition:

3.2.17 ***Polarization Identity.*** In any inner product space,

$$4\langle x, y \rangle = \|x + y\|^2 - \|x - y\|^2 + i\|x + iy\|^2 - i\|x - iy\|^2.$$

PROOF.

$$\|x + y\|^2 = \|x\|^2 + \|y\|^2 + 2\,\mathrm{Re}\langle x, y\rangle$$
$$\|x - y\|^2 = \|x\|^2 + \|y\|^2 - 2\,\mathrm{Re}\langle x, y\rangle$$
$$\|x + iy\|^2 = \|x\|^2 + \|y\|^2 + 2\,\mathrm{Re}\langle x, iy\rangle$$
$$\|x - iy\|^2 = \|x\|^2 + \|y\|^2 - 2\,\mathrm{Re}\langle x, iy\rangle$$

But $\mathrm{Re}\langle x, iy\rangle = \mathrm{Re}[-i\langle x, y\rangle] = \mathrm{Im}\langle x, y\rangle$, and the result follows. ▮

Problems

1. In the Hilbert space $l^2(S)$, show that the elements e_α, $\alpha \in S$, form an orthonormal basis, where

$$e_\alpha(s) = \begin{cases} 0, & s \neq \alpha, \\ 1, & s = \alpha. \end{cases}$$

2. (a) If A is an arbitrary subset of the Hilbert space H, show that $A^{\perp\perp} = S(A)$.
   (b) If M is a linear manifold of H, show that M is dense in H iff $M^\perp = \{0\}$.
3. Let $x_1, \ldots, x_n$ be elements of a Hilbert space. Show that the x_i are linearly dependent iff the Gramian (the determinant of the inner products $\langle x_i, x_j\rangle$, $i, j = 1, \ldots, n$) is 0.
4. (*Gram–Schmidt process*) Let $B = \{x_1, x_2, \ldots\}$ be a countable linearly independent subset of the Hilbert space H. Define $e_1 = x_1/\|x_1\|$; having chosen orthonormal elements $e_1, \ldots, e_n$, let y_{n+1} be the projection of x_{n+1} on the space spanned by $e_1, \ldots, e_n$:

$$y_{n+1} = \sum_{i=1}^{n} \langle x_{n+1}, e_i\rangle e_i.$$

Define

$$e_{n+1} = \frac{x_{n+1} - y_{n+1}}{\|x_{n+1} - y_{n+1}\|}.$$

   (a) Show that $L\{e_1, \ldots, e_n\} = L\{x_1, \ldots, x_n\}$ for all n, hence $x_{n+1} \neq y_{n+1}$ and the process is well defined.
   (b) Show that the e_n form an orthonormal basis for $S(B)$.

Comments. Consider the space $H = \mathbf{L}^2(-1, 1)$; if we take $x_n(t) = t^n$, $n = 0, 1, \ldots$, the Gram–Schmidt process yields the *Legendre polynomials* $e_n(t) = a_n\, d^n[(t^2 - 1)^n]/dt^n$, where a_n is chosen so that $\|e_n\| = 1$. Similarly, if in $\mathbf{L}^2(-\infty, \infty)$ we take $x_n(t) = t^n e^{-t^2/2}$, $n = 0, 1, \ldots$, we obtain the *Hermite polynomials* $e_n(t) = a_n(-1)^n e^{t^2}\, d^n(e^{-t^2})/dt^n$.

5. (a) Show that every Hilbert space has an orthonormal basis.
   (b) Show that any two orthonormal bases have the same cardinality.
6. Let U be an open subset of the complex plane, and let $H(U)$ be the collection of all functions f analytic on U such that

$$\|f\|^2 = \iint_U |f(x+iy)|^2 \, dx \, dy < \infty.$$

If we define

$$\langle f, g \rangle = \iint_U f(x+iy)\bar{g}(x+iy) \, dx \, dy, \qquad f, g \in H(U),$$

$H(U)$ becomes an inner product space.

(a) If K is a compact subset of U and $f \in H(U)$, show that

$$\sup\{|f(z)| : z \in K\} \le \|f\|/\sqrt{\pi} \, d_0$$

where d_0 is the Euclidean distance from K to the complement of U. Therefore convergence in $H(U)$ implies uniform convergence on compact subsets of U. (If $z \in K$, the Cauchy integral formula yields

$$f(z) = (2\pi)^{-1} \int_0^{2\pi} f(z + \mathrm{r}e^{i\theta}) \, d\theta, \qquad r < d_0.$$

Integrate this equation with respect to r, $0 \le r \le d < d_0$. Note also that if U is the entire plane, we may take $d_0 = \infty$, and it follows that $H(C) = \{0\}$.)

(b) Show that $H(U)$ is complete, and hence is a Hilbert space.

7. (a) If f is analytic on the unit disk $D = \{z\colon |z| < 1\}$ with Taylor expansion $f(z) = \sum_{n=0}^{\infty} a_n z^n$, show that

$$\sup_{0 \le r < 1} \frac{1}{2\pi} \int_0^{2\pi} |f(re^{i\theta})|^2 \, d\theta = \sum_{n=0}^{\infty} |a_n|^2.$$

It follows that if H^2 is the collection of all functions f analytic on D such that

$$N^2(f) = \sup_{0 \le r < 1} \frac{1}{2\pi} \int_0^{2\pi} |f(re^{i\theta})|^2 \, d\theta < \infty,$$

then H^2, with norm N, is a pre-Hilbert space.

(b) If $f \in H^2$, show that

$$\iint_D |f(x+iy)|^2 \, dx \, dy \le \pi N^2(f);$$

hence $H^2 \subset H(D)$ and convergence in H^2 implies convergence in $H(D)$.

(c) If $f_n(z) = z^n$, $n = 0, 1, \ldots$, show that $f_n \to 0$ in $H(D)$ but not in H^2.

(d) Show that H^2 is complete, and hence is a Hilbert space. [By (a), H^2 is isometrically isomorphic to a subspace of l^2.]

(e) If $e_n(z) = z^n$, $n = 0, 1, \ldots$, show that the e_n form an orthonormal basis for H^2.

(f) If $e_n(z) = [(2n+2)/2\pi]^{1/2} z^n$, $n = 0, 1, \ldots$, show that the e_n form an orthonormal basis for $H(D)$.

8. Let M be a closed convex subset of the Hilbert space H, and y_0 an element of M. If $x \in H$, show that

$$\|x - y_0\| = \inf\{\|x - y\| : y \in M\}$$

iff

$$\operatorname{Re}\langle x - y_0, y - y_0\rangle \le 0 \qquad \text{for all} \quad y \in M.$$

9. (a) If g is a continuous complex-valued function on $[0, 2\pi]$ with $g(0) = g(2\pi)$, use the Stone–Weierstrass theorem to show that g can be uniformly approximated by trigonometric polynomials $\sum_{k=-n}^{n} c_k e^{ikt}$. Conclude that the trigonometric polynomials are dense in $L^2[0, 2\pi]$.

(b) If $f \in L^2[0, 2\pi]$, show that the Fourier series $\sum_{n=-\infty}^{\infty} a_n e^{int}$, $a_n = (1/2\pi)\int_0^{2\pi} f(t)e^{-int}\,dt$, converges to f in L^2, that is,

$$\int_0^{2\pi} \left| f(t) - \sum_{k=-n}^{n} a_k e^{ikt} \right|^2 dt \to 0 \qquad \text{as} \quad n \to \infty.$$

(c) Show that $\{e^{int}/\sqrt{2\pi},\ n = 0, \pm 1, \pm 2, \ldots\}$ yields an orthonormal basis for $L^2[0, 2\pi]$.

10. (a) Give an example to show that if M is a nonempty, closed, but not convex subset of a Hilbert space H, there need not be an element of minimum norm in M. Thus the convexity hypothesis cannot be dropped from Theorem 3.2.9, even if we restrict ourselves to existence and forget about uniqueness.

(b) Show that convexity is not necessary in the existence part of 3.2.9 if H is finite-dimensional.

3.3 Linear Operators on Normed Linear Spaces

The idea of a linear transformation from one Euclidean space to another is familiar. If A is a linear map from R^n to R^m, then A is completely specified by giving its values on a basis $e_1, \ldots, e_n$: $A(\sum_{i=1}^{n} c_i e_i) = \sum_{i=1}^{n} c_i A(e_i)$;

furthermore, A is always continuous. If elements of R^n and R^m are represented by column vectors, A is represented by an $m \times n$ matrix. If $n = m$, so that A is a linear transformation on R^n, A is one-to-one iff it is onto, and if A^{-1} exists, it is always continuous (as well as linear).

Linear transformations on infinite-dimensional spaces have many features not found on the finite-dimensional case, as we shall see.

In this section, we study mappings A from one normed linear space L to another such space M. The mapping A will be a *linear operator*, that is, $A(ax + by) = aA(x) + bA(y)$ for all $x, y \in L$, $a, b \in C$. We use the symbol $\| \; \|$ for the norm on both spaces; no confusion should result. Linear operators can of course be defined on arbitrary vector spaces, but in this section, it is always understood that the domain and range are normed.

Linearity does not imply continuity; to study this idea, we introduce a new concept.

3.3.1 ***Definitions and Comments.*** If A is a linear operator, the *norm* of A is defined by:

(a) $\|A\| = \sup\{\|Ax\| : x \in L, \|x\| \le 1\}$. We may express $\|A\|$ in two other ways.

(b) $\|A\| = \sup\{\|Ax\| : x \in L, \|x\| = 1\}$.

(c) $\|A\| = \sup\{\|Ax\|/\|x\| : x \in L, x \ne 0\}$.

To see this, note that (b) $\le$ (a) is clear; if $x \ne 0$, then $\|Ax\|/\|x\| = \|A(x/\|x\|)\|$, and $x/\|x\|$ has norm 1; hence (c) $\le$ (b). Finally if $\|x\| \le 1$, $x \ne 0$, then $\|Ax\| \le \|Ax\|/\|x\|$, so (a) $\le$ (c).

It follows from (c) that $\|Ax\| \le \|A\| \, \|x\|$, and in fact $\|A\|$ is the smallest number k such that $\|Ax\| \le k\|x\|$ for all $x \in L$.

The linear operator A is said to be *bounded* iff $\|A\| < \infty$. Boundedness is often easy to check, a very fortunate circumstance because we can show that boundedness is equivalent to continuity.

3.3.2 ***Theorem.*** A linear operator A is continuous iff it is bounded.

PROOF. If A is bounded and $\{x_n\}$ is a sequence in L converging to 0, then $\|Ax_n\| \le \|A\| \, \|x_n\| \to 0$. (We use here the fact that the mapping $x \to \|x\|$ of L into the nonnegative reals is continuous; this follows because $| \, \|x\| - \|y\| \, | \le \|x - y\|$.) Thus A is continuous at 0, and therefore, by linearity, is continuous everywhere. On the other hand, if A is unbounded, we can find elements $x_n \in L$ with $\|x_n\| \le 1$ and $\|Ax_n\| \to \infty$. Let $y_n = x_n/\|Ax_n\|$; then $y_n \to 0$, but $\|Ay_n\| = 1$ for all n, hence Ay_n does not converge to 0. Consequently, A is discontinuous. ∎

We are going to show that the set of all bounded linear operators from L to M is itself a normed linear space, but first we consider some examples:

3.3.3 ***Examples.*** (a) Let $L = C[a, b]$, the set of all continuous complex-valued functions on the closed bounded interval $[a, b]$ of reals. Put the *sup norm* on L:

$$\|x\| = \sup\{|x(t)| : a \le t \le b\}, \qquad x \in C[a, b].$$

With this norm, L is a Banach space [see 3.1.2(a)]. Let $K = K(s, t)$ be a continuous complex-valued function on $[a, b] \times [a, b]$, and define a linear operator on L by

$$(Ax)(s) = \int_a^b K(s, t)x(t)\,dt, \qquad a \le s \le b.$$

(By the dominated convergence theorem, Ax actually belongs to L.) We show that A is bounded:

$$\begin{aligned}\|Ax\| &= \sup_{a \le s \le b} \left| \int_a^b K(s, t)x(t)\,dt \right| \\ &\le \sup_{a \le t \le b} |x(t)| \sup_{a \le s \le b} \int_a^b |K(s, t)|\,dt \\ &= \|x\| \max_{a \le s \le b} \int_a^b |K(s, t)|\,dt\end{aligned}$$

(note that $\int_a^b |K(s, t)|\,dt$ is a continuous function of s, by the dominated convergence theorem). Thus

$$\|A\| \le \max_{a \le s \le b} \int_a^b |K(s, t)|\,dt < \infty.$$

In fact $\|A\| = \max_{a \le s \le b} \int_a^b |K(s, t)|\,dt$ (see Problem 3).

(b) Let $L = l^p(S)$, where S is the set of all integers and $1 \le p < \infty$. Define a linear operator T on L by $(Tf)_n = f_{n+1}$, $n \in S$; T is called the *two-sided shift* or the *bilateral shift*. It follows from the definition that T is one-to-one onto, and $(T^{-1}f)_n = f_{n-1}$, $n \in S$. Also, $\|Tf\| = \|f\|$ for all $f \in L$; hence $\|T^{-1}f\| = \|f\|$ for all $f \in L$, so that T is an isometric isomorphism of L with itself. (In particular, $\|T\| = \|T^{-1}\| = 1$.)

If we replace S by the positive integers and define T as above, the resulting operator is called the *one-sided* or *unilateral* shift. The one-sided shift is onto with norm 1, but is not one-to-one; $T(f_1, f_2, \ldots) = (f_2, f_3, \ldots)$.

The shift operators we have defined are shifts to the left. We may also define shifts to the right; in the two-sided case we take $(Af)_n = f_{n-1}$, $n \in S$ (so that $A = T^{-1}$). In the one-sided case, we set $(Af)_n = f_{n-1}$, $n \ge 2$; $(Af)_1 = 0$;

thus $A(f_1, f_2, \ldots) = (0, f_1, f_2, \ldots)$. The operator A is one-to-one but not onto; $A(L)$ is the closed subspace of L consisting of those sequences whose first coordinate is 0.

(c) Assume that the Banach space L is the direct sum of the two closed subspaces M and N, in other words each $x \in L$ can be represented in a unique way as $y + z$ for some $y \in M$, $z \in N$. (We have already encountered this situation with L a Hilbert space, M a closed subspace, $N = M^\perp$.) We define a linear operator P on L by

$$Px = y.$$

P is called a *projection*; specifically, P is the projection of L on M. P has the following properties:

(1) P is *idempotent*; that is, $P^2 = P$, where P^2 is the composition of P with itself.

(2) P is continuous.

Property (1) follows from the definition of P; property (2) will be proved later, as a consequence of the closed graph theorem (see after 3.4.16).

Conversely, let P be a continuous idempotent linear operator on L. Define

$$M = \{x \in L \colon Px = x\}, \qquad N = \{x \in L \colon Px = 0\}.$$

Then we can prove that M and N are closed subspaces, L is the direct sum of M and N, and P is the projection of L on M.

By continuity of P, M and N are closed subspaces. If $x \in L$, then $x = Px + (I - P)x = y + z$ where $y \in M$, $z \in N$. Since $M \cap N = \{0\}$, L is the direct sum of M and N. Furthermore, $Px = y$ by definition of y, so that P is the projection of L on M.

If f is a linear operator from a vector space L to the scalar field, f is called a *linear functional*. (The norm of a scalar b is taken as $|b|$.) Considerable insight is gained about normed linear spaces by studying ways of representing continuous linear functionals on such spaces. We give some examples:

3.3.4 ***Representations of Continuous Linear Functionals.*** (a) Let f be a continuous linear functional on the Hilbert space H. We show that there is a unique element $y \in H$ such that

$$f(x) = \langle x, y \rangle \qquad \text{for all} \quad x \in H.$$

This is one of several results called the *Riesz representation theorem*.

If the desired y exists, it must be unique, for if $\langle x, y \rangle = \langle x, z \rangle$ for all x, then $y - z$ is orthogonal to everything in H (including itself), so $y = z$.

To prove existence, let N be the *null space* of f, that is, $N = \{x \in H: f(x) = 0\}$. If $N^\perp = \{0\}$, then $N = H$ by the projection theorem; hence $f \equiv 0$, and we may take $y = 0$. Thus assume we have an element $u \in N^\perp$ with $u \neq 0$. Then $u \notin N$, and if we define $z = u/f(u)$, we have $z \in N^\perp$ and $f(z) = 1$.

If $x \in H$ and $f(x) = a$, then

$$x = (x - az) + az, \qquad \text{with} \quad x - az \in N, \quad az \perp N.$$

If $y = z/\|z\|^2$, then

$$\begin{aligned} \langle x, y \rangle &= \langle x - az, y \rangle + a\langle z, y \rangle \\ &= a\langle z, y \rangle \qquad \text{since} \quad y \perp N \\ &= a = f(x) \end{aligned}$$

as desired.

The above argument shows that if f is not identically 0, then $N^\perp$ is one-dimensional. For if $x \in N^\perp$ and $f(x) = a$, then $x - az \in N \cap N^\perp$, hence $x = az$. Therefore $N^\perp = \{az: a \in C\}$.

Notice also that if $\|f\|$ is the norm of f, considered as a linear operator, then

$$\|f\| = \|y\|.$$

For $|f(x)| = |\langle x; y \rangle| \leq \|x\| \, \|y\|$ for all x, hence $\|f\| \leq \|y\|$; but $|f(y)| = |\langle y, y \rangle| = \|y\| \, \|y\|$, so $\|f\| \geq \|y\|$.

Now consider the space H^* of all continuous linear functionals on H; H^* is a vector space under the usual operations of addition and scalar multiplication. If to each $f \in H^*$ we associate the element of H given by the Riesz representation theorem, we obtain a map $\psi: H^* \to H$ that is one-to-one onto, norm-preserving, and *conjugate linear*; that is,

$$\psi(af + bg) = \bar{a}\psi(f) + \bar{b}\psi(g), \qquad f, g \in H^*, \quad a, b \in C.$$

[Note that if $f(x) = \langle x, y \rangle$ for all x, then $af(x) = \langle x, \bar{a}y \rangle$.] Such a map is called a *conjugate isometry*.

(b) Let f be a continuous linear functional on l^p $(= l^p(S)$, where S is the set of positive integers), $1 < p < \infty$. We show that if q is defined by $(1/p) + (1/q) = 1$, there is a unique element $y = (y_1, y_2, \ldots) \in l^q$ such that

$$f(x) = \sum_{k=1}^{\infty} x_k y_k \qquad \text{for all} \quad x \in l^p.$$

Furthermore,

$$\|f\| = \|y\| = \left(\sum_{k=1}^{\infty} |y_k|^q \right)^{1/q}.$$

To prove this, let e_n be the sequence in l^p defined by $e_n(j) = 0$, $j \neq n$; $e_n(n) = 1$. If $x \in l^p$, then $\|x - \sum_{k=1}^{n} x_k e_k\|^p = \sum_{k=n+1}^{\infty} |x_k|^p \to 0$; hence $x = \sum_{k=1}^{\infty} x_k e_k$,where the series converges in l^p. By continuity of f,

$$f(x) = \sum_{k=1}^{\infty} x_k y_k, \qquad \text{where} \quad y_k = f(e_k). \tag{1}$$

Now write y_k in polar form, that is, $y_k = r_k e^{i\theta_k}$, $r_k \geq 0$. Let

$$z_n = (r_1^{q-1} e^{-i\theta_1}, \ldots, r_n^{q-1} e^{-i\theta_n}, 0, 0, \ldots);$$

by (1),

$$f(z_n) = \sum_{k=1}^{n} r_k^{q-1} e^{-i\theta_k} r_k e^{i\theta_k} = \sum_{k=1}^{n} |y_k|^q. \tag{2}$$

But

$$\begin{aligned} |f(z_n)| &\leq \|f\| \, \|z_n\| \\ &= \|f\| \left(\sum_{k=1}^{n} |y_k|^{[q-1]p} \right)^{1/p} \\ &= \|f\| \left(\sum_{k=1}^{n} |y_k|^q \right)^{1/p}. \end{aligned}$$

By Eq. (2),

$$\left(\sum_{k=1}^{n} |y_k|^q \right)^{1/q} \leq \|f\|;$$

hence $y \in l^q$ and $\|y\| \leq \|f\|$. To prove that $\|f\| \leq \|y\|$, observe that Eq. (1) is of the form $f(x) = \int_\Omega xy \, d\mu$ where Ω is the positive integers and μ is counting measure. By Hölder's inequality, $|f(x)| \leq \|x\| \, \|y\|$, so that $\|f\| \leq \|y\|$.

Finally, we prove uniqueness. If $y, z \in l^q$ and $f(x) = \sum_{k=1}^{\infty} x_k y_k = \sum_{k=1}^{\infty} x_k z_k$ for all $x \in l^p$, then $g(x) = \sum_{k=1}^{\infty} x_k(y_k - z_k) = 0$ for all $x \in l^p$. The above argument with f replaced by g shows that $\|g\| = \|y - z\|$; hence $\|y - z\| = 0$, and thus $y = z$.

(c) Let f be a continuous linear functional on l^1. We show that there is a unique element $y \in l^\infty$ such that

$$f(x) = \sum_{k=1}^{\infty} x_k y_k \qquad \text{for all} \quad x \in l^1.$$

The argument of (b) may be repeated up to Eq. (1). In this case, however, if $y_k = r_k e^{i\theta_k}$, $k = 1, 2, \ldots$, we take

$$z_n = (0, \ldots, 0, e^{-i\theta_n}, 0, \ldots), \qquad \text{with } e^{-i\theta_n} \text{ in position } n.$$

Thus by (1),

$$f(z_n) = e^{-i\theta_n} r_n e^{i\theta_n} = |y_n|.$$

But

$$|f(z_n)| \le \|f\| \, \|z_n\| = \|f\|.$$

Therefore $|y_n| \le \|f\|$ for all n, so that $y \in l^\infty$ and $\|y\| \le \|f\|$. But by (1),

$$|f(x)| \le \left(\sup_k |y_k|\right) \sum_{k=1}^{\infty} |x_k| = \|y\| \, \|x\|;$$

hence $\|f\| \le \|y\|$. Uniqueness is proved as in (b).

In 3.3.4(b) and (c), the map $f \to y$ of $(l^p)^*$ to l^q [$q = \infty$ in 3.3.4(c)] is linear, and is one-to-one by uniqueness of y. To show that it is onto, observe that any linear functional of the form $f(x) = \sum_{k=1}^{\infty} x_k y_k$, $x \in l^p$, $y \in l^q$, satisfies $\|f\| \le \|y\|$ by the analysis of 3.3.4(b) and (c). Therefore f is continuous, so that every $y \in l^q$ is the image of some $f \in (l^p)^*$. Since $\|f\| = \|y\|$, we have an isometric isomorphism of $(l^p)^*$ and l^q.

If we replace y_k by $\bar{y}_k$, we obtain the result that there is a unique $y \in l^q$ such that $f(x) = \sum_{k=1}^{\infty} x_k \bar{y}_k$ for all $x \in l^p$. This makes the map $f \to y$ a conjugate isometry rather than an isometric isomorphism. If $p = 2$, then $q = 2$ also, and thus we have another proof of the Riesz representation theorem for separable Hilbert spaces (see 3.2.16). In fact, essentially the same argument may be used in an arbitrary Hilbert space if the e_n are replaced by an arbitrary orthonormal basis. (Other examples of representation of continuous linear functionals are given in Problems 10 and 11.)

We now show that the set of bounded linear operators from one normed linear space to another can be made into a normed linear space.

3.3.5 ***Theorem.*** Let L and M be normed linear spaces, and let $[L, M]$ be the collection of all bounded linear operators from L to M. The operator norm defined by 3.3.1 is a norm on $[L, M]$, and if M is complete, then $[L, M]$ is complete. In particular, the set L^* of all continuous linear functionals on L is a Banach space (whether or not L is complete).

PROOF. It follows from 3.3.1 that $\|A\| \ge 0$ and $\|aA\| = |a| \, \|A\|$ for all $A \in [L, M]$ and $a \in C$. Also by 3.3.1, if $\|A\| = 0$, then $Ax = 0$ for all $x \in L$, hence $A = 0$. If $A, B \in [L, M]$, then again by 3.3.1,

$$\|A + B\| = \sup\{\|(A + B)x\| : x \in L, \quad \|x\| \le 1\}.$$

Since $\|(A + B)x\| \le \|Ax\| + \|Bx\|$ for all x,

$$\|A + B\| \le \|A\| + \|B\|$$

and it follows that $[L, M]$ is a vector space and the operator norm is in fact a norm on $[L, M]$.

Now let $A_1, A_2, \ldots$ be a Cauchy sequence in $[L, M]$. Then

$$\|(A_n - A_m)x\| \leq \|A_n - A_m\| \, \|x\| \to 0 \quad \text{as} \quad n, m \to \infty. \tag{1}$$

Therefore $\{A_n x\}$ is a Cauchy sequence in M for each $x \in L$, hence A_n converges pointwise on L to an operator A. Since the A_n are linear, so is A (observe that $A_n(ax + by) = aA_n x + bA_n y$, and let $n \to \infty$). Now given $\varepsilon > 0$, choose N such that $\|A_n - A_m\| \leq \varepsilon$ for $n, m \geq N$. Fix $n \geq N$ and let $m \to \infty$ in Eq. (1) to conclude that $\|(A_n - A)x\| \leq \varepsilon \|x\|$ for $n \geq N$; therefore $\|A_n - A\| \to 0$ as $n \to \infty$. Since $\|A\| \leq \|A - A_n\| + \|A_n\|$, we have $A \in [L, M]$ and $A_n \to A$ in the operator norm. ∎

In the above proof we have talked about two different types of convergence of sequences of operators.

3.3.6 ***Definitions and Comments.*** Let $A, A_1, A_2, \ldots \in [L, M]$. We say that A_n converges *uniformly* to A iff $\|A_n - A\| \to 0$ (notation: $A_n \xrightarrow{u} A$). Since $\|(A_n - A)x\| \leq \|A_n - A\| \, \|x\|$, uniform operator convergence means that $A_n x \to Ax$, uniformly for $\|x\| \leq 1$ (or equally well for $\|x\| \leq k$, k any positive real number).

We say that A_n converges *strongly* to A (notation: $A_n \xrightarrow{s} A$) iff $A_n x \to Ax$ for each $x \in L$. Thus strong operator convergence is pointwise convergence on all of L.

Uniform convergence implies strong convergence, but not conversely. For example, let $\{e_1, e_2, \ldots\}$ be an orthonormal set in a Hilbert space, and let $A_n x = \langle x, e_n \rangle$, $n = 1, 2, \ldots$. Then $A_n \xrightarrow{s} 0$ by Bessel's inequality, but A_n does not converge uniformly to 0. In fact $\|A_n e_n\| = 1$ for all n, hence $\|A_n\| \equiv 1$.

There is an important property of finite-dimensional spaces that we are now in a position to discuss. In the previous section, we regarded C^n as a Hilbert space, so that if $x \in C^n$, the norm of x was taken as the Euclidean norm $\|x\| = (\sum_{i=1}^n |x_i|^2)^{1/2}$. The metric associated with this norm yields the standard topology on C^n. However, we may put various other norms on C^n, for example, the L^p norm $\|x\|_p = (\sum_{i=1}^n |x_i|^p)^{1/p}$, $1 \leq p < \infty$, or the sup norm $\|x\|_\infty = \max(|x_1|, \ldots, |x_n|)$. Since the space is finite-dimensional, there are no convergence difficulties and all elements have finite norm. Fortunately, the proliferation of norms causes no confusion because *all norms on a given finite-dimensional space induce the same topology*. The proof of this result is outlined in Problems 6 and 7.

Problems

1. Let f be a linear functional on the normed linear space L. If f is not identically 0, show that the following are equivalent:

   (a) f is continuous.
   (b) The null space $N = f^{-1}\{0\}$ is closed.
   (c) N is not dense in L.
   (d) f is bounded on some neighborhood of 0.

   [To prove that (c) implies (d), show that if $B(x, \varepsilon) \cap N = \varnothing$, and f is unbounded on $B(0, \varepsilon)$, then $f(B(0, \varepsilon)) = C$. In particular, there is a point $z \in B(0, \varepsilon)$ such that $f(z) = -f(x)$, and this leads to a contradiction.]
2. Show that any infinite-dimensional normed linear space had a discontinuous linear functional. (Let $e_1, e_2, \ldots$ be an infinite sequence of linearly independent elements such that $\|e_n\| = 1$ for all n. Define f appropriately on the e_n and extend f to the whole space using linearity.)
3. In Example 3.3.3(a), show that $\|A\| = \max_{a \le s \le b} \int_b^a |K(s, t)|\, dt$. [If $\int_a^b |K(s, t)|\, dt$ assumes a maximum at the point u, and $K(u, t) = r(t)e^{i\theta(t)}$, let $z(t) = r(t)e^{-i\theta(t)}$. Let $x_1, x_2, \ldots$ be continuous functions such that $|x_n(t)| \le 1$ for all n and t, and $\int_a^b |x_n(t) - z(t)|\, dt \to 0$ as $n \to \infty$. Since $\|Ax_n\| \le \|A\|$ and $(Ax_n)(s) \to \int_a^b K(s, t)z(t)\, dt$ as $n \to \infty$, it follows that $\int_a^b |K(u, t)|\, dt \le \|A\|$.]

   The same argument, with integrals replaced by sums, shows that if A is a matrix operator on C^n, with sup norm, $\|A\| = \max_{1 \le i \le n} \sum_{j=1}^{n} |a_{ij}|$.
4. Let M be a linear manifold in the normed linear space L. Denote by $[x]$ the coset of x modulo M, that is, $\{x + y : y \in M\}$. Define

   $$\|[x]\| = \inf\{\|y\| : y \in [x]\};$$

   note that $\|[x]\| = \inf\{\|x - z\| : z \in M\} = \text{dist}(x, M)$. Show that the above formula defines a seminorm on the quotient space L/M, a norm if M is closed.
5. If L is a Banach space and M is a closed subspace, show that L/M is a Banach space.
6. (a) Let A be a linear operator from L to M. Show that A is one-to-one with A^{-1} continuous on its domain $A(L)$, iff there is a finite number $m > 0$ such that $\|Ax\| \ge m\|x\|$ for all $x \in L$.

   (b) Let $\|\ \|_1$ and $\|\ \|_2$ be norms on the linear space L. Show that the norms induce the same topology iff there are finite numbers $m, M > 0$ such that

   $$m\|x\|_1 \le \|x\|_2 \le M\|x\|_1 \qquad \text{for all} \quad x \in L.$$

(This may be done using part (a), or it may be shown directly that, for example, if $\|x\|_1 \le (1/m)\|x\|_2$ for all x, then the topology induced by $\| \ \|_1$ is weaker than (that is, included in) the topology induced by $\| \ \|_2$.)

7. Let L be a finite-dimensional normed linear space, with basis $e_1, \ldots, e_n$. Let $\| \ \|_1$ be any norm on L, and define

$$\|x\|_2 = \left(\sum_{i=1}^{n} |x_i|^2 \right)^{1/2}, \qquad \text{where} \quad x = \sum_{i=1}^{n} x_i e_i .$$

   (a) Show that for some positive real number k, $\|x\|_1 \le k\,\|x\|_2$ for all $x \in L$.
   (b) Show that for some positive real number m, $\|x\|_1 \ge m$ on $\{x\colon \|x\|_2 = 1\}$. [By (a), the map $x \to \|x\|_1$ is continuous in the topology induced by $\| \ \|_2$.]
   (c) Show that $\|x\|_1 \ge m\,\|x_2\|$ for all $x \in L$, and conclude that all norms on a finite-dimensional space induce the same topology.
   (d) Let M be an *arbitrary* normed linear space, and let L be a finite-dimensional subspace of M. Show that L is closed in M.

8. (*Riesz lemma*) Let M be a closed proper subspace of the normed linear space L. Show that if $0 < \delta < 1$, there is an element $x_\delta \in L$ such that $\|x_\delta\| = 1$ and $\|x - x_\delta\| \ge \delta$ for all $x \in M$. [Choose $x_1 \notin M$, and let $d = \text{dist}(x_1, M) > 0$ since M is closed. Choose $x_0 \in M$ such that $\|x_1 - x_0\| \le d/\delta$, and set $x_\delta = (x_1 - x_0)/\|x_1 - x_0\|$.]

9. Let L be a normed linear space. Show that the following are equivalent:
   (a) L is finite-dimensional.
   (b) L is topologically isomorphic to a Euclidean space C^n (or R^n if L is a real vector space), that is, there is a one-to-one, onto, linear, bicontinuous map between the two spaces.
   (c) L is locally compact (every point of L has a neighborhood whose closure is compact).
   (d) Every closed bounded subset of L is compact.
   (e) The set $\{x \in L\colon \|x\| = 1\}$ is compact.
   (f) The set $\{x \in L\colon \|x\| = 1\}$ is totally bounded, that is, can be covered by a finite number of open balls of any preassigned radius.

   [Problem 7 shows that (a) implies (b); (b) implies (c), (d) implies (e), and (e) implies (f) are obvious, and (c) implies (d) is easy. To prove that (f) implies (a), use Problem 8.]

10. Let c be the space of convergent sequences of complex numbers [see 3.1.2(b)]. If f is a continuous linear functional on c, show that there is a unique element $y = (y_0, y_1, \ldots) \in l^1$ such that for all $x \in c$,

$$f(x) = \left(\lim_{n \to \infty} x_n \right) \left(y_0 - \sum_{k=1}^{\infty} y_k \right) + \sum_{k=1}^{\infty} x_k y_k .$$

Furthermore, $\|f\| = |y_0 - \sum_{k=1}^{\infty} y_k| + \sum_{k=1}^{\infty} |y_k|$. If c_0 is the closed subspace of c consisting of those sequences converging to 0, the representation of a continuous linear functional on c_0 is simpler: $f(x) = \sum_{k=1}^{\infty} x_k y_k$, $\|f\| = \sum_{k=1}^{\infty} |y_k|$. Thus $c_0{}^*$ is isometrically isomorphic to l^1.

11. Let $(\Omega, \mathscr{F}, \mu)$ be a measure space.

(a) Assume μ finite, $1 < p < \infty$, $(1/p) + (1/q) = 1$. If f is a continuous linear functional on $L^p = L^p(\Omega, \mathscr{F}, \mu)$ show that there is an element $y \in L^q$ such that

$$f(x) = \int_\Omega xy \, d\mu \qquad \text{for all} \quad x \in L^p.$$

Furthermore, $\|f\| = \|y\|_q$. If y_1 is another such function, show that $y = y_1$ a.e. $[\mu]$. [Define $\lambda(A) = f(I_A)$, $A \in \mathscr{F}$, and apply the Radon–Nikodym theorem.]

(b) Drop the finiteness assumption on μ. If $A \in \mathscr{F}$ and $\mu(A) < \infty$, part (a) applied to $(A, \mathscr{F} \cap A, \mu)$ provides an essentially unique $y_A \in L^q$ such that $y_A = 0$ on A^c and

$$f(xI_A) = \int_\Omega xy_A \, d\mu \qquad \text{for all} \quad x \in L^p;$$

also, $\|y_A\|_q$ is the norm of the restriction of f to $L^p(A)$, so

$$\|y_A\|_q \le \|f\|.$$

(i) If $\mu(A) < \infty$ and $\mu(B) < \infty$, show that $y_A = y_B$ a.e. $[\mu]$ on $A \cap B$. Thus $y_{A \cup B}$ may be obtained by piecing together y_A and y_B.

(ii) Let A_n, $n = 1, 2, \ldots$, be sets with $\|y_{A_n}\|_q \to k = \sup\{\|y_A\|_q : A \in \mathscr{F}, \mu(A) < \infty\} \le \|f\|$. Show that y_{A_n} converges in L_q to a limit function y, and y is essentially independent of the particular sequence $\{A_n\}$. Furthermore, $\|y\|_q \le \|f\|$.

(iii) Show that $f(x) = \int_\Omega xy \, d\mu$ for all $x \in L^p$. Since $\|y\|_q \le \|f\|$ by (ii) and $\|f\| \le \|y\|_q$ by Hölder's inequality, we have $\|f\| = \|y\|_q$. Thus the result of (a) holds for arbitrary μ.

(c) Prove (a) with μ finite, $p = 1$, $q = \infty$.

(d) Prove (a) with μ σ-finite, $p = 1$, $q = \infty$. [For an extension of this result, see Kelley and Namioka (1963, Problem 14M).]

It follows that there is an isometric isomorphism of $(L^p)^*$ and L^q if $1 < p < \infty$, $1 < q < \infty$, $(1/p) + (1/q) = 1$; if μ is σ-finite, this is true also for $p = 1$, $q = \infty$.

3.4 Basic Theorems of Functional Analysis

Almost every area of functional analysis leans heavily on at least one of the three basic results of this section: the Hahn–Banach theorem, the uniform boundedness principle, and the open mapping theorem. We are going to establish these results and discuss applications.

We first consider an extension problem. If f is a linear functional defined on a subspace M of a vector space L, there is no difficulty in extending f to a linear functional on all of L; simply extend a Hamel basis of M to a Hamel basis for L, define f arbitrarily on the basis vectors not belonging to M, and extend by linearity. However, if L is normed and we require that the extension of f have the same norm as the original functional, the problem becomes more difficult. We first prove a preliminary result.

3.4.1 ***Lemma.*** Let L be a real vector space, not necessarily normed, and let p be a map from L to R satisfying

$$\begin{aligned} p(x+y) &\le p(x) + p(y) \qquad \text{for all} \quad x, y \in L \\ p(ax) &= ap(x) \qquad \text{for all} \quad x \in L \quad \text{and} \quad \text{all } a > 0. \end{aligned}$$

[The first property is called *subadditivity*, the second *positive-homogeneity*. Note that positive-homogeneity implies that $p(0) = 0$ [set $x = 0$ to obtain $p(0) = ap(0)$ for all $a > 0$]. A subadditive, positive-homogeneous map is sometimes called a *sublinear functional*.]

Let M be a subspace of L and g a linear functional defined on M such that $g(x) \le p(x)$ for all $x \in M$. Let x_0 be a fixed element of L. For any real number c, the following are equivalent:

(1) $g(x) + \lambda c \le p(x + \lambda x_0)$ for all $x \in M$ and all $\lambda \in R$.
(2) $-p(-x - x_0) - g(x) \le c \le p(x + x_0) - g(x)$ for all $x \in M$.

Furthermore, there is a real number c satisfying (2), and hence (1).

PROOF. To prove that (1) implies (2), first set $\lambda = 1$, and then set $\lambda = -1$ and replace x by $-x$ [note $g(-x) = -g(x)$ by linearity]. Conversely, if (2) holds and $\lambda > 0$, replace x by x/λ in the right-hand inequality of (2); if $\lambda < 0$, replace x by x/λ in the left-hand inequality. In either case, the positive homogeneity of p yields (1). If $\lambda = 0$, (1) is true by hypothesis.

To produce the desired c, let x and y be arbitrary elements of M. Then

$$\begin{aligned} g(x) - g(y) = g(x - y) &\le p(x - y) \\ &\le p(x + x_0) + p(-y - x_0) \qquad \text{by subadditivity.} \end{aligned}$$

It follows that

$$\sup_{y \in M}[-p(-y - x_0) - g(y)] \le \inf_{x \in M}[p(x + x_0) - g(x)].$$

Any c between the sup and the inf will work. ∎

We may now prove the main extension theorem.

3.4.2 *Hahn–Banach Theorem.* Let p be a subadditive, positive-homogeneous functional on the real linear space L, and g a linear functional on the subspace M, with $g \le p$ on M. There is a linear functional f on L such that $f = g$ on M and $f \le p$ on all of L.

PROOF. If $x_0 \notin M$, consider the subspace $M_1 = L(M \cup \{x_0\})$, consisting of all elements $x + \lambda x_0$, $x \in M$, $\lambda \in R$. We may extend g to a linear functional on M_1 by defining $g_1(x + \lambda x_0) = g(x) + \lambda c$, where c is any real number. If we choose c to satisfy (1) of 3.4.1, then $g_1 \le p$ on M_1.

Now let $\mathscr{C}$ be the collection of all pairs (h, H) where h is an extension of g to the subspace $H \supset M$, and $h \le p$ on H. Partially order $\mathscr{C}$ by $(h_1, H_1) \le (h_2, H_2)$ iff $H_1 \subset H_2$ and $h_1 = h_2$ on H_1; then every chain in $\mathscr{C}$ has an upper bound (consider the union of all subspaces in the chain). By Zorn's lemma, $\mathscr{C}$ has a maximal element (f, F). If $F \ne L$, the first part of the proof yields an extension of f to a larger subspace, contradicting maximality. ∎

There is a version of the Hahn–Banach theorem for complex spaces. First, we observe that if L is a vector space over C, L is automatically a vector space over R, since we may restrict scalar multiplication to real scalars. For example, C^n is an n-dimensional space over C, with basis vectors $(1, 0, \dots, 0)$, $\dots$, $(0, \dots, 0, 1)$. If C^n is regarded as a vector space over R, it becomes $2n$-dimensional, with basis vectors

$$(1, 0, \dots, 0), \dots, (0, \dots, 0, 1), \qquad (i, 0, \dots, 0), \dots, (0, \dots, 0, i).$$

Now if f is a linear functional on L, with $f_1 = \operatorname{Re} f$, $f_2 = \operatorname{Im} f$, then f_1 and f_2 are linear functionals on L', where L' is L regarded as a vector space over R. Also, for all $x \in L$,

$$f(ix) = f_1(ix) + if_2(ix).$$

But

$$f(ix) = if(x) = -f_2(x) + if_1(x).$$

Thus $f_1(ix) = -f_2(x), f_2(ix) = f_1(x)$; consequently

$$f(x) = f_1(x) - if_1(ix) = f_2(ix) + if_2(x).$$

Therefore f is determined by f_1 (or by f_2). Conversely, let f_1 be a linear functional on L'. Then f_1 is a map from L to R such that $f_1(ax + by) = af_1(x) + bf_1(y)$ for all $x, y \in L$ and all $a, b \in R$. Define $f(x) = f_1(x) - if_1(ix)$, $x \in L$. It follows that f is a linear functional on L (and $f_1 = \operatorname{Re} f$). For f_1 is additive, hence so is f, and if $a, b \in R$, we have

$$\begin{aligned} f((a + ib)x) &= f_1(ax + ibx) - if_1(-bx + iax) \\ &= af_1(x) + bf_1(ix) + ibf_1(x) - iaf_1(ix) \\ &= (a + ib)[f_1(x) - if_1(ix)] \\ &= (a + ib)f(x). \end{aligned}$$

Note that homogeneity of f_1 does not immediately imply homogeneity of f. For $f_1(\lambda x) = \lambda f_1(x)$ for real λ but not in general for complex λ. [For example, let $L = C$, $f(x) = x$, $f_1(x) = \operatorname{Re} x$.]

We now prove the complex version of the Hahn–Banach theorem.

3.4.3 ***Theorem.*** Let L be a vector space over C, and p a seminorm on L. If g is a linear functional on the subspace M, and $|g| \le p$ on M, there is a linear functional f on L such that $f = g$ on M and $|f| \le p$ on L.

PROOF. Since p is a seminorm, it is subadditive and absolutely homogeneous, and hence positive-homogeneous. If $g_1 = \operatorname{Re} g$, then $g_1 \le |g| \le p$; so by 3.4.2, there is an extension of g_1 to a linear map f_1 of L to R such that $f_1 = g_1$ on M and $f_1 \le p$ on L. Define $f(x) = f_1(x) - if_1(ix)$, $x \in L$. Then f is a linear functional on L and $f = g$ on M. Fix $x \in L$, and let $f(x) = re^{i\theta}$, $r \ge 0$. Then

$$\begin{aligned} |f(x)| &= r = f(e^{-i\theta}x) \\ &= f_1(e^{-i\theta}x) && \text{since } r \text{ is real} \\ &\le p(e^{-i\theta}x) && \text{since } f_1 \le p \text{ on } L \\ &= p(x) && \text{by absolute homogeneity. ∎} \end{aligned}$$

3.4.4 ***Corollary.*** Let g be a continuous linear functional on the subspace M of the normed linear space L. There is an extension of g to a continuous linear functional f on L such that $\|f\| = \|g\|$.

PROOF. Let $p(x) = \|g\| \, \|x\|$; then p is a seminorm on L and $|g| \le p$ on M by definition of $\|g\|$. The result follows from 3.4.3. ∎

A direct application of the Hahn–Banach theorem is the result that in a normed linear space, there are enough continuous linear functionals to distinguish points; in other words, if $x \neq y$, there is a continuous linear functional f such that $f(x) \neq f(y)$. We now prove this, along with other related results.

3.4.5 ***Theorem.*** Let M be a subspace of the normed linear space L, and let L^* be the collection of all continuous linear functionals on L.

(a) If $x_0 \notin \overline{M}$, there is an $f \in L^*$ such that $f = 0$ on M, $f(x_0) = 1$, and $\|f\| = 1/d$, where d is the distance from x_0 to M.

(b) $x_0 \in \overline{M}$ iff every $f \in L^*$ that vanishes on M also vanishes at x_0.

(c) If $x_0 \neq 0$, there is an $f \in L^*$ such that $\|f\| = 1$ and $f(x_0) = \|x_0\|$; thus the maximum value of $|f(x)|/\|x\|$, $x \neq 0$, is achieved at x_0. In particular, if $x \neq y$, there is an $f \in L^*$ such that $f(x) \neq f(y)$.

PROOF. (a) First note that $L(M \cup \{x_0\})$ is the set of all elements $y = x + ax_0$, $x \in M$, $a \in C$, and since $x_0 \notin M$, a is uniquely determined by y. Define f on $N = L(M \cup \{x_0\})$ by $f(x + ax_0) = a$; f is linear, and furthermore, $\|f\| = 1/d$, as we now prove. By 3.3.1 we have

$$\begin{aligned}\|f\| &= \sup\left\{\frac{|f(y)|}{\|y\|} : y \in N, \quad y \neq 0\right\}\\ &= \sup\left\{\frac{|a|}{\|x + ax_0\|} : x \in M, \quad a \in C, \quad x \neq 0 \quad \text{or} \quad a \neq 0\right\}\\ &= \sup\left\{\frac{|a|}{\|x + ax_0\|} : x \in M, \quad a \in C, \quad a \neq 0\right\}\end{aligned}$$

since $f(y) = 0$ when $a = 0$. Now

$$\frac{|a|}{\|x + ax_0\|} = \frac{1}{\left\|x_0 + \dfrac{x}{a}\right\|} = \frac{1}{\|x_0 - z\|} \qquad \text{for some} \quad z \in M;$$

hence $\|f\| = (\inf\{\|x_0 - z\| : z \in M\})^{-1} = 1/d < \infty$. The result now follows from 3.4.4.

(b) This is immediate from (a).

(c) Apply (a) with $M = \{0\}$, to obtain $g \in L^*$ with $g(x_0) = 1$ and $\|g\| = 1/\|x_0\|$; set $f = \|x_0\| g$. ∎

The Hahn–Banach theorem is basic in the study of the concept of reflexivity, which we now discuss. Let L be a normed linear space, and L^* the set of continuous linear functionals on L; L^* is sometimes called the *conjugate space* of L. By 3.3.5, L^* is a Banach space, so that we may talk about L^{**}, the conjugate space of L^*, or the *second conjugate space* of L. We may identify L with a subspace of L^{**} as follows: If $x \in L$, we define $x^{**} \in L^{**}$ by

$$x^{**}(f) = f(x), \qquad f \in L^*.$$

If $\|f_n - f\| \to 0$, then $f_n(x) \to f(x)$; hence x^{**} is in fact a continuous linear functional on L^*. Let us examine the map $x \to x^{**}$ of L into L^{**}.

3.4.6 ***Theorem.*** Define $h: L \to L^{**}$ by $h(x) = x^{**}$. Then h is an isometric isomorphism of L and the subspace $h(L)$ of L^{**}; therefore, if $x \in L$, we have, by 3.3.1, $\|x\| = \|x^{**}\| = \sup\{|f(x)| : f \in L^*, \|f\| \le 1\}$.

PROOF: To show that h is linear, we write

$$[h(ax + by)](f) = f(ax + by) = af(x) + bf(y) = [ah(x)](f) + [bh(y)](f).$$

We now prove that h is norm-preserving ($\|h(x)\| = \|x\|$ for all $x \in L$); consequently, h is one-to-one. If $x \in L$, $|[h(x)](f)| = |f(x)| \le \|x\| \; \|f\|$, and hence $\|h(x)\| \le \|x\|$. On the other hand, by 3.4.5(c), there is an $f \in L^*$ such that $\|f\| = 1$ and $|f(x)| = \|x\|$. Thus

$$\sup\{|[h(x)](f)| : f \in L^*, \quad \|f\| = 1\} \ge \|x\|$$

so that $\|h(x)\| \ge \|x\|$, and consequently $\|h(x)\| = \|x\|$. ∎

If $h(L) = L^{**}$, L is said to be *reflexive*. Note that L^{**} is complete by 3.3.5 so by 3.4.6, a reflexive normed linear space is necessarily complete. We shall now consider some examples.

3.4.7 ***Examples.*** (a) Every Hilbert space is reflexive. For if ψ is the conjugate isometry of 3.3.4(a), H^* becomes a Hilbert space if we take $\langle f, g \rangle = \langle \psi(g), \psi(f) \rangle$. Thus if $q \in H^{**}$, we have, for some $g \in H^*$, $q(f) = \langle f, g \rangle = \langle \psi(g), \psi(f) \rangle = f(x)$, where $x = \psi(g)$. Therefore $q = h(x)$.

(b) If $1 < p < \infty$, l^p is reflexive. For by 3.3.4(b), $(l^p)^*$ is isometrically isomorphic to l^q, where $(1/p) + (1/q) = 1$. Thus if $t \in (l^p)^{**}$ we have $t(y) = \sum_{k=1}^{\infty} y_k z_k$, $y \in l^q$, where z is an element of $(l^q)^* = l^p$. But then $t = h(z)$.

Essentially the same argument, with the aid of Problem 11 of Section 3.3, shows that if $(\Omega, \mathscr{F}, \mu)$ is an arbitrary measure space $\mathbf{L}^p(\Omega, \mathscr{F}, \mu)$ is reflexive for $1 < p < \infty$.

(c) The space l^1 is not reflexive. This depends on the following result:

3.4.8 ***Theorem.*** If L is a normed linear space and L^* is separable, so is L. Thus if L is reflexive and separable (so that L^{**} is separable), then so is L^*.

PROOF. Let $f_1, f_2, \ldots$ form a countable dense subset of $\{f \in L^*: \|f\| = 1\}$ (note that any subset of a separable metric space is separable). Since $\|f_n\| = 1$, we can find points $x_n \in L$ with $\|x_n\| = 1$ and $|f_n(x_n)| \geq \frac{1}{2}$ for all n. Let M be the space spanned by the x_n; we claim that $M = L$. If not, 3.4.5(a) yields an $f \in L^*$ with $f = 0$ on M and $\|f\| = 1$. But then $\frac{1}{2} \leq |f_n(x_n)| = |f_n(x_n) - f(x_n)| \leq \|f_n - f\|$ for all n, contradicting the assumption that $\{f_1, f_2, \ldots\}$ is dense. ∎

To return to 3.4.7(c), we note that l^1 is separable since $\{x \in l^1: x_k = 0$ for all but finitely many $k\}$ is dense. [If $x \in l^1$ and $x^{(n)} = (x_1, \ldots, x_n, 0, 0, \ldots)$ then $x^{(n)} \to x$ in l^1.] But $(l^1)^* = l^\infty$ by 3.3.4(c), and this space is not separable. For if $S = \{x \in l^\infty: x_k = 0$ or 1 for all $k\}$, then S is uncountable and $\|x - y\| = 1$ for all $x, y \in S$, $x \neq y$. Thus the sets $B_x = \{y \in l^\infty: \|y - x\| < \frac{1}{2}\}$, $x \in S$, form an uncountable family of disjoint open sets. If there were a countable dense set D, there would be at least one point of D in each B_x, a contradiction. Thus l^1 is not reflexive.

We now consider the second basic result of this section. Suppose that the A_i, i belonging to the arbitrary index set I, are bounded linear operators from L to M, where L and M are normed linear spaces. The uniform boundedness principle asserts that if L is complete and the A_i are pointwise bounded, that is, $\sup\{\|A_i x\|: i \in I\} < \infty$ for each $x \in L$, then the A_i are uniformly bounded, that is, $\sup\{\|A_i\|: i \in I\} < \infty$. Completeness of L is essential; to see this, let L be the set of all sequences $x = (x_1, x_2, \ldots)$ of complex numbers such that $x_k = 0$ for all but finitely many k, with the l^p norm, $1 \leq p \leq \infty$. Take $M = C$, and $A_n x = n x_n$, $n = 1, 2, \ldots$. For any x, $A_n x = 0$ for sufficiently large n, so the A_n are pointwise bounded, although $\|A_n\| = n \to \infty$.

The proof that we shall give uses the Baire category theorem, one version of which states if a complete metric space is a countable union of closed sets, one of the sets must have a nonempty interior. (See the appendix on general topology, Theorem A9.2.)

3.4.9 ***Principle of Uniform Boundedness.*** Let A_i, $i \in I$, be bounded linear operators from the Banach space L to the normed linear space M. If the A_i are pointwise bounded, they are uniformly bounded.

PROOF. Let $C_n = \{x \in L: \sup_i \|A_i x\| \leq n\}$, $n = 1, 2, \ldots$. Since the A_i are pointwise bounded, $\bigcup_{n=1}^{\infty} C_n = L$, and since the A_i are continuous, each C_n is closed. By the Baire category theorem, for some n there is a closed ball $\bar{B} = \{x \in L\colon \|x - x_0\| \le r\} \subset C_n$. Now if $\|y\| \le 1$ and $i \in I$, we have

$$\begin{aligned} \|A_i y\| &= \frac{1}{r}\|A_i z\| \qquad \text{where} \quad z = ry \\ &\le \frac{1}{r}\|A_i(x_0 + z)\| + \frac{1}{r}\|A_i x_0\| \\ &\le \frac{2n}{r} \qquad \text{since} \quad x_0 + z \quad \text{and} \quad x_0 \quad \text{belong to} \quad \bar{B}. \end{aligned}$$

Thus $\|A_i\| \le 2n/r$. ∎

The uniform boundedness principle is used in an important way in the study of weak convergence, a concept that we now describe.

3.4.10 ***Definitions and Comments.*** The sequence (or net) $\{x_n\}$ in the normed linear space L is said to converge *weakly* to $x \in L$ iff $f(x_n) \to f(x)$ for every $f \in L^*$ (notation: $x_n \xrightarrow{w} x$). Convergence in the metric of L ($\|x_n - x\| \to 0$) will be called *strong convergence* and will be written simply as $x_n \to x$. [In this terminology, strong convergence of the sequence of linear operators A_n to the linear operator A means strong convergence of $A_n x$ to Ax for each x (see 3.3.6).]

For each $x_0 \in L$, consider the collection of sets

$$U(x_0) = \{x \in L\colon |f_i(x) - f_i(x_0)| < \varepsilon_i, \quad i = 1, \ldots, n\},$$

where $\varepsilon_1, \ldots, \varepsilon_n > 0, f_1, \ldots, f_n \in L^*$, $n = 1, 2, \ldots$.

We form a topology by taking the $U(x_0)$ as a base for the neighborhood system at x_0. (There is no difficulty checking that the required axioms for a neighborhood system are satisfied.) The resulting (Hausdorff) topology is called the *weak topology* on L; it follows from the definition that convergence relative to the weak topology coincides with weak convergence. (The topology induced by the norm of L will be called the *strong topology*.)

It follows from the definitions that strong convergence implies weak convergence, hence every set that is open in the weak topology is open in the strong topology (and similarly for closed sets). To see that the converse does not hold, let $\{e_1, e_2, \ldots\}$ be an infinite orthonormal sequence in a Hilbert space H. If $x \in H$, then $\langle e_n, x\rangle \to 0$ as $n \to \infty$ by Bessel's inequality, hence $e_n \xrightarrow{w} 0$. But $\|e_n\| \equiv 1$, so e_n does not converge strongly to 0.

In a Euclidean space, strong and weak convergence coincide. For if $x_n = a_{1n}e_1 + \cdots + a_{kn}e_k$ converges weakly to $x = a_1 e_1 + \cdots + a_k e_k$ (where the e_i

are basis vectors for C^k), let f be a continuous linear functional that is 1 at e_i and 0 at e_j, $j \neq i$. Then $f(x_n) \to f(x)$, so that $a_{in} \to a_i$ as $n \to \infty$ $(i = 1, \ldots, k)$. Thus $x_n \to x$ strongly.

The weak topology is an important example of a product topology (see the appendix on general topology, Section A3). For L may be identified (see 3.4.6) with a subset $h(L)$ of the set C^{L^*} of all complex-valued functions on L^*, and the weak topology is the product topology, also called the "topology of pointwise convergence," on C^{L^*}, relativized to $h(L)$.

We give a few properties of weak convergence.

3.4.11 ***Theorem.*** (a) A weakly convergent sequence $\{x_n\}$ is bounded, that is, $\sup_n \|x_n\| < \infty$. In fact if $x_n \xrightarrow{w} x_0$, then $\|x_0\| \leq \liminf_{n\to\infty} \|x_n\|$.

(b) If A is a bounded linear operator from L to M and the net $\{x_i\}$ converges weakly to x_0 in L, then Ax_i converges weakly to Ax_0 in M.

(c) If the net $\{x_i\}$ converges weakly to x_0, then x_0 belongs to the subspace spanned by the x_i.

(d) If M is a linear manifold of L, the closures of M relative to the weak and strong topologies coincide. Consequently if M is closed in the strong topology, it is closed in the weak topology.

PROOF. (a) The x_n may be regarded as continuous linear functionals on L^* with $x_n(f) = f(x_n)$ (see 3.4.6). The x_n are pointwise bounded on L^* since $f(x_n) \to f(x_0)$; hence by the uniform boundedness principle, $\sup_n \|x_n\| < \infty$. Also, if $f \in L^*$,

$$|f(x_0)| = \lim_{n\to\infty} |f(x_n)| = \liminf_{n\to\infty} |f(x_n)| \leq \|f\| \liminf_{n\to\infty} \|x_n\|.$$

Since $\|x_0\| = \sup\{|f(x_0)| : f \in L^*, \|f\| \leq 1\}$, we may conclude that $\|x_0\| \leq \liminf_{n\to\infty} \|x_n\|$.

(b) If $g \in M^*$, define $f = g \circ A$; since A is continuous, we have $f \in L^*$, so that $f(x_i) \to f(x_0)$, that is, $g(Ax_i) \to g(Ax_0)$. But g is arbitrary; hence $Ax_i \xrightarrow{w} Ax_0$.

(c) If this is false, 3.4.5(a) yields an $f \in L^*$ with $f(x_0) = 1$ and $f(x_i) \equiv 0$, contradicting $x_i \xrightarrow{w} x_0$.

(d) The strong closure is always a subset of the weak closure, so assume x belongs to the weak closure of M. Let $\{x_i\}$ be a net in M with $x_i \xrightarrow{w} x$. By (c), x is a strong limit of a net of finite linear combinations of the x_i. But these finite linear combinations belong to M since M is a subspace; hence x belongs to the strong closure. ▌

In order to characterize weak convergence in specific spaces, the following result is useful.

3.4.12 ***Theorem.*** Let E be a subset of L^* such that $S(E) = L^*$, and let $\{x_n\}$ be a sequence in L. If $x_0 \in L$, then $x_n \xrightarrow{w} x_0$ iff $\sup_n \|x_n\| < \infty$ and $f(x_n) \to f(x_0)$ for all $f \in E$.

PROOF. The "only if" part follows from 3.4.11(a) and the definition of weak convergence. For the "if" part, let $f \in L^*$, and choose elements $f_k \in L(E)$ with $\|f - f_k\| \to 0$. Then

$$\begin{aligned} |f(x_n) - f(x_0)| &\le |f(x_n) - f_k(x_n)| + |f_k(x_n) - f_k(x_0)| + |f_k(x_0) - f(x_0)| \\ &\le \|f - f_k\| \, \|x_n\| + |f_k(x_n) - f_k(x_0)| + \|f_k - f\| \, \|x_0\|. \end{aligned}$$

Since the x_n are bounded in norm, given $\varepsilon > 0$, we may choose k such that the right-hand side is at most $|f_k(x_n) - f_k(x_0)| + \varepsilon$; but $f_k(x_n) \to f_k(x_0)$ as $n \to \infty$ since $f_k \in L(E)$, and since ε is arbitrary, the result follows. ∎

We now describe weak convergence in l^p and $\mathbf{L}^p$.

3.4.13 ***Theorem.*** Assume $1 < p < \infty$.

(a) Let $x_n = (x_{n1}, x_{n2}, \ldots) \in l^p$, $n = 1, 2, \ldots$. If $z \in l^p$, then $x_n \xrightarrow{w} z$ iff $\sup_n \|x_n\| < \infty$ and $x_{nk} \xrightarrow{w} z_k$ as $n \to \infty$ for each k.

(b) Let $x_n \in \mathbf{L}^p(\Omega, \mathscr{F}, \mu)$, $n = 1, 2, \ldots$, where μ is assumed finite. If $z \in \mathbf{L}^p(\Omega, \mathscr{F}, \mu)$, then $x_n \to z$ iff $\sup_n \|x_n\| < \infty$ and $\int_A x_n \, d\mu \to \int_A z \, d\mu$ for each $A \in \mathscr{F}$. (It will often be convenient to blur the distinction between L^p and $\mathbf{L}^p$, and treat the elements of $\mathbf{L}^p$ as functions rather than equivalence classes.)

PROOF. (a) Define $f_k \in (l^p)^*$ by $f_k(x) = x_k$; then f_k corresponds to the sequence in l^q with a one in position k and zeros elsewhere [see 3.3.4(b)]. Take $E = \{f_1, f_2, \ldots\}$ and apply 3.4.12.

(b) For each $A \in \mathscr{F}$, define $f_A \in (\mathbf{L}^p)^*$ by $f_A(x) = \int_A x \, d\mu$; f_A corresponds to the indicator function $I_A \in \mathbf{L}^q$ (see Problem 11, Section 3.3). Take E as the set of all f_A, $A \in \mathscr{F}$; E spans $(\mathbf{L}^p)^*$ because the simple functions are dense in $\mathbf{L}^q$, and the result follows from 3.4.12. ∎

We now consider the third basic result, the open mapping theorem. This will allow us to conclude that under certain conditions, the inverse of a one-to-one continuous linear operator is continuous. We cannot make this assertion in general, as the following example shows: Let L be the set of all continuous complex-valued functions x on $[0, 1]$ such that $x(0) = 0$, $M = \{x \in L\colon x$ has a continuous derivative on $[0, 1]\}$; put the sup norm on L and M. If A is defined by $(Ax)(t) = \int_0^t x(s) \, ds$, $0 \le t \le 1$, then A is a one-to-one, bounded, linear operator from L onto M. But A^{-1} is discontinuous; for example, if $x_n(t) = \sin nt$, then $y_n(t) = (Ax_n)(t) = (1 - \cos nt)/n$, so that $y_n \to 0$ in M, but x_n has no limit in L. A hypothesis under which continuity of the inverse holds is the completeness of both spaces L and M. In the above example, M is not complete; for example, a sequence of polynomials may converge uniformly to a continuous function without a continuous derivative (in fact to a continuous nowhere differentiable function).

We now state the third basic theorem.

3.4.14 ***Open Mapping Theorem.*** Let A be a bounded linear operator from the Banach space L onto the Banach space M. Then A is an open map, that is, if D is an open subset of L, then $A(D)$ is an open subset of M. Consequently if A is also one-to-one, then A^{-1} is bounded.

PROOF. We defer this until the next section (see 3.5.11) since it will not be much more difficult to prove a more general statement about mappings of topological vector spaces. ▮

The open mapping theorem allows us to prove the closed graph theorem, which is often useful in proving that a particular linear operator is bounded.

First we observe that if L and M are normed linear spaces, we may define a norm on the product space $L \times M$ by $\|(x, y)\| = (\|x\|^p + \|y\|^p)^{1/p}$, $x \in L$, $y \in M$, where p is any fixed real number in $[1, \infty)$. For if $x, x' \in L$, $y, y' \in M$,

$$
\begin{aligned}
(\|x + x'\|^p + \|y + y'\|^p)^{1/p} &\le [(\|x\| + \|x'\|)^p + (\|y\| + \|y'\|)^p]^{1/p} \\
&\le (\|x\|^p + \|y\|^p)^{1/p} + (\|x'\|^p + \|y'\|^p)^{1/p} \\
&\qquad \text{by Minkowski's inequality applied to } R^2.
\end{aligned}
$$

Therefore the triangle inequality is satisfied and we have defined a norm on $L \times M$ (the other requirements for a norm are immediate). Furthermore, $(x_n, y_n) \to (x, y)$ iff $x_n \to x$ and $y_n \to y$; thus regardless of the value of p, the norm on $L \times M$ induces the product topology; also, if L and M are complete, so is $L \times M$. [The same result is obtained using the analog of the L^∞ norm, that is, $\|(x, y)\| = \max(\|x\|, \|y\|)$.]

3.4.15 ***Definition.*** Let A be a linear operator from L to M, where L and M are normed linear spaces. We say that A is *closed* iff the graph $G(A) = \{(x, Ax): x \in L\}$ is a closed subset of $L \times M$. Equivalently, A is closed iff the following condition holds:

If $x_n \in L$, $x_n \to x$, and $Ax_n \to y$, then $(x, y) \in G(A)$; in other words, $y = Ax$. This formulation shows that every bounded linear operator is closed. The converse holds if L and M are Banach spaces.

3.4.16 ***Closed Graph Theorem.*** If A is a closed linear operator from the Banach space L to the Banach space M, then A is bounded.

PROOF. Since $G(A)$ is a closed subspace of $L \times M$, it is a Banach space. Define $P\colon G(A) \to L$ by $P(x, Ax) = x$. Then P is linear and maps onto L, and $\|P(x, Ax)\| = \|x\| \le \|(x, Ax)\|$; hence $\|P\| \le 1$ so that P is bounded. [Alternatively, if $(x_n, Ax_n) \to (x, y)$, then $x_n \to x$, proving continuity of P.] Similarly, the linear operator $Q\colon G(A) \to M$ given by $Q(x, Ax) = Ax$ is bounded. If $P(x, Ax) = 0$, then $x = Ax = 0$, so P is one-to-one. By 3.4.14, P^{-1} is bounded, and since $A = Q \circ P^{-1}$, A is bounded. ▌

As an application of the closed graph theorem, we show that if P is the projection of a Banach space L on a closed subspace M, then P is continuous [see 3.3.3(c)]. Let $\{x_n\}$ be a sequence of points in L with $x_n \to x$, and assume Px_n converges to the element $y \in M$. Recall that in defining a projection operator it is assumed that L is the direct sum of closed subspaces M and N; thus

$$x_n = y_n + z_n \qquad \text{where} \quad y_n = Px_n \in M, \quad z_n \in N.$$

Since $x_n \to x$ and $y_n \to y$, it follows that $z_n \to z = x - y$, necessarily in N. Therefore $x = y + z$, $y \in M$, $z \in N$, so that $y = Px$, proving P closed. By 3.4.16, P is continuous.

Problems

1. Show that a subadditive, absolutely homogeneous functional on a vector space must be nonnegative, and hence a seminorm. Give an example of a subadditive, positive-homogeneous functional that fails to be nonnegative.
2. Let $(\Omega, \mathscr{F}, \mu)$ be a measure space, and assume $\mathscr{F}$ is countably generated, that is, there is a countable set $\mathscr{C} \subset \mathscr{F}$ such that $\sigma(\mathscr{C}) = \mathscr{F}$. (Note that the minimal field $\mathscr{F}_0$ over $\mathscr{C}$ is also countable; see Problem 9, Section 1.2.) If μ is σ-finite on $\mathscr{F}_0$ and $1 \le p < \infty$, show that $\mathbf{L}^p(\Omega, \mathscr{F}, \mu)$ is separable. If in addition there is an infinite collection of disjoint sets $A \in \mathscr{F}$ with $\mu(A) > 0$, show that $\mathbf{L}^1(\Omega, \mathscr{F}, \mu)$ is not reflexive.

3. If L and M are normed linear spaces and $[L, M]$ is complete, show that M must be complete.
4. Let $A \in [L, M]$, where L and M are normed linear spaces. The *adjoint* of A is an operator $A^* : M^* \to L^*$, defined as follows: If $f \in M^*$ we take $(A^*f)(x) = f(Ax)$, $x \in L$. Establish the following results:

   (a) $\|A^*\| = \|A\|$.
   (b) $(aA + bB)^* = aA^* + bB^*$ for all $a, b \in C$, $A, B \in [L, M]$.
   (c) If $A \in [L, M]$, $B \in [M, N]$, then $(BA)^* = A^*B^*$, where BA is the composition of A and B.
   (d) If $A \in [L, M]$, A maps onto M, and A^{-1} exists and belongs to $[M, L]$, then $(A^{-1})^* = (A^*)^{-1}$.
5. Define the *annihilator* of the subset K of the normed linear space L as $K^\perp = \{f \in L^* : f(x) = 0 \text{ for all } x \in K\}$. Similarly, if $J \subset L^*$, define $J^\perp = \{x \in L : f(x) = 0 \text{ for all } f \in J\}$. If $A \in [L, M]$, we denote by $N(A)$ the null space $\{x \in L : Ax = 0\}$, and by $R(A)$ the closure of the range of A, that is, the closure of $\{Ax : x \in L\}$. Establish the following:

   (a) For any $K \subset L$, $K^{\perp\perp} = S(K)$, the space spanned by K.
   (b) $R(A)^\perp = N(A^*)$ and $R(A) = N(A^*)^\perp$.
   (c) $R(A) = M$ iff A^* is one-to-one.
   (d) $R(A^*)^\perp = N(A)$.
   (e) For any $J \subset L^*$, $S(J) \subset J^{\perp\perp}$; $S(J) = J^{\perp\perp}$ if L is reflexive.
   (f) $R(A^*) \subset N(A)^\perp$; $R(A^*) = N(A)^\perp$ if L is reflexive.
   (g) If $R(A^*) = L^*$, then A is one-to-one; the converse holds if L is reflexive.
6. Consider the Hahn–Banach theorem 3.4.2, with the additional assumption that L is a normed linear space (or more generally, a topological vector space) and p is continuous at 0; hence continuous on all of L since $|p(x) - p(y)| \le p(x - y)$. Show that if L is separable, the theorem may be proved without Zorn's lemma. It follows that 3.4.3 and 3.4.4 do not require Zorn's lemma under the above hypothesis.
7. If μ_0 is a finitely additive, nonnegative real-valued set function on a field $\mathscr{F}_0$ of subsets of a set Ω, use the Hahn–Banach theorem to show that μ_0 has an extension to a finitely additive, nonnegative real-valued set function on the class of *all* subsets of Ω. Thus in one respect, at least, finite additivity is superior to countable additivity.
8. If the sequence (or net) of bounded linear operators A_n on a Banach space converges strongly to the (necessarily linear) operator A, show that A is bounded; in fact

$$\|A\| \le \liminf_{n\to\infty} \|A_n\| \le \sup_n \|A_n\| < \infty.$$

9. Let $\{A_i,\ i \in I\}$ be a family of continuous linear operators from the Banach space L to the normed linear space M. Assume the A_i are weakly bounded, that is, $\sup_i |f(A_i x)| < \infty$ for all $x \in L$ and all $f \in M^*$. Show that the A_i are uniformly bounded, that is, $\sup_i \|A_i\| < \infty$.
10. (a) If the elements x_i, $i \in I$, belong to the normed linear space L, and $\sup_i |f(x_i)| < \infty$ for each $f \in L^*$, show that $\sup_i \|x_i\| < \infty$.
    (b) If A is a linear operator from the normed linear space L to the normed linear space M, and $f \circ A$ is continuous for each $f \in M^*$, show that A is continuous.
11. Let L, M, and N be normed linear spaces, with L or M complete, and let $B: L \times M \to N$ be a bilinear form, that is, $B(x, y)$ is linear in x for each fixed y, and linear in y for each fixed x. If for each $f \in N^*, f(B(x, y))$ is continuous in x for each fixed y, and continuous in y for each fixed x, show that B is bounded, that is,

$$\sup\{|B(x, y)| : \|x\|, \quad \|y\| \leq 1\} < \infty.$$

    Equivalently, for some positive constant k we have $|B(x, y)| \leq k\|x\| \|y\|$ for all x, y.
12. Give an example of a closed unbounded operator from one normed linear space to another.
13. Let A be a bounded linear operator from the Banach space L onto the Banach space M. Show that there is a positive number k such that for each $y \in M$ there is an $x \in L$ with $y = Ax$ and $\|x\| \leq k\|y\|$. This result is sometimes called the *solvability theorem.*

3.5 Some Properties of Topological Vector Spaces

Recall that a topological vector space is a vector space L with a topology that makes addition and scalar multiplication continuous; that is, the maps $(x, y) \to x + y$ of $L \times L \to L$ and $(a, x) \to ax$ of $C \times L \to L$ are continuous, with the product topology on $L \times L$ and $C \times L$. In this section we look at some of the important properties of such spaces.

One cannot put an arbitrary topology on a vector space L and obtain a topological vector space. For example, consider the discrete topology (all subsets of L are open); with this topology, L is not a topological vector space (exclude the trivial case $L = \{0\}$). To see this, pick a nonzero $x \in L$, and let $a_n = 1/n$, $n = 1, 2, \ldots$. If scalar multiplication were continuous, we would have $a_n x \to 0 \cdot x = 0$; since $\{0\}$ is open, this implies that $a_n x = 0$ for sufficiently large n, a contradiction.

The following result is helpful in checking whether a given topology makes L into a topological vector space. [Throughout this section, we use the following notation: If A and B are subsets of L,

$$A + B = \{x + y \colon x \in A, \quad y \in B\},$$
$$A - B = \{x - y \colon x \in A, \quad y \in B\}.$$

If $x \in L$,

$$x + B = \{x + y \colon y \in B\}$$
$$x - B = \{x - y \colon y \in B\}.]$$

3.5.1 ***Theorem.*** Let L be a topological vector space, and let $\mathscr{U}$ be a base of overneighborhoods at 0 (the sets in $\mathscr{U}$ are overneighborhoods of 0, and for each neighborhood V of 0 there is a set $U \in \mathscr{U}$ with $U \subset V$). Then

(a) if $U, V \in \mathscr{U}$, there is a set $W \in \mathscr{U}$ such that $W \subset U \cap V$;
(b) if $U \in \mathscr{U}$ then for each $x \in L$ there is a $\delta > 0$ such that $ax \in U$ for all complex numbers a with $|a| \le \delta$;
(c) for each $U \in \mathscr{U}$, there is a set $V \in \mathscr{U}$ with $V + V \subset U$.

Conversely, if $\mathscr{U}$ is a collection of subsets of L satisfying (a)–(c), and in addition,

(d) each $U \in \mathscr{U}$ is *circled*, in other words, if $x \in U$, then $ax \in U$ whenever $|a| \le 1$,

there is a unique topology that makes L a topological vector space with $\mathscr{U}$ a base of overneighborhoods at 0.

PROOF. (a) This follows since the intersection of two neighborhoods of 0 is a neighborhood of 0.

(b) By continuity of scalar multiplication, $ax \to 0$ as $a \to 0$, with x fixed; hence $ax \in U$ for sufficiently small $|a|$.

(c) By continuity of the map $(x, y) \to x + y$ at $x = y = 0$, there are sets $V_1, V_2 \in \mathscr{U}$ such that if $x \in V_1$ and $y \in V_2$, then $x + y \in U$; in other words, $V_1 + V_2 \subset U$. By (a), there is a set $V \in \mathscr{U}$ with $V \subset V_1 \cap V_2$; then $V + V \subset V_1 + V_2 \subset U$.

Conversely, let $\mathscr{U}$ satisfy (a)–(d). Take $\mathscr{U}$ as an overneighborhood base at 0, and, for each $x \in L$, $x + \mathscr{U} = \{x + U \colon U \in \mathscr{U}\}$ as an overneighborhhood base at x. Thus $\mathscr{V}(x) = \{V \subset L \colon V \supset x + U$ for some $U \in \mathscr{U}\}$ is the proposed class of overneighborhoods of x. Now:

(i) If $V \in \mathscr{V}(x)$, then $x \in V$. This follows because 0 belongs to each $U \in \mathscr{U}$, by (*b*).

(ii) If V_1, $V_2 \in \mathscr{V}(x)$, then $V_1 \cap V_2 \in \mathscr{V}(x)$. For if $x + U_i \subset V_i$, $i = 1, 2$, by (a) there is a set $W \in \mathscr{U}$ with $W \subset U_1 \cap U_2$. Then $x + W \subset (x + U_1) \cap (x + U_2) \subset V_1 \cap V_2$.

(iii) If $V \in \mathscr{V}(x)$ and $V \subset W$, then $W \in \mathscr{V}(x)$. This is immediate from the definition of $\mathscr{V}(x)$.

(iv) If $V \in \mathscr{V}(x)$, there is a set $W \in \mathscr{V}(x)$ such that $W \subset V$ and $V \in \mathscr{V}(y)$ for each $y \in W$. For if $x + U \subset V$, by (c) there is a set $T \in \mathscr{U}$ with $T + T \subset U$. If $W = x + T$ and $y = x + t \in W$, then $y + T = x + t + T \subset x + T + T \subset x + U \subset V$, hence $V \in \mathscr{V}(y)$.

Thus the axioms of a system of overneighborhoods are satisfied, so there is a topology on L with $\mathscr{U}$ an overneighborhood base at 0. By definition of the $\mathscr{V}(x)$, we have $x_n \to x$ iff $x_n - x \to 0$; hence if the addition operation $(x, y) \to x + y$ is continuous at $(0, 0)$, it is continuous everywhere. But continuity of addition at $(0, 0)$ is immediate from (c).

Finally, we look at continuity of the multiplication operation $(a, x) \to ax$. By (b), the map is continuous at $a = 0$ for each fixed x, hence, as above, continuous at any a for each fixed x. If $U \in \mathscr{U}$ and $a \in C$, choose a positive integer $n \geq |a|$ and apply (c) inductively to obtain $V \in \mathscr{U}$ such that $V + \cdots + V$ (n times) is a subset of U. (Note, for example, that $V + V + V \subset V + V + V + V$ since $0 \in V$.) If $nV = \{ny\colon y \in V\}$, then $nV \subset V + \cdots + V \subset U$; hence by (d), $aV \subset U$, so that multiplication is continuous at $x = 0$ for each fixed a, and thus continuous at any x for each fixed a. Now if $U \in \mathscr{U}$, then by (d), if $|a| \leq 1$ and $x \in U$, then $ax \in U$, so that multiplication is jointly continuous at $(0, 0)$. Thus if $x_n \to x$ and $a_n \to a$, then $(a_n - a)(x_n - x) \to 0$, and by the above analysis, $a_n x \to ax$, $ax_n \to ax$. It follows that $a_n x_n \to ax$.

Finally, to prove uniqueness of the topology, observe that if $\mathscr{U}$ is a base of overneighborhoods at 0, continuity of addition and subtraction makes the map $y \to x + y$, $y \in L$, a homeomorphism, and therefore $\mathscr{V}(x)$ is a system of overneighborhoods of x. ∎

We have seen that a seminorm on a vector space L induces a topology under which L becomes a topological vector space. We now show that a similar result holds if instead of a single seminorm we use an entire family.

3.5.2 ***Theorem.*** Let $\{p_i, i \in I\}$ be a family of seminorms on the vector space L. Let $\mathscr{U}$ be the class of all finite intersections of sets $\{x \in L\colon p_i(x) < \delta_i\}$ where $i \in I$, $\delta_i > 0$. Then $\mathscr{U}$ is a neighborhood base for a topology that makes L a topological vector space. This topology is the weakest, making all the p_i continuous, and for a net $\{x_n\}$ in L, $x_n \to x$ iff $p_i(x_n - x) \to 0$ for each $i \in I$.

PROOF. We show that the conditions of 3.5.1 are satisfied. Condition (a) follows because an intersection of two finite intersections of sets $\{x: p_i(x) < \delta_i\}$ is a finite intersection of such sets. If $x \in L$, then $p_i(ax) = |a| p_i(x)$, which is less than δ_i if $|a|$ is sufficiently small; this proves (b). To prove (c), suppose that $U = \bigcap_{i=1}^n \{x: p_i(x) < \delta_i\}$, and let $V = \bigcap_{i=1}^n \{x: p_i(x) < \delta\}$ where $0 < \delta \leq \frac{1}{2} \min_i \delta_i$. If $y, z \in V$, then $p_i(y + z) \leq p_i(y) + p_i(z) < \delta_i$, hence $y + z \in U$. Finally, each $U \in \mathscr{U}$ is circled since $p_i(ax) = |a| p_i(x)$, proving (d). If $x \in U = \bigcap_{i=1}^n \{z: p_i(z) < \delta_i\}$ and $y \in V = \bigcap_{i=1}^n \{z: p_i(z) < \min_j[\delta_j - p_j(x)]\}$, then $p_i(x + y) \leq p_i(x) + p_i(y) < p_i(x) + \delta_i - p_i(x) = \delta_i$; hence $x + V \subset U$. Thus U is an overneighborhood of each of its points, so the sets in $\mathscr{U}$ are open and $\mathscr{U}$ is a neighborhood base, not merely an overneighborhood base.

Now continuity of p_i at 0 is equivalent to continuity everywhere $[|p_i(x) - p_i(y)| \leq p_i(x - y)]$ and the sets $U \in \mathscr{U}$ are open in any topology for which the p_i are continuous at 0. This proves that the given topology is the weakest, making the p_i continuous. Finally, $x_n \to x$ iff $x_n - x \to 0$ (see the proof of 3.5.1), and $x_n - x \to 0$ iff for each i, $p_i(x_n - x) \to 0$, by definition of $\mathscr{U}$. ∎

The topology induced by a family of seminorms is *locally convex*; that is, for each $x \in L$, the family of convex overneighborhoods forms a base of overneighborhoods at x. This follows because the sets $\{x: p_i(x) < \delta_i\}$ are convex. We shall prove later (see 3.5.7) that every locally convex topology is generated by a family of seminorms.

If for each $x \neq 0$ there is an $i \in I$ such that $p_i(x) \neq 0$, the topology is Hausdorff. For if $x \neq y$ and $p_i(x - y) = \delta > 0$, then $\{z: p_i(z - x) < \delta/2\}$ and $\{z: p_i(z - y) < \delta/2\}$ are disjoint neighborhoods of x and y.

3.5.3 ***Examples.*** (a) Let L be a vector space of complex-valued functions on a topological space Ω. For each compact subset K of Ω, define $p_K(x) = \sup\{|x(t)| : t \in K\}$. In the (Hausdorff) topology induced by the seminorms p_K, convergence means uniform convergence on all compact subsets of Ω. If we restrict the K to finite subsets of Ω, we obtain the topology of pointwise convergence. In general, if the K are restricted to a class $\mathscr{C}$ of subsets of Ω, we obtain the topology of uniform convergence on sets in $\mathscr{C}$.

(b) Let $L = C^\infty[a, b]$, the collection of all infinitely differentiable complex-valued functions on the closed bounded interval $[a, b] \subset R$. For each n, let $p_n(x) = \sup\{|x^{(n)}(t)| : a \leq t \leq b\}$ where $x^{(n)}$ is the nth derivative of x. In the topology induced by the p_n, convergence means uniform convergence of all derivatives.

We now examine convex sets in more detail.

3.5.4 ***Definitions.*** Let K be a subset of the vector space L. Then K is said to be *radial at* x iff K contains a line segment through x in each direction, in other words, if $y \in L$, there is a $\delta > 0$ such that $x + \lambda y \in K$ for $0 \le \lambda < \delta$. (If K is radial at x, x is sometimes called an *internal point* of K.) If K is convex and radial at 0, the *Minkowski functional* of K is defined as

$$p(x) = \inf\left\{r > 0 : \frac{x}{r} \in K\right\}.$$

Intuitively, $p(x)$ is the factor by which x has to be shrunk in order to reach the boundary of K. The Minkowski functional has the following properties.

3.5.5 ***Lemma.*** Let K be a convex subset of L, radial at 0, and let p be the Minkowski functional of K.

(a) The functional p is sublinear, that is, subadditive and positive-homogeneous.

(b) $\{x \in L : p(x) < 1\}$ is the radial kernel of K, defined by rad ker $K = \{x \in K : K \text{ is radial at } x\}$; also, $K \subset \{x \in L : p(x) \le 1\}$.

(c) If K is circled, then p is a seminorm.

(d) If L is a topological vector space and 0 belongs to the interior K^0 of K, then p is continuous, $\overline{K} = \{x \in L : p(x) \le 1\}$, and $K^0 = \{x \in L : p(x) < 1\}$; hence $\{x \in L : p(x) = 1\}$ is the boundary of K.

PROOF. (a) If $x/r \in K$ and $y/s \in K$, then

$$\frac{x+y}{r+s} = \frac{r}{r+s}\frac{x}{r} + \frac{s}{r+s}\frac{y}{s} \in K \qquad \text{by convexity.}$$

Thus $p(x + y) \le r + s$; take the inf over r, then over s, to obtain $p(x + y) \le p(x) + p(y)$, proving subadditivity. Positive homogeneity follows from the definition of p.

(b) If $p(x) < 1$, then $sx \in K$ for some $s > 1$; hence $x \in K$ by convexity. [Write $x = (1/s)(sx) + [1 - (1/s)]0$.] Now if $p(x) < 1$ and $y \in L$, then $p(x + \lambda y) \le p(x) + \lambda p(y) < 1$ for λ sufficiently small and nonnegative; hence K is radial at x. Conversely, if K is radial at x, then $x + \lambda x \in K$ for some $\lambda > 0$; hence $p(x + \lambda x) \le 1$ by definition of p. Thus $p(x) \le (1 + \lambda)^{-1} < 1$. The last statement follows from the definition of p.

(c) If $x/r \in K$ and $a \in C$, $a \ne 0$, then $ax/|a|r \in K$ since K is circled; consequently $p(ax) \le |a|r$. Take the inf over r to obtain $p(ax) \le |a|p(x)$. Replace x by x/a to obtain $p(x) \le |a|p(x/a)$ or, with $b = 1/a$, $p(bx) \ge |b|p(x)$. Now $p(0) = 0$ since $0 \in K$, and the result follows.

(d) Since $0 \in K^0$ there is a neighborhood U of 0 such that $U \subset K$. If $\lambda > 0$ and $y \in \lambda U$, that is, $y = \lambda x$ for some $x \in U$, then $p(y) = \lambda p(x) \le \lambda$.

[Note $x \in K$ implies $p(x) \leq 1$, by (b).] Thus p is continuous at the origin, and therefore continuous everywhere by subadditivity.

Since p is continuous, $\{x: p(x) \leq 1\}$ is closed, so by (*b*), $\overline{K} \subset \{x: p(x) \leq 1\}$. But if $0 < \lambda < 1$ and $p(x) \leq 1$, then $p(\lambda x) < 1$; hence $\lambda x \in K$. If $\lambda \to 1$, then $\lambda x \to x$; therefore $p(x) \leq 1$ implies $x \in \overline{K}$.

Again, by continuity of p, $\{x: p(x) < 1\}$ is open, and hence is a subset of K^0. But if $p(x) \geq 1$, then by considering λx with $\lambda > 1$ we see that x is a limit of a sequence of points not in K, hence $x \notin K^0$. ▮

Before characterizing locally convex spaces, we need the following result.

3.5.6 ***Lemma.*** If U is a neighborhood of 0 in the topological vector space L, there is a circled neighborhood V of 0 with $V \subset U$, and a closed circled overneighborhood W of 0 with $W \subset U$. If L is locally convex, V and W can be taken as convex.

PROOF. Choose $T \in \mathscr{U}$ and $\delta > 0$ such that $aT \subset U$ for $|a| \leq \delta$, and take $V = \bigcup\{aT: |a| \leq \delta\}$. Now if $A \subset L$, we claim that

$$\overline{A} = \bigcap\{A + N: N \in \mathscr{N}\}, \tag{1}$$

where $\mathscr{N}$ is the family of all neighborhoods of 0. [This may be written as $\overline{A} = \bigcap\{A - N: N \in \mathscr{N}\}$; note that $N \in \mathscr{N}$ iff $-N \in \mathscr{N}$ since the map $y \to -y$ is a homeomorphism.] For if $x \in \overline{A}$ and $N \in \mathscr{N}$, then $x + N$ is a neighborhood of x; hence $(x + N) \cap A \neq \varnothing$. If $y \in (x + N) \cap A$, then $x \in y - N$. If $x \notin \overline{A}$, then $(x + N) \cap A = \varnothing$ for some $N \in \mathscr{N}$; hence $x \notin A - N$.

Now if U is a neighborhood of 0, let V_1 be a circled neighborhood of 0 with $V_1 + V_1 \subset U$. By (1), $\overline{V_1} \subset V_1 + V_1$, and since V_1 is circled, so is $\overline{V_1}$. Thus we may take $W = \overline{V_1}$.

In the locally convex case, we may as well assume U convex [the interior of a convex set is convex, by 3.5.5(d) and the fact that a translation of a convex set is convex]. If V_2 is a circled neighborhood included in U, the convex hull

$$\hat{V}_2 = \left\{\sum_{i=1}^{n} \lambda_i x_i : x_i \in V_2, \quad \lambda_i \geq 0, \quad \sum_{i=1}^{n} \lambda_i = 1, \quad n = 1, 2, \ldots\right\},$$

the smallest convex overset of V_2, is also included in U. Since V_2 is circled so is $\hat{V}_2$, and therefore so is $(\hat{V}_2)^0$. (If $x + N \subset \hat{V}_2$, $N \in \mathscr{N}$, and $|a| \leq 1$, then $ax + aN \subset \hat{V}_2$.) Thus we may take $V = (\hat{V}_2)^0$.

Finally, if V_3 is a circled convex neighborhood of 0 with $V_3 + V_3 \subset U$, then $W = \overline{V_3}$ is closed, circled, and convex, and by (1), $W \subset V_3 + V_3 \subset U$. ▮

3.5.7 ***Theorem.*** If L is a locally convex topological vector space, the topology of L is generated, in the sense of 3.5.2, by a family of seminorms. Specifically, if $\mathscr{U}$ is the collection of all circled convex neighborhoods of 0, the Minkowski functionals p_U of the sets $U \in \mathscr{U}$ are the desired seminorms.

PROOF. By 3.5.5(c), the p_U are seminorms, and 3.5.6, $\mathscr{U}$ is a base at 0 for the topology of L. By 3.5.5(d), for each $U \in \mathscr{U}$ we have $U = \{x\colon p_U(x) < 1\}$, and it follows that the topology of L is the same as the topology induced by the p_U. ∎

The fact that the Minkowski functional is sublinear suggests that the Hahn–Banach theorem may provide useful information. The next result illustrates this idea.

3.5.8 ***Theorem.*** Let K be a convex subset of the real vector space L; assume K is radial at 0 and has Minkowski functional p.

(a) If f is a linear functional on L, then $f \le 1$ on K iff $f \le$ p on L.

(b) If g is a linear functional on the subspace M, and $g \le 1$ on $K \cap M$, then g may be extended to a linear functional f on L such that $f \le 1$ on K.

(c) If in addition K is circled and L is a complex vector space, and g is a linear functional on the subspace M with $|g| \le 1$ on $K \cap M$, then g may be extended to a linear functional f on L such that $|f| \le 1$ on K.

(d) A continuous linear functional g on a subspace M of a locally convex topological vector space L may be extended to a continuous linear functional on L.

PROOF. (a) If $f \le p$ on L, then $f \le 1$ on K by 3.5.5(b). Conversely, assume $f \le 1$ on K. If $x/r \in K$, then $f(x/r) \le 1$, so $f(x) \le r$. Take the inf over r to obtain $f(x) \le p(x)$.

(b) By (*a*), $g \le p$ on M, so by the Hahn–Banach theorem, g extends to a linear functional f with $f \le p$ on L. By (a), $f \le 1$ on K.

(c) Let L' be L regarded as a real vector space; by (b), $g_1 = \operatorname{Re} g$ may be extended to a linear function f_1 on L' with $f_1 \le 1$ on K. If $f(x) = f_1(x) - if_1(ix)$, then (as in 3.4.3) f is a linear functional on L, and on M we have $f(x) = g_1(x) - ig_1(ix) = g(x)$. Finally, if $f(x) = re^{i\theta}$, $r \ge 0$, then $|f(x)| = r = f(e^{-i\theta}x) = f_1(e^{-i\theta}x)$ since r is real. If $x \in K$, then $e^{-i\theta}x \in K$ since K is circled, so $|f(x)| \le 1$.

(d) By continuity, g must be bounded on some neighborhood of 0 in M, so that by 3.5.6 there is a convex circled neighborhood K of 0 in L such that $|g|$ is bounded on $K \cap M$. By (c), g has an extension to a linear functional f on L with $|f|$ bounded on K. But if $|f| \le r$ on K, then, given $\delta > 0$, we have $|f| \le \delta$ on the neighborhood $(\delta/r)K$, proving continuity. ∎

In 3.4.14 we promised a general version of the open mapping theorem, and we now consider this question. The following technical lemma is the key step.

3.5.9 ***Lemma.*** Let A be a continuous linear operator from L to M, where L is a complete, metrizable topological vector space and M is a Hausdorff topological vector space. Assume that for every neighborhood W of 0 in L, $\overline{A(W)}$ is an overneighborhood of 0 in M. Then in fact $A(W)$ is an overneighborhood of 0 in M.

PROOF. Denote by $\mathscr{U}$ the collection of overneighborhoods of 0 in L, and by $\mathscr{V}$ the corresponding collection in M. It suffices to show that if U and V are neighborhoods of 0 in L, then $\overline{A(U)} \subset A(U + V)$. For then given the neighborhood W we may find a closed circled neighborhood U with $U + U + U \subset W$; then [see 3.5.6, Eq. (1)] $\overline{U + U} \subset U + U + U \subset W$. Now $\overline{A(U)}$, which belongs to $\mathscr{V}$ by hypothesis, is a subset of $A(U + U)$ by the result to be established, and this in turn is a subset of $A(W)$, as desired.

Thus let U and V be neighborhoods of 0 in L. Since L is metrizable it has a countable base of neighborhoods at 0, say $U_1, U_2, \ldots$. It may be assumed that $U_{n+1} + U_{n+1} \subset U_n$ for all n [3.5.1(c)], and that $U_1 = U$, $U_2 + U_2 \subset U \cap V$.

Let $y \in \overline{A(U)}$; we try to find elements $x_n \in U_n$ such that if $y_n = Ax_n$, then, for all $n = 1, 2, \ldots$,

$$y_1 + \cdots + y_n - y \in \overline{A(U_{n+1})}. \tag{1}$$

Since $y \in \overline{A(U_1)}$ $(=\overline{A(U)})$ and $y + \overline{A(U_2)}$ is an overneighborhood of y, there is an element $y_1 \in A(U_1) \cap (y + \overline{A(U_2)})$. Thus $y_1 = Ax_1$ for some $x_1 \in U_1$, and $y_1 - y \in \overline{A(U_2)}$. If $x_1, \ldots, x_n$ have been found such that $y_1 + \cdots + y_n - y \in \overline{A(U_{n+1})}$, then since $y - (y_1 + \cdots + y_n) + \overline{A(U_{n+2})}$ is an overneighborhood of $y - (y_1 + \cdots + y_n)$, there is an element $y_{n+1} \in A(U_{n+1}) \cap (y - (y_1 + \cdots + y_n) + \overline{A(U_{n+2})})$, so that we have $x_{n+1} \in U_{n+1}$, $y_{n+1} = Ax_{n+1}$, and $y_1 + \cdots + y_{n+1} - y \in \overline{A(U_{n+2})}$, completing the induction.

Let $z_n = \sum_{i=1}^n x_i$; then $z_{n+k} - z_n = x_{n+1} + \cdots + x_{n+k} \in U_{n+1} + \cdots + U_{n+k} \subset U_n$. (Note $U_{n+k-1} + U_{n+k} \subset U_{n+k-1} + U_{n+k-1} \subset U_{n+k-2}$, and proceed backward.) Since the U_n form a base at 0, $\{z_n\}$ is a Cauchy sequence in L; hence z_n converges to some $x \in L$ by completeness. [By the appendix on general topology, Theorem A10.9, the metric d of L may be assumed invariant, that is, $d(x, y) = d(x + z, y + z)$ for all z. We know that $z_m - z_n \to 0$ as $n, m \to \infty$;

hence by invariance, $d(z_m, z_n) = d(z_m - z_n, 0) \to 0$, so that $\{z_n\}$ is Cauchy.] Since

$$x_1 + \cdots + x_n \in U_1 + \cdots + U_n \subset U_1 + (U_2 + U_2) \subset U + (U \cap V) \subset U + V$$

it follows that $x \in \overline{U + V}$. If we can show that $y = Ax$, then $y \in A(\overline{U + V})$ and we have finished.

If $y \neq Ax$, the Hausdorff hypothesis for M yields a closed circled $W' \in \mathscr{V}$ such that $Ax + W'$ and $y + W'$ are disjoint; let W be a closed set in $\mathscr{V}$ such that $W + W \subset W'$. By continuity of A there is a set $W_1 \in \mathscr{U}$ such that $A(W_1) \subset W$. Since the U_n form a base at 0, $U_n \subset W_1$ for large n; hence $A(U_n) \subset A(W_1) \subset W$; since W is closed, we have

$$\overline{A(U_n)} \subset W \qquad \text{for sufficiently large } n. \tag{2}$$

If $n < m$,

$$A\left(\sum_{j=1}^{m} x_j\right) - y = \sum_{j=1}^{n} y_j - y + \sum_{j=n+1}^{m} y_j$$

$$\in \overline{A(U_{n+1})} + A(U_{n+1} + \cdots + U_m) \qquad \text{by (1).}$$

By (2), $A(\sum_{j=1}^{m} x_j) - y \in W + A(U_{n+1} + \cdots + U_m) \subset W + A(U_n) \subset W + W$. Let $m \to \infty$; by continuity of A we find $Ax - y \in \overline{W + W} \subset \overline{W'} = W' \subset W' + W'$, contradicting the disjointness of $Ax + W'$ and $y + W'$. ∎

Under the hypothesis of 3.5.9, if $W \in \mathscr{V}$, then by continuity of A there is a set U_n in the countable base at 0 in L such that $A(U_n) \subset V$. By 3.5.9, $\overline{A(U_n)} \in \mathscr{V}$ for all n; hence M also has a countable base at 0. It is shown in the appendix on general topology (Theorem A10.8) that the existence of a countable base at 0 in a topological vector space is equivalent to pseudo-metrizability. Thus under the hypothesis of 3.5.9, M, being Hausdorff, is metrizable.

One more preliminary result is needed.

3.5.10 ***Lemma.*** Let A be a linear operator from the topological vector space L to the topological vector space M. Assume $A(L)$ is of the second category in M; that is, $A(L)$ cannot be expressed as a countable union of nowhere dense sets. If U is an overneighborhood of 0 in L, then $\overline{A(U)}$ is an overneighborhood of 0 in M.

PROOF. Let $\mathscr{U}$ and $\mathscr{V}$ be the overneighborhoods of 0 in L and M, respectively. If $U \in \mathscr{U}$, choose a circled $V \in \mathscr{U}$ with $V + V \subset U$. By 3.5.1(b) applied to V

we have $L = \bigcup_{n=1}^{\infty} nV$, hence $A(L) = \bigcup_{n=1}^{\infty} nA(V)$. Since $A(L)$ is of the second category, $\overline{nA(V)}$ $[= n\overline{A(V)}]$ has a nonempty interior for some n, hence $\overline{A(V)}$ has a nonempty interior. If B is a nonempty open subset of $\overline{A(V)}$, then

$$\begin{aligned} 0 \in B - B &\subset \overline{A(V)} - \overline{A(V)} \subset \overline{A(V) - A(V)} \\ &= \overline{A(V) + A(V)} \qquad \text{since } V, \text{ hence } A(V), \text{ is circled} \\ &\subset \overline{A(U)}. \end{aligned}$$

But $B - B = \bigcup\{x - B\colon x \in B\}$, an open set; hence $\overline{A(U)} \in \mathscr{V}$. ∎

3.5.11 ***Open Mapping Theorem.*** Let L and M be metrizable topological vector spaces, with L complete. If A is a continuous linear operator from L to M and $A(L)$ is of the second category in M, then $A(L) = M$ and A is an open map.

PROOF. Take $\mathscr{U}$ and $\mathscr{V}$ as in 3.5.9 and 3.5.10. Let U be any set in $\mathscr{U}$; by 3.5.9 and 3.5.10, $A(U) \in \mathscr{V}$. If $y \in M$, then $y \in nA(U)$ for some n by 3.5.1(b); hence $y = A(nx)$ for some $x \in U$, proving that $A(L) = M$. If G is open in L and $x_0 \in G$, choose $V \in \mathscr{U}$ such that $x_0 + V \subset G$. Then $A(V) \in \mathscr{V}$ and $Ax_0 + A(V) \subset A(G)$, proving that $A(G)$ is open. ∎

3.5.12 ***Corollary.*** If A is a continuous linear operator from L onto M, where L and M are complete metrizable topological vector spaces, then A is an open map.

PROOF. By completeness of M, $A(L)$ $(=M)$ is of category 2 in M, and the result follows from 3.5.11. ∎

We conclude this section with a discussion of separation theorems and their applications. If K_1 and K_2 are disjoint convex subsets of R^3, there is a plane P such that K_1 is on one side of P and K_2 is on the other side. Now P can be described as $\{x \in R^3\colon f(x) = c\}$ for some linear functional f and real number c; hence we have $f(x) \le c$ for all x in one of the two convex sets, and $f(x) \ge c$ for all x in the other set. We are going to consider generalizations of this idea.

The following theorem is the fundamental result of this type.

3.5.13 ***Basic Separation Theorem.*** Let K_1 and K_2 be disjoint, nonempty convex subsets of the real vector space L, and assume that K_1 has at least one internal point. There is a linear functional f on L separating K_1 and K_2, that is, $f \not\equiv 0$ and $f(x) \leq f(y)$ for all $x \in K_1$ and $y \in K_2$ [for short, $f(K_1) \leq f(K_2)$].

PROOF. First assume that 0 is an internal point of K_1. Pick an element $z \in K_2$; then $-z$ is an internal point of $-z + K_1 \subset K_1 - K_2$; hence 0 is an internal point of the convex set $K = z + K_1 - K_2$. If $z \in K$, then $K_1 \cap K_2 \neq \varnothing$, contradicting the hypothesis; therefore $z \notin K$; so if p is the Minkowski functional of K, we have $p(z) \geq 1$ by 3.5.5(b). Define a linear functional g on the subspace $M = \{\lambda z : \lambda \in R\}$ by $g(\lambda z) = \lambda$. If $a > 0$, then $g(az) = a = a(1) \leq ap(z) = p(az)$, and if $a \leq 0$, then $g(az) = a \leq 0 \leq p(az)$. By the Hahn–Banach theorem, g extends to a (nontrivial) linear functional f with $f \leq p$ on L. By 3.5.8(a), $f \leq 1$ on K. Since $f(z) = g(z) = 1$, $f(K_1) \leq f(K_2)$.

If x is an internal point of K_1, then 0 is an internal point of $-x + K_1$, and $-x + K_1$ is disjoint from $-x + K_2$. If $f(-x + K_1) \leq f(-x + K_2)$, then $f(K_1) \leq f(K_2)$. ∎

If L is a complex vector space, the discussion after 3.4.2 shows that there is a nontrivial linear functional f on L such that $\operatorname{Re} f(x) \leq \operatorname{Re} f(y)$ for all $x \in K_1$, $y \in K_2$; this is what we shall mean by "f separates K_1 and K_2" in the complex case.

3.5.14 ***Corollaries.*** Assume the hypothesis of 3.5.13.

(a) If L is a topological vector space and K_1 has nonempty interior, the linear functional f constructed in 3.5.13 is continuous.

(b) If L is locally convex, and K_1 and K_2 disjoint convex subsets of L, with K_1 closed and K_2 compact, there is a continuous linear functional f on L strongly separating K_1 and K_2, that is, $f(x) \leq c_1 < c_2 \leq f(y)$ for some real numbers c_1 and c_2 and all $x \in K_1$, $y \in K_2$. In particular if L is Hausdorff and $x, y \in L$, $x \neq y$, there is a continuous linear functional f on L with $f(x) \neq f(y)$.

(c) If L is locally convex, M a closed subspace of L, and $x_0 \notin M$, there is a continuous linear functional f on L such that $f = 0$ on M and $f(x_0) \neq 0$.

PROOF. (a) Since ${K_1}^0 \neq \varnothing$, *the interior and internal points of* K_1 *coincide*, by 3.5.5(b) and (d). Thus in the proof of 3.5.13, 0 is an interior point of K, so that $K^0 = \{x : p(x) < 1\}$ by 3.5.5(d). Since $f \leq p$ on L it follows that $f \leq 1$ on some neighborhood U of 0 (for example, $U = K^0$); since U may be assumed circled, $f(-x) \leq 1$ for all $x \in U$; hence $|f| \leq 1$ on U. But then $|f| \leq \delta$ on the neighborhood δU, proving continuity.

(b) Let $K = K_2 - K_1$; K is convex, and is also closed since the sum of a compact set A and a closed set B in a topological vector space is closed. (If

$x_n \in A$, $y_n \in B$, $x_n + y_n \to z$, find a subnet $x_{n_i} \to x \in A$; then $y_{n_i} \to z - x$, hence $z - x \in B$ since B is closed.) Now $0 \notin K$ by disjointness of K_1 and K_2; so by local convexity, there is a convex circled neighborhood V of 0 such that $V \cap K = \varnothing$. By 3.5.13 and part (a) above there is an $f \in L^*$ separating V and K, say $f(V) \le c \le f(K)$. Now f cannot be identically 0 on V, for if so $f \equiv 0$ on L by 3.5.1(b). Thus there is an $x \in V$ with $f(x) \neq 0$, and since V is circled, we may assume $f(x) > 0$. Thus c must be greater than 0; hence $f(K_2) \ge f(K_1) + c$, $c > 0$, as desired.

(c) Let $K_1 = M$, $K_2 = \{x_0\}$, and apply (b) to obtain $f \in L^*$ with $f(M) \le c_1 < c_2 \le f(x_0)$. Since f is bounded above on the subspace M, f must be identically 0 on M. [If $f(x) \neq 0$, consider $f(\lambda x)$ for arbitrarily large λ.] But then $f(x_0) \ge c_2 > 0$. ∎

If we adopt the above definition of separation in complex vector spaces, all parts of 3.5.14 extend immediately to the complex case.

Separation theorems may be applied effectively in the study of weak topologies. In 3.4.10 we defined the weak topology on a normed linear space; the definition is identical for an arbitrary topological vector space L. Specifically, for each $f \in L^*$, $p_f(x) = |f(x)|$ defines a seminorm on L. The locally convex topology induced by the seminorms p_f, $f \in L^*$, is called the *weak topology* on L. By 3.5.1 and 3.5.2, a base at x_0 for the weak topology consists of finite intersections of sets of the form $\{x: p_f(x - x_0) < \varepsilon\}$, so in the case of a normed linear space we obtain the topology defined in 3.4.10.

There is a dual topology defined on L^*; if $x \in L$, then $p_x(f) = |f(x)|$ defines a seminorm on L^*. The locally convex topology induced by the seminorms p_x is called the *weak* topology* on L^*.

By 3.5.2, the weak topology is the weakest topology on L making each $f \in L^*$ continuous, so the weak topology is weaker than the original topology of L. Convergence of x_n to x in the weak topology means $f(x_n) \to f(x)$ for each $f \in L^*$. The weak* topology on L^* is the weakest topology making all evaluation maps $f \to f(x)$ continuous. Convergence of f_n to f in the weak* topology means $f_n(x) \to f(x)$ for all $x \in L$; thus weak* convergence is simply pointwise convergence, so if L is a normed linear space, the weak* topology is weaker than the norm topology on L^*.

We have observed in 3.4.10 that the weak topology is an example of a product topology. Since weak* convergence is pointwise convergence, the weak* topology is the product topology on the set C^L of all complex-valued functions on L, relativized to L^*.

In distinguishing between the weak topology and the original topology on L, it will be convenient to call the original topology the *strong* topology.

By the above discussion, a weakly closed subset of L is closed in the original topology. Under certain conditions there is a converse statement:

3.5.15 ***Theorem.*** Let L be a locally convex topological vector space. If K is a convex subset of L, then K is strongly closed in L iff it is weakly closed.

PROOF. Assume K strongly closed. If $y \notin K$, then by 3.5.14(b) there are real numbers c_1 and c_2 and an $f \in L^*$ with $\operatorname{Re} f(x) \le c_1 < c_2 \le \operatorname{Re} f(y)$ for all $x \in K$. But then $W = \{x \in L: |f(x) - f(y)| < c_2 - c_1\}$ is a weak neighborhood of y, and if $x \in K$, we have $|f(x - y)| \ge |\operatorname{Re} f(y) - \operatorname{Re} f(x)| \ge c_2 - c_1$; therefore $W \cap K = \varnothing$, proving K^c weakly open. ∎

For the remainder of this section we consider normed linear spaces. The closed unit ball $\{f: \|f\| \le 1\}$ of L^* is never compact (in the strong topology) unless L (hence L^*) is finite-dimensional; see Problem 9, Section 3.3. However, it is always compact in the weak* topology, as we now prove:

3.5.16 ***Banach–Alaoglu Theorem.*** If L is a normed linear space and $B = \{f \in L^*: \|f\| \le 1\}$, then B is compact in the weak* topology.

PROOF. If $f \in B$, then $|f(x)| \le \|f\| \, \|x\| \le \|x\|$ for all $x \in L$. If $I(x)$ is the set $\{z \in C: |z| \le \|x\|\}$, then $B \subset \prod\{I(x): x \in L\}$, the set of all functions defined on L such that $f(x) \in I(x)$ for each $x \in L$. By the Tychonoff theorem (see the appendix on general topology, Theorem A5.4), $\prod\{I(x): x \in L\}$ is compact in the product topology (the topology of pointwise convergence).

Let $\{f_n\}$ be a net in B; by the above discussion there is a subnet converging pointwise to a complex-valued function f on L, and since the f_n are linear so is f. Now $|f_n(x)| \le \|x\|$ for each x, and it follows that $|f(x)| \le \|x\|$, hence $f \in B$, proving compactness. ∎

If L is reflexive, then L and L^{**} may be identified, and the weak topology on L corresponds to the weak* topology on L^{**}. By 3.5.16, the closed unit ball of L is weakly compact. Conversely, weak compactness of the closed unit ball implies reflexivity. To prove this we need the following auxiliary results:

3.5.17 ***Lemma.*** Let L be a topological vector space.

(a) If f is a linear functional on L, f is continuous relative to the weak topology iff f is continuous relative to the strong topology, that is, iff $f \in L^*$.

(b) If g is a linear functional on L^*, g is continuous relative to the weak* topology iff there is an $x \in L$ such that $g(f) = f(x)$ for all $f \in L^*$.

PROOF. (a) If $f \in L^*$ and x_n converges weakly to x, then $f(x_n) \to f(x)$ by definition of the weak topology, so f is weakly continuous. Conversely, if f is continuous relative to the weak topology and x_n converges strongly to x, then x_n converges weakly to x, hence $f(x_n) \to f(x)$. Thus $f \in L^*$.

(b) The "if" part follows from the definition of the weak* topology, so assume g weak* continuous. A basic neighborhood of 0 in the weak* topology is of the form $U = \bigcap_{i=1}^n \{f \in L^*: |f(x_i)| < \delta_i, i = 1, \ldots, n\}$ for some $x_1, \ldots, x_n \in L$, $\delta_1, \ldots, \delta_n > 0$, $n = 1, 2, \ldots$; thus there is a U of this type such that $|g(f)| < 1$ for all $f \in U$. By linearity, $|f(x_i)| < \delta_i/M$ for all i implies $|g(f)| < 1/M$. In particular, if $f(x_i) = 0$ for all $i = 1, \ldots, n$, then $|g(f)| < 1/M$, and since M is arbitrary, $g(f) = 0$. But this implies, by a standard algebraic result (see Problem 15) that $g(f) = \sum_{i=1}^n c_i f(x_i)$ for some complex numbers $c_1, \ldots, c_n$. Set $x = \sum_{i=1}^n c_i x_i$. ∎

3.5.18 ***Lemma.*** Let L be a normed linear space, and let h be the isometric isomorphism of L into L^{**} (see 3.4.6). Let B be the closed unit ball in L, B^{**} the closed unit ball in L^{**}. Then $h(B)$ is dense in B^{**} relative to the weak* topology. Furthermore, $h(L)$ is dense in L^{**}.

PROOF. Let M be the weak* closure of $h(B)$, and let $x^{**} \in B^{**}$, $x^{**} \notin M$. By 3.5.16, B^{**} is weak* compact (and convex); so by 3.5.14(b), there is a linear functional g on L^{**} such that g is continuous relative to the weak* topology and $\operatorname{Re} g(M) \le c_1 < c_2 \le \operatorname{Re} g(x^{**})$. Since M is a subspace (in particular is circled), $|g(y^{**})| \le c_1$ for all $y^{**} \in M$. [If $y^{**} \in M$ and $g(y^{**}) = re^{i\theta}$, then $|g(y^{**})| = g(e^{-i\theta}y^{**}) = r = \operatorname{Re} g(e^{-i\theta}y^{**}) \le c_1$.] By 3.5.17(b), g is of the form $g(y^{**}) = y^{**}(f)$ for some $f \in L^*$.

If $y \in B$, and $y^{**} = h(y)$ is the corresponding element of $h(B) \subset M$, then $|f(y)| = |y^{**}(f)| \le c_1$; hence

$$\|f\| \le c_1 < c_2 \le \operatorname{Re} g(x^{**}) \le |x^{**}(f)| \le \|x^{**}\| \, \|f\| \le \|f\|,$$

a contradiction. Consequently $M = B^{**}$.

Finally, if $y^{**} \in L^{**}$, $y^{**} \ne 0$, then $y^{**}/\|y^{**}\| \in B^{**}$, which we have just seen is the weak* closure of $h(B)$. Thus $y^{**}/\|y^{**}\|$ belongs to the weak* closure of $h(L)$, which is a subspace. The result follows. ∎

We now characterize reflexive normed linear spaces.

3.5.19 ***Theorem.*** A normed linear space L is reflexive iff its closed unit ball B is weakly compact.

PROOF. The "only if" part has been pointed out after 3.5.16, so assume B weakly compact. The isometric isomorphism h of L into L^{**} is a homeomorphism of the weak topology of L and the weak* topology of $h(L)$ [note that $x_n \xrightarrow{w} x$ iff $f(x_n) \to f(x)$ for all $f \in L^*$; in other words, iff $h(x_n) \xrightarrow{w^*} h(x)$]. Thus $h(B)$ is a weak* compact (hence closed) subset of B^{**}, and is dense in B^{**} by 3.5.18. It follows that $h(B) = B^{**}$, so as in 3.5.18, $h(L) = L^{**}$, proving reflexivity. ∎

3.5.20 ***Corollaries.*** (a) If M is a closed subspace of the reflexive normed linear space L, then M is reflexive.

(b) If L is a Banach space, L is reflexive iff L^* is reflexive.

PROOF. (a) If B is the closed unit ball in L, then $B \cap M$ is the closed unit ball in M. By hypothesis, $B \cap M$ is strongly closed, hence weakly closed by 3.5.15. But B is weakly compact by 3.5.19, so $B \cap M$ is weakly compact. Again by 3.5.19, M is reflexive.

(b) Assume L reflexive. The weak* topology on L^* then coincides with its weak topology, hence by 3.5.16, the closed unit ball of L^* is weakly compact, and consequently L^* is reflexive by 3.5.19. (Completeness of L is not used in this part.)

If L^* is reflexive, so is L^{**} by the first part of the proof. Since L is complete, so is $h(L)$; thus $h(L)$ is a closed subspace of L^{**}, and the result follows from (a). ∎

Problems

1. Let $\mathscr{U}$ be the family of overneighborhoods of 0 in a topological vector space L. Show that L is Hausdorff iff $\bigcap\{U: U \in \mathscr{U}\} = \{0\}$.
2. If f is a linear functional on the topological vector space L, with f not identically 0, show that the four conditions of Problem 1, Section 3.3, are still equivalent.
3. If L is a finite-dimensional Hausdorff topological vector space, show that L is topologically isomorphic to C^n for some n. Show also that a finite-dimensional subspace of an arbitrary Hausdorff topological vector space is always closed. [It can be shown that for a Hausdorff topological vector space, finite-dimensionality is equivalent to local compactness; see Kelley and Namioka (1963, Theorem 7.8).]
4. (a) Let K be a convex set, radial at 0, in the vector space L. If g is any nonnegative sublinear functional on L such that $\{x \in L: g(x) < 1\} \subset K \subset \{x \in L: g(x) \leq 1\}$, show that g is the Minkowski functional of K. [Thus by 3.5.5(d), if $0 \in K^0$, then the Minkowski functionals of K^0, K, and $\overline{K}$ coincide.]

(b) Use part (a) to find the Minkowski functionals of each of the following subsets of R^2:

(i) $\{(x, y): |x| + |y| \le 1\}$.
(ii) $\{(x, y): -1 \le x < 1, -1 \le y < 1\}$
(iii) $\{(x, y): x^2 + y^2 < 1\}$.

(c) If K_1 and K_2 are convex and radial at 0, with Minkowski functionals p_1 and p_2, show that the Minkowski functional of $K_1 \cap K_2$ is $\max(p_1, p_2)$.

5. Let L be the set of all complex-valued functions on $[0, 1]$, with the topology of pointwise convergence. Then L is a locally convex topological vector space, with topology induced by the seminorms $p_x(f) = |f(x)|$, $f \in L$, where x ranges over $[0, 1]$; see 3.5.3(a). Show that L is not metrizable. (The same result holds even if we restrict L to the continuous complex-valued functions on $[0, 1]$.)
6. Let L be a locally convex topological vector space. Show (without using the theory of uniform spaces given in the appendix on general topology, Section A10) that the following conditions are equivalent:

(a) L is pseudometrizable, that is, there is a pseudometric inducing the topology of L.
(b) L has a countable base at 0 (hence at every point of L).
(c) The topology of L is generated by a countable family of seminorms.

7. Let L be the collection of analytic functions on the unit disk $D = \{z: |z| < 1\}$, with the topology of uniform convergence on compact subsets [see 3.5.3(a)]. Show that L is metrizable but not normable.
8. Let $L = L^p[0, 1]$ where $0 < p < 1$; with the pseudometric $d(x, y) = \int_0^1 |x(t) - y(t)|^p \, dt$ (see 2.4.12), L is a topological vector space. [Apply 3.5.1, using the sets $U = \{x: \int_0^1 |x(t)|^p \, dt < \varepsilon\}$, $\varepsilon > 0$.]

(a) Let f be a continuous linear functional on L. If $f(x) = 1$, show that x can be written as $y + z$ where $d(y, 0) = d(z, 0) = \frac{1}{2} d(x, 0)$.
(b) With f as in part (a), show that there are functions $x_1, x_2, \ldots \in L$ such that $d(x_n, 0) \to 0$ as $n \to \infty$, but $|f(x_n)| \ge 1$ for all n. Conclude that the only continuous linear functional on L is the zero functional. [In part (a), either $|f(y)| \ge \frac{1}{2}$ or $|f(z)| \ge \frac{1}{2}$; if, say, $|f(y)| \ge \frac{1}{2}$, take $x_1 = 2y$ and proceed inductively.]

9. Let L be a topological vector space. Show that there is a nontrivial continuous linear functional on L iff there is a proper convex subset of L with nonempty interior.
10. Let B_1 and B_2 be (not necessarily convex) nonempty subsets of the topological vector space L, with $B_1{}^0 \ne \varnothing$. If f is a linear functional on

L that separates B_1 and B_2, show that f is continuous. [This gives another proof of 3.5.14(a).]

11. Let K be a convex subset of a locally convex topological vector space L. Show that the closure of K in the weak topology coincides with the closure in the strong topology.
12. Let K be a convex set, radial at 0, in the real vector space L. Let x be a point of L and M a subspace of L such that $x + M$ does not contain any internal points of K. Show that there is a hyperplane H (that is, H is the null space of a nontrivial linear functional on L) such that $M \subset H$ and $x + H$ contains no internal points of K. [Define g on $L(M \cup \{x\})$ by $g(y + ax) = a$, and extend using the Hahn–Banach theorem.]
13. (a) If L is a topological vector space and $\mathscr{T}_1$, $\mathscr{T}_2$ are topologies on L such that $\mathscr{T}_1 \subset \mathscr{T}_2$ and L is complete and metrizable under both $\mathscr{T}_1$ and $\mathscr{T}_2$, show that $\mathscr{T}_1 = \mathscr{T}_2$.
    (b) If L is a Banach space under two norms $\| \;\|_1$ and $\| \;\|_2$, and, for some $k > 0$, $\|x\|_1 \le k\|x\|_2$ for all $x \in L$, show that for some $r > 0$, $\|x\|_2 \le r\|x\|_1$ for all $x \in L$; in other words (Problem 6, Section 3.3) the norms induce the same topology.
14. (a) (*Closed graph theorem*) Let A be a linear map from L to M, where L and M are complete metrizable topological vector spaces. Assume A is closed; in other words, the graph of A is a closed subset of $L \times M$, with the product topology. Show that A is continuous.
    (b) If A is a continuous linear operator from L to M where L and M are topological vector spaces and M is Hausdorff, show that A is closed.
15. Let $g, f_1, \ldots, f_n$ be linear functionals on the vector space L. If N denotes null space and $\bigcap_{i=1}^n N(f_i) \subset N(g)$, show that g is a linear combination of the f_i.
16. Let L be a normed linear space. If the weak and strong topologies coincide on L, show that L is finite-dimensional, as follows:
    (a) The unit ball $B = \{x\colon \|x\| < 1\}$ is strongly open, hence weakly open by hypothesis. Thus we can find $f_1, \ldots, f_n \in L^*$ and $\delta_1, \ldots, \delta_n > 0$ such that $\{x\colon |f_i(x)| < \delta_i,\ i = 1, \ldots, n\} \subset B$. Show that [with $N(f)$ denoting the null space of f] $\bigcap_{i=1}^n N(f_i) = \{0\}$.
    (b) Define $T(x) = (f_1(x), \ldots, f_n(x))$; by (a), T is a one-to-one map of L onto a subspace M of C^n. If $\{y_1, \ldots, y_k\}$ is a basis for M and $x_j = T^{-1}(y_j)$, $j = 1, \ldots, k$, show that $\{x_1, \ldots, x_k\}$ is a basis for L, hence L is finite-dimensional.
17. Let L be a separable normed linear space. If B is the closed unit ball in L^*, show that B is metrizable in the weak* topology. Thus since B is compact by 3.5.16, every sequence in B (hence every norm-bounded

sequence in L^*) has a weak* convergent subsequence. Similarly, if L is reflexive, the closed unit ball of L is metrizable in the weak topology. (Let $\{x_1, x_2, \ldots\}$ be a countable dense subset of L, and set

$$d(f, g) = \sum_{n=1}^{\infty} \frac{1}{2^n} \frac{|f(x_n) - g(x_n)|}{1 + |f(x_n) - g(x_n)|}, \qquad f, g \in L^*.$$

Show that weak* convergence implies d-convergence.)

18. Show that a reflexive Banach space is weakly complete; in other words every weak Cauchy sequence [$\{f(x_n)\}$ is Cauchy for each $f \in L^*$] converges weakly.
19. A subset B of a topological vector space L is said to be *absorbed* by a subset A iff $B \subset aA$ for sufficiently large $|a|$; B is said to be *bounded* iff it is absorbed by every neighborhood of 0.
    (a) Show that B is bounded iff for each sequence of points $x_n \in B$ and each sequence of complex numbers $\lambda_n \to 0$, we have $\lambda_n x_n \to 0$.
    (b) A *bornivore* in a topological vector space L is a convex circled set that absorbs every bounded set; L is *bornological* iff L is locally convex and every bornivore is an overneighborhood of 0. Show that every metrizable locally convex space is bornological.
    (c) Let T be a linear operator from L to M, where L is bornological and M is locally convex. If T maps bounded sets into bounded sets, show that T is continuous. (The converse is true for arbitrary L and M.)
    (d) If L is a Hausdorff topological vector space, show that there is no bounded subspace of L except $\{0\}$.

3.6 References

There is a vast literature on functional analysis, and we only give a few representative titles. Readable introductory treatments are given in Liusternik and Sobolev (1961), Taylor (1958), Bachman and Narici (1966), and Halmos (1951); the last deals exclusively with Hilbert spaces. Among the more advanced treatments, Dunford and Schwartz (1958, 1963, 1970) emphasize normed spaces, Kelley and Namioka (1963) and Schaefer (1966) emphasize topological vector spaces. Yosida (1968) gives a broad survey of applications to differential equations, semigroup theory, and other areas of analysis.

4

The Interplay between Measure Theory and Topology

4.1 Introduction

A connection between measure theory and topology is established when a σ-field $\mathscr{F}$ is defined in terms of topological properties. In the most common situation, we have a topological space Ω, and $\mathscr{F}$ is taken as the smallest σ-field containing all open sets of Ω. If this is done, there is a natural connection between measure-theoretic and topological questions. For example, if μ is a measure on $\mathscr{F}$ and $A \in \mathscr{F}$, we may ask whether A can be approximated by a compact subset. In other words, we wish to know if $\mu(A) = \sup\{\mu(K)$: K a compact subset of $A\}$. As another example, we may ask whether a function in $L^p(\Omega, \mathscr{F}, \mu)$ can be approximated by a continuous function. One formulation of this is to ask whether the continuous functions are dense in L^p.

In this chapter we investigate questions of this type. The results in the first two sections are not topological, but they serve as basic tools in the later development. We first consider a result that is a companion to the monotone class theorem:

4.1.1 ***Definition.*** Let $\mathscr{D}$ be a class of subsets of a set Ω. Then $\mathscr{D}$ is said to be a *Dynkin system* (D-system for short) iff the following conditions hold.

(a) $\Omega \in \mathscr{D}$.

(b) If $A, B \in \mathscr{D}$, $B \subset A$, then $A - B \in \mathscr{D}$. Thus $\mathscr{D}$ is closed under proper differences.

(c) If $A_1, A_2, \ldots \in \mathscr{D}$ and $A_n \uparrow A$, then $A \in \mathscr{D}$.

Note that by (a) and (b), $\mathscr{D}$ is closed under complementation; hence by (c), $\mathscr{D}$ is a monotone class. If $\mathscr{D}$ is closed under finite union (or closed under finite intersection), then $\mathscr{D}$ is a field, and hence a σ-field (see 1.2.1).

4.1.2 ***Dynkin System Theorem.*** Let $\mathscr{S}$ be a class of subsets of Ω, and assume $\mathscr{S}$ closed under finite intersection. If $\mathscr{D}$ is a Dynkin system and $\mathscr{D} \supset \mathscr{S}$, then $\mathscr{D}$ includes the minimal σ-field $\mathscr{F} = \sigma(\mathscr{S})$.

PROOF. Let $\mathscr{D}_0$ be the smallest D-system including $\mathscr{S}$. We show that $\mathscr{D}_0 = \mathscr{F}$, in other words *the smallest D-system and the smallest σ-field over a class closed under finite intersection coincide*. Since $\mathscr{D}_0 \subset \mathscr{D}$, the result will follow.

Now $\mathscr{D}_0 \subset \mathscr{F}$ since $\mathscr{F}$ is a D-system. To show that $\mathscr{F} \subset \mathscr{D}_0$, let $\mathscr{C} = \{A \in \mathscr{D}_0 : A \cap B \in \mathscr{D}_0$ for all $B \in \mathscr{S}\}$. Then $\mathscr{S} \subset \mathscr{C}$ since $\mathscr{S}$ is closed under finite intersection, and since $\mathscr{D}_0$ is a D-system, so is $\mathscr{C}$. Thus $\mathscr{D}_0 \subset \mathscr{C}$, hence $\mathscr{D}_0 = \mathscr{C}$.

Now let $\mathscr{C}' = \{C \in \mathscr{D}_0 : C \cap D \in \mathscr{D}_0$ for all $D \in \mathscr{D}_0\}$. The result $\mathscr{D}_0 = \mathscr{C}$ implies that $\mathscr{S} \subset \mathscr{C}'$, and since $\mathscr{C}'$ is a D-system we have $\mathscr{D}_0 \subset \mathscr{C}'$, hence $\mathscr{D}_0 = \mathscr{C}'$. It follows that $\mathscr{D}_0$ is closed under finite intersection; by 4.1.1, $\mathscr{D}_0$ is a σ-field, so that $\mathscr{F} \subset \mathscr{D}_0$. ▮

If $\mathscr{S}$ is a field and $\mathscr{M}$ is the smallest monotone class including $\mathscr{S}$, then $\mathscr{M} = \sigma(\mathscr{S})$ (see 1.3.9). In the Dynkin system theorem, we have a weaker hypothesis on $\mathscr{S}$ (it is closed under finite intersection but need not be a field) but a stronger hypothesis on the class of sets including $\mathscr{S}$ ($\mathscr{D}$ is a Dynkin system, not merely a monotone class).

4.1.3 ***Corollary.*** Let $\mathscr{S}$ be a class of subsets of Ω, and let μ_1 and μ_2 be finite measures on $\sigma(\mathscr{S})$. Assume $\Omega \in \mathscr{S}$ and $\mathscr{S}$ is closed under finite intersection. If $\mu_1 = \mu_2$ on $\mathscr{S}$, then $\mu_1 = \mu_2$ on $\sigma(\mathscr{S})$.

PROOF. Let $\mathscr{D}$ be the collection of sets $A \in \sigma(\mathscr{S})$ such that $\mu_1(A) = \mu_2(A)$. Then $\mathscr{D}$ is a D-system and $\mathscr{S} \subset \mathscr{D}$, hence $\mathscr{D} = \sigma(\mathscr{S})$ by 4.1.2. ▮

4.1.4 ***Corollary.*** Let $\mathscr{S}$ be a class of subsets of Ω; assume that $\Omega \in \mathscr{S}$ and $\mathscr{S}$ is closed under finite intersection. Let H be a vector space of real-valued functions on Ω, such that $I_A \in H$ for each $A \in \mathscr{S}$.

Suppose that whenever $f_1, f_2, \ldots$ are nonnegative functions in H, $|f_n| \le M < \infty$ for all n, and $f_n \uparrow f$, the limit function f belongs to H. Then $I_A \in H$ for all $A \in \sigma(\mathscr{S})$.

Proof. Let $\mathscr{D} = \{A \subset \Omega \colon I_A \in H\}$; then $\mathscr{S} \subset \mathscr{D}$ and $\mathscr{D}$ is a D-system. For if $A, B \in \mathscr{D}$, $B \subset A$, then $I_{A-B} = I_A - I_B \in H$, so that $A - B \in \mathscr{D}$. If $A_n \in \mathscr{D}$ and $A_n \uparrow A$, then $I_{A_n} \uparrow I_A$, hence $I_A \in H$ by hypothesis; thus $A \in \mathscr{D}$. The result now follows from 4.1.2. ∎

4.2 The Daniell Integral

One of the basic properties of the integral is linearity; if f and g are μ-integrable functions and a and b are real or complex numbers, then

$$\int_\Omega (af + bg)\, d\mu = a \int_\Omega f\, d\mu + b \int_\Omega g\, d\mu.$$

Thus the integral may be regarded as a linear functional on the vector space of integrable functions. This idea may be used as the basis for a different approach to integration theory. Instead of beginning with a measure and constructing the corresponding integral, we start with a given linear functional E on a vector space. Under appropriate hypotheses, we extend E to a larger space, and finally we show that there is a measure μ such that E is in fact the integral with respect to μ.

We first fix the notation to be used.

4.2.1 ***Notation.*** In this section, L will denote a vector space of real-valued functions on a set Ω; L is assumed *closed under the lattice operations*, in other words if $f, g \in L$ and $f \vee g = \max(f, g), f \wedge g = \min(f, g)$, then $f \vee g, f \wedge g \in L$. There are several familiar examples of such spaces; L can be the class of continuous real-valued functions on a given topological space, or equally well, the bounded continuous real-valued functions. Another possibility is to take L as the collection of all real-valued functions on a given set.

The letter E will denote a *positive linear functional* on L, that is, a linear map E from L to R such that $f \ge 0$ implies $E(f) \ge 0$. This implies that E is monotone, that is, $f \le g$ implies $E(f) \le E(g)$.

If H is any class of functions from Ω to R (or $\overline{R}$), H^+ will denote $\{f \in H : f \ge 0\}$. The collection of functions $f\colon \Omega \to \overline{R}$ of the form $\lim_n f_n$ where the f_n form an increasing sequence of functions in L^+, will be denoted by L'; if the f_n are allowed to form an increasing net in L^+, the resulting class is denoted by L''. (See the appendix on general topology, Section A1, for a discussion of nets.)

If H is as above, $\sigma(H)$ is defined as the smallest σ-field of subsets of Ω making every function in H Borel measurable, that is, the minimal σ-field containing all sets $f^{-1}(B)$, where f ranges over H and B over the Borel sets. If $\mathscr{G}$ is a class of subsets of Ω, $\sigma(\mathscr{G})$ as usual denotes the minimal σ-field over $\mathscr{G}$.

The following will be assumed throughout:

Hypothesis A: If the functions f_n form a sequence in L decreasing to 0, then $E(f_n)$ decreases to 0. Equivalently, if the f_n belong to L and increase to $f \in L$, then $E(f_n)$ increases to $E(f)$.

We are also going to carry through a parallel development under the following assumption:

Hypothesis B: If the functions f_n form a net in L decreasing to 0, then $E(f_n)$ decreases to 0 (with the equivalent statement just as in hypothesis A).

Hypothesis A is always assumed in the statements of theorems. Corresponding results under hypothesis B will be added in brackets.

4.2.2 ***Lemma.*** Let $\{f_m\}$ and $\{f_n'\}$ be sequences in L increasing to f and f', respectively, with $f \le f'$ (f and f' need not belong to L). Then

$$\lim_m E(f_m) \le \lim_n E(f_n').$$

Hence E may be extended to L' by defining $E(\lim_n f_n) = \lim_n E(f_n)$. [Under hypothesis B (and with "sequence" replaced by "net" in the above statement), E may be extended to L'' in the same fashion.]

PROOF. As $n \to \infty, f_m \wedge f_n' \uparrow f_m \wedge f' = f_m$; hence $\lim_n E(f_n') \ge \lim_n E(f_m \wedge f_n') = E(f_m)$. Let $m \to \infty$ to finish the proof. [Under hypothesis B, the proof is the same, with sequences replaced by nets.] ∎

We now study the extension of E.

4.2.3 ***Lemma.*** The extension of E to L' has the following properties:

(a) $0 \le E(f) \le \infty$ for all $f \in L'$.
(b) If $f, g \in L', f \le g$, then $E(f) \le E(g)$.
(c) If $f \in L'$ and $0 \le c < \infty$, then $cf \in L'$ and $E(cf) = cE(f)$.
(d) If $f, g \in L'$, then $f + g, f \vee g, f \wedge g \in L'$ and

$$E(f + g) = E(f \vee g) + E(f \wedge g) = E(f) + E(g).$$

(e) If $\{f_n\}$ is a sequence in L' increasing to f, then $f \in L'$ and $E(f_n)$ increases to $E(f)$.

[Under hypothesis B, the extension of E to L'' has exactly the same properties; "sequence" is replaced by "net" in (e).]

PROOF. Properties (a)–(c) follow from 4.2.2. To prove (d), let $f_n, g_n \in L^+$, with $f_n \uparrow f$, $g_n \uparrow g$. Then

$$\begin{aligned} E(f_n \vee g_n) + E(f_n \wedge g_n) &= E[(f_n \vee g_n) + (f_n \wedge g_n)] \\ &\quad \text{by linearity} \\ &= E(f_n + g_n) \\ &\quad \text{(the sum of two numbers is} \\ &\quad \text{the larger plus the smaller)} \\ &= E(f_n) + E(g_n) \\ &\quad \text{by linearity.} \end{aligned}$$

Let $n \to \infty$ to obtain (d). To prove (e), let $f_{nm} \in L^+, f_{nm} \uparrow f_n$ as $m \to \infty$. If $g_m = f_{1m} \vee \cdots \vee f_{mm}$, the g_m form an increasing sequence in L^+, and

$$f_{nm} \leq g_m \leq f_m \qquad \text{for} \quad n \leq m; \tag{1}$$

hence

$$E(f_{nm}) \leq E(g_m) \leq E(f_m), \qquad n \leq m. \tag{2}$$

In (1) let $m \to \infty$ to obtain $f_n \leq \lim_m g_m \leq \lim_m f_m = f$; then let $n \to \infty$ to obtain $g_m \uparrow f$, hence $E(g_m) \uparrow E(f)$. Now let $m \to \infty$ in (2) to find that $E(f_n) \leq E(f) \leq \lim_m E(f_m)$; then let $n \to \infty$ to establish that $\lim_n E(f_n) = E(f)$.

[Under hypothesis B, (a)–(d) are proved as above, but (e) is more complicated. Let $\{f_n\}$ be a net in L'' with $f_n \uparrow f$; for each n there is a net $\{f_{nm}\}$ in L^+ with $f_{nm} \uparrow f_n$, so that $f = \sup_{n,m} f_{nm}$. Let D consist of all finite sets of pairs (n, m). Direct D by inclusion, and define $g_M = \max\{f_{nm} : (n, m) \in M\}$, $M \in D$. The g_M form a monotone net in L^+ and $g_M \uparrow f$; therefore $f \in L''$ and, by 4.2.2, $E(g_M) \uparrow E(f)$. Now for each M, $E(g_M) \leq \sup_{n,m} E(f_{n,m})$, and since $f_{nm} \leq f_n$ for all m, it follows that $E(g_M) \leq \sup_n E(f_n) = \lim_n E(f_n)$. But M is arbitrary, hence $E(f) \leq \lim_n E(f_n)$. The reverse inclusion holds because $f_n \uparrow f$, and the proof is complete.] ▮

We now begin the construction of a measure derived from the linear functional E. Fron now on we assume that all constant functions belong to L and, as a normalization, $E(1) = 1$ (hence $E(c) = c$ for all c).

4.2.4 ***Lemma.*** Let $\mathscr{G} = \{G \subset \Omega: I_G \in L'\}$ and define $\mu(G) = E(I_G), G \in \mathscr{G}$. Then $\mathscr{G}$ satisfies the four conditions of 1.3.2, namely:

(a) $\varnothing, \Omega \in \mathscr{G}$, $\mu(\varnothing) = 0$, $\mu(\Omega) = 1$, $0 \le \mu(A) \le 1$ for all $A \in \mathscr{G}$.
(b) If $G_1, G_2 \in \mathscr{G}$, then $G_1 \cup G_2, G_1 \cap G_2 \in \mathscr{G}$, and

$$\mu(G_1 \cup G_2) + \mu(G_1 \cap G_2) = \mu(G_1) + \mu(G_2).$$

(c) If $G_1, G_2 \in \mathscr{G}$ and $G_1 \subset G_2$, then $\mu(G_1) \le \mu(G_2)$.
(d) If $G_n \in \mathscr{G}$, $n = 1, 2, \ldots$ and $G_n \uparrow G$, then $G \in \mathscr{G}$ and $\mu(G_n) \uparrow \mu(G)$.

Thus by 1.3.3 and 1.3.5, $\mu^*(A) = \inf\{\mu(G): G \in \mathscr{G}, G \supset A\}$ is a probability measure on the σ-field $\mathscr{H} = \{H \subset \Omega: \mu^*(H) + \mu^*(H^c) = 1\}$, and $\mu^* = \mu$ on $\mathscr{G}$.

[Under hypothesis B, we take $\mathscr{G} = \{G \subset \Omega: I_G \in L''\}$ and replace sequences by nets in 4.2.4(d). The class $\mathscr{G}$ then has exactly the same properties.]

Proof. Part (a) follows since L contains the constant functions and $E(c) = c$. Part (b) follows from 4.2.3(d) with $f = I_{G_1}, g = I_{G_2}$. Part (c) follows from 4.2.3(b), and part (d) from 4.2.3(e). [The proof under hypothesis B is identical.] ▮

We now investigate Borel measurability relative to the σ-field $\sigma(\mathscr{G})$.

4.2.5 ***Lemma.*** If $f \in L'$ and $a \in R$, then $\{\omega: f(\omega) > a\} \in \mathscr{G}$. Hence

$$f: (\Omega, \sigma(\mathscr{G})) \to (\bar{R}, \mathscr{B}(\bar{R})).$$

[The same result holds for $f \in L''$ under hypothesis B.]

Proof. Let $f_n \in L^+, f_n \uparrow f$. Then $(f_n - a)^+ = (f_n - a) \vee 0 \in L^+ \subset L'$, and $(f_n - a)^+ \uparrow (f - a)^+$. By 4.2.3(e) (or the definition of L'), $(f - a)^+ \in L'$; hence $k(f - a)^+ \in L'$ for each positive integer k, by 4.2.3(c). But as $k \to \infty$,

$$1 \wedge k(f - a)^+ \uparrow I_{\{f > a\}};$$

so by 4.2.3(e), $I_{\{f > a\}} \in L'$, and therefore $\{f > a\} \in \mathscr{G}$. [Under hypothesis B, the proof is the same, with $\{f_n\}$ a net instead of a sequence.] ▮

4.2.6 ***Lemma.*** The σ-fields $\sigma(L)$, $\sigma(L')$, and $\sigma(\mathscr{G})$ are identical. [Under hypothesis B we only have $\sigma(L'') = \sigma(\mathscr{G})$ and $\sigma(L) \subset \sigma(L'')$.]

Proof. By 4.2.5, $\sigma(\mathscr{G})$ makes every function in L' measurable; hence $\sigma(L') \subset \sigma(\mathscr{G})$. If $G \in \mathscr{G}$, then $I_G \in L'$; hence $G = \{I_G = 1\} \in \sigma(L')$; therefore $\sigma(\mathscr{G}) \subset \sigma(L')$. [Under hypothesis B, $\sigma(L'') = \sigma(\mathscr{G})$ by the same argument.]

If $f \in L$, then $f = f^+ - f^-$, where $f^+, f^- \in L^+ \subset L'$. Since f^+ and f^- are $\sigma(L')$-measurable, so is f. In other words, $\sigma(L')$ makes every function in L Borel measurable, so that $\sigma(L) \subset \sigma(L')$. [Under hypothesis B, $\sigma(L) \subset \sigma(L'')$ by the same argument.] Now if $f \in L'$, then f is the limit of a sequence $f_n \in L$, and since the f_n are $\sigma(L)$-measurable, so is f. [This fails under hypothesis B because the limit of a net of measurable functions need not be measurable; see Problem 1.] Thus $\sigma(L') \subset \sigma(L)$. ∎

Now by definition of the set function μ^* (see 4.2.4) we have, for all $A \subset \Omega$,

$$\begin{aligned}\mu^*(A) &= \inf\{E(I_G)\colon G \in \mathscr{G},\quad G \supset A\}\\ &= \inf\{E(f)\colon f = I_G \in L',\quad f \geq I_A\}\\ &\geq \inf\{E(f)\colon f \in L',\quad f \geq I_A\}.\end{aligned}$$

In fact equality holds.

4.2.7 ***Lemma.*** For any $A \subset \Omega$, $\mu^*(A) = \inf\{E(f)\colon f \in L', f \geq I_A\}$. [The result is the same under hypothesis B, with L' replaced by L''.]

PROOF. Let $f \in L', f \geq I_A$. If $0 < a < 1$, then $A \subset \{f > a\}$, which belongs to $\mathscr{G}$ by 4.2.5. Thus $\mu^*(A) \leq \mu\{f > a\} = E(I_{\{f>a\}})$. But since $f \geq 0$ we have $f \geq aI_{\{f>a\}}$; hence $E(I_{\{f>a\}}) \leq E(f)/a$. Let $a \to 1$ to conclude that $\mu^*(A) \leq E(f)$. [The proof is the same under hypothesis B.] ∎

We may now prove that μ^* is a measure on $\sigma(\mathscr{G})$:

4.2.8 ***Lemma.*** If $\mathscr{H} = \{H \subset \Omega\colon \mu^*(H) + \mu^*(H^c) = 1\}$, then $\mathscr{G} \subset \mathscr{H}$, hence $\sigma(\mathscr{G}) \subset \mathscr{H}$. [The result is the same under hypothesis B.]

PROOF. Let $G \in \mathscr{G}$; since $I_G \in L'$, there are functions $f_n \in L^+$ with $f_n \uparrow I_G$. Then

$$\mu^*(G) = \mu(G) = E(I_G) = \lim_n E(f_n) \qquad \text{by 4.2.3(e).}$$

By 4.2.7, $\mu^*(G^c) = \inf\{E(f)\colon f \in L', f \geq I_{G^c}\}$. But if $f_n \leq I_G$, then $1 - f_n \geq I_{G^c}$; since $1 - f_n \geq 0$, we have $1 - f_n \in L^+ \subset L'$; hence

$$\mu^*(G^c) \leq \inf_n E(1 - f_n) = 1 - \lim_n E(f_n) = 1 - E(I_G).$$

Thus $\mu^*(G) + \mu^*(G^c) \leq 1$; since $\mu^*(G) + \mu^*(G^c)$ is always at least 1 by 1.3.3(b), we have $G \in \mathscr{H}$. [Under hypothesis B, the proof is the same, with $\{f_n\}$ a net instead of a sequence.] ∎

We now prove the main theorem; for clarity we state the results under hypotheses A and B separately (in 4.2.9 and 4.2.10).

4.2.9 ***Daniell Representation Theorem.*** Let L be a vector space of real-valued functions on the set Ω; assume that L contains the constant functions and is closed under the lattice operations. Let E be a *Daniell integral* on L, that is, a positive linear functional on L such that $E(f_n) \downarrow 0$ for each sequence of functions $f_n \in L$ with $f_n \downarrow 0$; assume that $E(1) = 1$.

Then there is a unique probability measure P on $\sigma(L)$ $(=\sigma(L') = \sigma(\mathscr{G})$ by 4.2.6) such that each $f \in L$ is P-integrable and $E(f) = \int_\Omega f\, dP$.

PROOF. Let P be the restriction of μ^* to $\sigma(L)$. By 4.2.6 and 4.2.8, P is a probability measure. If $G \in \mathscr{G}$, then

$$E(I_G) = \mu(G) = \mu^*(G) = P(G) = \int_\Omega I_G\, dP.$$

Now if $f \in L'$, we define

$$h_n = \sum_{k=1}^{n2^n} \frac{k-1}{2^n} I_{\{(k-1)/2^n < f \le k/2^n\}} + nI_{\{f > n\}}.$$

Since $I_{\{a < f \le b\}} = I_{\{f > a\}} - I_{\{f > b\}}$ for $a < b$, and $\{f > a\}, \{f > b\} \in \mathscr{G}$ by 4.2.5, it follows that $E(h_n) = \int_\Omega h_n\, dP$. But the h_n form a sequence of nonnegative simple functions increasing to f [see 1.5.5(a)], so by the monotone convergence theorem, $E(f) = \int_\Omega f\, dP$.

Now let $f \in L$; $f = f^+ - f^-$, where $f^+, f^- \in L^+ \subset L'$. Then $E(f) = E(f^+) - E(f^-) = \int_\Omega f^+\, dP - \int_\Omega f^-\, dP = \int_\Omega f\, dP$. (Since $f^+, f^- \in L$, the integrals are finite.)

This establishes the existence of the desired probability measure P. If P' is another such measure, then $\int_\Omega f\, dP = \int_\Omega f\, dP'$ for all $f \in L$, and hence for all $f \in L'$, by the monotone convergence theorem. Set $f = I_G$, $G \in \mathscr{G}$, to show that $P = P'$ on $\mathscr{G}$. Now $\mathscr{G}$ is closed under finite intersection by 4.2.4(b); hence by 4.1.3, $P = P'$ on $\sigma(\mathscr{G})$, proving uniqueness. ∎

4.2.10 ***Theorem.*** Let L be a vector space of real-valued functions on the set Ω; assume that L contains the constant functions and is closed under the lattice operations. Let E be a positive linear functional on L such that $E(f_n) \downarrow 0$ for each net of functions $f_n \in L$ with $f_n \downarrow 0$; assume that $E(1) = 1$.

Then there is a unique probability measure P on $\sigma(L'')$ $(=\sigma(\mathscr{G})$ by 4.2.6) such that:

(a) Each $f \in L$ is P-integrable and $E(f) = \int_\Omega f\, dP$.
(b) If $\{G_n\}$ is a net of sets in $\mathscr{G}$ and $G_n \uparrow G$, then $G \in \mathscr{G}$ and $P(G_n) \uparrow P(G)$.

PROOF. Let P be the restriction of μ^* to $\sigma(L'')$. The proof that P satisfies (a) is done exactly as in 4.2.9, with L' replaced by L''; P satisfies (b) by 4.2.4(d).

The uniqueness part cannot be done exactly as in 4.2.9 because the monotone convergence theorem fails in general for nets (see Problem 2). Let P' be a probability measure satisfying (a) and (b). If $f \in L''$, there is a net of functions $f_\alpha \in L^+$ with $f_\alpha \uparrow f$. We define

$$h_{n\alpha} = \frac{1}{2^n} \sum_{j=1}^{n2^n} I_{\{f_\alpha > j2^{-n}\}}, \qquad n = 1, 2, \ldots.$$

If $(k-1)/2^n < f_\alpha(\omega) \le k/2^n$, $k = 1, 2, \ldots, n2^n$, then $h_{n\alpha}(\omega) = (k-1)/2^n$, and if $f_\alpha(\omega) > n$, then $h_{n\alpha}(\omega) = n$. Thus the $h_{n\alpha}$, $n = 1, 2, \ldots$, are in fact the standard sequence of nonnegative simple functions increasing to f_α [see 1.5.5(a)]. Similarly, if

$$h_n = \frac{1}{2^n} \sum_{j=1}^{n2^n} I_{\{f > j2^{-n}\}},$$

the h_n are nonnegative simple functions increasing to f. Now

$$\begin{aligned}
\int_\Omega h_n \, dP' &= \frac{1}{2^n} \sum_{j=1}^{n2^n} P'\{f > j2^{-n}\} \\
&= \frac{1}{2^n} \sum_{j=1}^{n2^n} \lim_\alpha P'\{f_\alpha > j2^{-n}\} \qquad \text{by (b)} \\
&= \lim_\alpha \frac{1}{2^n} \sum_{j=1}^{n2^n} P'\{f_\alpha > j2^{-n}\} \\
&\qquad \text{since the sum on } j \text{ is finite} \\
&= \lim_\alpha \int_\Omega h_{n\alpha} \, dP';
\end{aligned}$$

but

$$\begin{aligned}
\int_\Omega f \, dP' &= \lim_n \int_\Omega h_n \, dP' \\
&\qquad \text{by the monotone convergence theorem} \\
&= \lim_n \lim_\alpha \int_\Omega h_{n\alpha} \, dP' \\
&= \lim_\alpha \lim_n \int_\Omega h_{n\alpha} \, dP' \\
&\qquad \text{since } h_{n\alpha} \text{ is monotone in each variable,} \\
&\qquad \text{so that "lim" may be replaced by "sup"} \\
&= \lim_\alpha \int_\Omega f_\alpha \, dP' \\
&\qquad \text{by the monotone convergence theorem}
\end{aligned}$$

Equation continues

$$= \lim_{\alpha} \int_{\Omega} f_\alpha \, dP \qquad \text{by (a)}$$

$$= \int_{\Omega} f \, dP$$

by the above argument with P' replaced by P.

Set $f = I_G$, $G \in \mathscr{G}$, to show that $P = P'$ on $\mathscr{G}$; hence, as in 4.2.9, on $\sigma(\mathscr{G})$. ▌

The following approximation theorem will be helpful in the next section.

4.2.11 ***Theorem.*** Assume the hypothesis of 4.2.9, and in addition assume that L is closed under limits of uniformly convergent sequences. Let

$$\mathscr{G}' = \{G \subset \Omega \colon G = \{f > 0\} \quad \text{for some} \quad f \in L^+\}.$$

Then:

(a) $\mathscr{G}' = \mathscr{G}$.
(b) If $A \in \sigma(L)$, then $P(A) = \inf\{P(G) \colon G \in \mathscr{G}', G \supset A\}$.
(c) If $G \in \mathscr{G}$, then $P(G) = \sup\{E(f) \colon f \in L^+, f \leq I_G\}$.

PROOF. (a) We have $\mathscr{G}' \subset \mathscr{G}$ by 4.2.5. Conversely, suppose $G \in \mathscr{G}$, and let $f_n \in L^+$ with $f_n \uparrow I_G$ $(\in L')$. Set $f = \sum_{n=1}^{\infty} 2^{-n} f_n$. Since $0 \leq f_n \leq 1$, the series is uniformly convergent, hence $f \in L^+$. But

$$\{f > 0\} = \bigcup_{n=1}^{\infty} \{f_n > 0\} = \{I_G = 1\} = G.$$

Consequently, $G \in \mathscr{G}'$.

(b) This is immediate from (a) and the fact that $P = \mu^*$ on $\sigma(L)$.

(c) If $f \in L^+, f \leq I_G$, then $E(f) \leq E(I_G) = P(G)$. Conversely, let $G \in \mathscr{G}$, with $f_n \in L^+, f_n \uparrow I_G$. Then $P(G) = E(I_G) = \lim_n E(f_n) = \sup_n E(f_n)$, hence $P(G) \leq \sup\{E(f) \colon f \in L^+, f \leq I_G\}$. ▌

Problems

1. Give an example to show that the limit of a net of measurable functions need not be measurable.
2. Give an example of a net of nonnegative Borel measurable functions f_α increasing to a Borel measurable function f, with $\lim_\alpha \int f_\alpha \, d\mu \neq \int f \, d\mu$.
3. Let L be the class of real-valued continuous functions on $[0, 1]$, and let $E(f)$ be the Riemann integral of f. Show that E is a Daniell integral on L, and show that $\sigma(L) = \mathscr{B}[0, 1]$ and P is Lebesgue measure.

4.3 Measures on Topological Spaces

We are now in a position to obtain precise results on the interplay between measure theory and topology.

4.3.1 ***Definitions and Comments.*** Let Ω be a normal topological space (Ω is Hausdorff, and if A and B are disjoint closed subsets of Ω, there are disjoint open sets U and V with $A \subset U$ and $B \subset V$). The basic property of normal spaces that we need is *Urysohn's lemma*: If A and B are disjoint closed subsets of Ω, there is a continuous function $f: \Omega \to [0, 1]$ such that $f = 0$ on A and $f = 1$ on B. Other standard results are that every compact Hausdorff space is normal, and every metric space is normal.

The class of *Borel sets* of Ω, denoted by $\mathscr{B}(\Omega)$ or simply by $\mathscr{B}$, is defined as the smallest σ-field of subsets of Ω containing the open (or equally well the closed) sets. The class of *Baire sets* of Ω, denoted by $\mathscr{A}(\Omega)$ or simply by $\mathscr{A}$, is defined as the smallest σ-field of subsets of Ω making all continuous real-valued functions (Borel) measurable, that is, $\mathscr{A}$ is the minimal σ-field containing all sets $f^{-1}(B)$ where B ranges over $\mathscr{B}(R)$ and f ranges over the class $C(\Omega)$ of continuous maps from Ω to R. Note that $\mathscr{A}$ is the smallest σ-field making all bounded continuous functions measurable. For let $\mathscr{F}$ be a σ-field that makes all bounded continuous functions measurable. If $f \in C(\Omega)$, then $f^+ \wedge n$ is a bounded continuous function and $f^+ \wedge n \uparrow f^+$ as $n \to \infty$. Thus f^+ (and similarly f^-) is $\mathscr{F}$-measurable, hence $f = f^+ - f^-$ is $\mathscr{F}$-measurable. Thus $\mathscr{A} \subset \mathscr{F}$, as desired.

The class of bounded continuous real-valued functions on Ω will be denoted by $C_b(\Omega)$.

If V is an open subset of R and $f \in C(\Omega)$, then $f^{-1}(V)$ is open in Ω, hence $f^{-1}(V) \in \mathscr{B}(\Omega)$. But the sets $f^{-1}(V)$ generate $\mathscr{A}(\Omega)$, since any σ-field containing the sets $f^{-1}(V)$ for all open sets V must contain the sets $f^{-1}(B)$ for all Borel sets B. (Problem 6 of Section 1.2 may be used to give a formal proof, with $\mathscr{C}$ taken as the class of open sets.) It follows that $\mathscr{A}(\Omega) \subset \mathscr{B}(\Omega)$.

An F_σ *set* in Ω is a countable union of closed sets, and a G_σ *set* is a countable intersection of open sets.

4.3.2 ***Theorem.*** Let Ω be a normal topological space. Then $\mathscr{A}(\Omega)$ is the minimal σ-field containing the open F_σ sets (or equally well, the minimal σ-field containing the closed G_δ sets).

PROOF. Let $f \in C(\Omega)$; then $\{f > a\} = \bigcup_{n=1}^{\infty} \{f \geq a + (1/n)\}$ is an open F_σ set. As above, the sets $\{f > a\}$, $a \in R, f \in C(\Omega)$, generate $\mathscr{A}$; hence $\mathscr{A}$ is included in the minimal σ-field $\mathscr{H}$ over the F_σ sets. Conversely, let $H = \bigcup_{n=1}^\infty F_n$, F_n closed, be an open F_σ set. By Urysohn's lemma, there are functions $f_n \in C(\Omega)$ with $0 \le f_n \le 1$, $f_n = 0$ on H^c, $f_n = 1$ on F_n. If $f = \sum_{n=1}^\infty 2^{-n} f_n$, then $f \in C(\Omega)$, $0 \le f \le 1$, and $\{f > 0\} = \bigcup_{n=1}^\infty \{f_n > 0\} = H$. Thus $H \in \mathscr{A}$, so that $\mathscr{H} \subset \mathscr{A}$. ▌

4.3.3 ***Corollary.*** If Ω is a normal topological space, the open F_σ sets are precisely the sets $\{f > 0\}$ where $f \in C_b(\Omega)$, $f \ge 0$.

PROOF. By the argument of 4.3.2. ▌

4.3.4 ***Corollary.*** If Ω is a metric space, then $\mathscr{A}(\Omega) = \mathscr{B}(\Omega)$.

PROOF. If F is a closed subset of Ω, then F is a G_δ

$$\left(F = \bigcap_{n=1}^\infty \{\omega \colon \operatorname{dist}(\omega, F) < \frac{1}{n}\right);$$

hence every open subset of Ω is an F_σ. The result now follows from 4.3.2. ▌

Corollary 4.3.4 has a direct proof that avoids use of 4.3.2; see Problem 1.

We may obtain some additional information about the open F_σ sets of a normal space.

4.3.5 ***Lemma.*** Let A be an open F_σ set in the normal space Ω, so that by 4.3.3, $A = \{f > 0\}$ where $f \in C_b(\Omega)$, $f \ge 0$. Then I_A is the limit of an increasing sequence of continuous functions.

PROOF. We have $\{f > 0\} = \bigcup_{n=1}^\infty \{f \ge 1/n\}$, and by Urysohn's lemma there are functions $f_n \in C(\Omega)$ with $0 \le f_n \le 1$, $f_n = 0$ on $\{f = 0\}$, $f_n = 1$ on $\{f \ge 1/n\}$. If $g_n = \max(f_1, \ldots, f_n)$, then $g_n \uparrow I_{\{f>0\}}$. ▌

The Daniell theory now gives us a basic approximation theorem.

4.3.6 ***Theorem.*** Let P be any probability measure on $\mathscr{A}(\Omega)$, where Ω is a normal topological space. If $A \in \mathscr{A}$, then

(a) $P(A) = \inf\{P(V) \colon V \supset A,\ V \text{ an open } F_\sigma \text{ set}\}$,
(b) $P(A) = \sup\{P(C) \colon C \subset A,\ C \text{ a closed } G_\delta \text{ set}\}$.

PROOF. Let $L = C_b(\Omega)$ and define $E(f) = \int_\Omega f\,dP, f \in L$. [Note that $\sigma(L) = \mathscr{A}$, so each $f \in L$ is $\mathscr{A}$-measurable; furthermore, since f is bounded, $\int_\Omega f\,dP$ is finite. Thus E is well-defined.] Now E is a positive linear functional on L, and by the dominated convergence theorem, E is a Daniell integral. By 4.2.11(b),

$$P(A) = \inf\{P(G): G \in \mathscr{G}', \quad G \supset A\},$$

where

$$\mathscr{G}' = \{G \subset \Omega: G = \{f > 0\} \quad \text{for some } f \in L^+\}.$$

By 4.3.3, $\mathscr{G}'$ is the class of open F_σ sets, proving (a). Part (b) follows upon applying (a) to the complement of A. ∎

4.3.7 ***Corollary.*** If Ω is a metric space, and P is a probability measure on $\mathscr{B}(\Omega)$, then for each $A \in \mathscr{B}(\Omega)$,

(a) $P(A) = \inf\{P(V): V \supset A, V \text{ open}\}$,
(b) $P(A) = \sup\{P(C): C \subset A, C \text{ closed}\}$.

PROOF. In a metric space, every closed set is a G_δ and every open set is an F_σ (see 4.3.4); the result follows from 4.3.6. ∎

Under additional hypotheses on Ω, we obtain approximations by compact subsets.

4.3.8 ***Theorem.*** Let Ω be a complete separable metric space (sometimes called a "Polish space"). If P is a probability measure on $\mathscr{B}(\Omega)$, then for each $A \in \mathscr{B}(\Omega)$,

$$P(A) = \sup\{P(K): K \text{ compact subset of } A\}.$$

PROOF. By 4.3.7, the approximation property holds with "compact" replaced by "closed." We are going to show that if $\varepsilon > 0$, there is a compact set K_ε such that $P(K_\varepsilon) \geq 1 - \varepsilon$. This implies the theorem, for if C is closed, then $C \cap K_\varepsilon$ is compact, and $P(C) - P(C \cap K_\varepsilon) = P(C - K_\varepsilon) \leq P(\Omega - K_\varepsilon) \leq \varepsilon$.

Since Ω is separable, there is a countable dense set $\{\omega_1, \omega_2, \ldots\}$. Let $B(\omega_n, r\}$ (respectively, $\overline{B}(\omega_n, r))$ be the open (respectively, closed) ball with center at ω_n and radius r. Then for every $r > 0$, $\Omega = \bigcup_{n=1}^\infty \overline{B}(\omega_n, r)$ so that $\bigcup_{k=1}^m \overline{B}(\omega_k, 1/n) \uparrow \Omega$ as $m \to \infty$ (n fixed). Thus given $\varepsilon > 0$ and a positive integer n, there is a positive integer $m(n)$ such that $P(\bigcup_{k=1}^m \overline{B}(\omega_k, 1/n)) \geq 1 - \varepsilon 2^{-n}$ for all $m \geq m(n)$.

Let $K_\varepsilon = \bigcap_{n=1}^{\infty} \bigcup_{k=1}^{m(n)} \bar{B}(\omega_k, 1/n)$. Then K_ε is closed, and

$$P(K_\varepsilon{}^c) \le \sum_{n=1}^{\infty} P\left(\bigcup_{k=1}^{m(n)} \bar{B}\left[\omega_k, \frac{1}{n}\right]\right)^c \le \sum_{n=1}^{\infty} \varepsilon 2^{-n} = \varepsilon.$$

Therefore $P(K_\varepsilon) \ge 1 - \varepsilon$.

It remains to show that K_ε is compact. Let $\{x_1, x_2, \ldots\}$ be a sequence in K_ε. Then $x_p \in \bigcap_{n=1}^{\infty} \bigcup_{k=1}^{m(n)} \bar{B}(\omega_k, 1/n)$ for all p; hence $x_p \in \bigcup_{k=1}^{m(1)} \bar{B}(\omega_k, 1)$ for all p. We conclude that for some k_1, $x_p \in \bar{B}(\omega_{k_1}, 1)$ for infinitely many p, say, for $p \in T_1$, an infinite subset of the positive integers. But $x_p \in \bigcup_{k=1}^{m(2)} \bar{B}(\omega_k, \frac{1}{2})$ for all p, in particular for all $p \in T_1$; hence for some k_2, $x_p \in \bar{B}(\omega_{k_1}, 1) \cap \bar{B}(\omega_{k_2}, \frac{1}{2})$ for infinitely many $p \in T_1$, say, for $p \in T_2 \subset T_1$. Continue inductively to obtain integers $k_1, k_2, \ldots$ and infinite sets $T_1 \supset T_2 \supset \cdots$ such that

$$x_p \in \bigcap_{j=1}^{i} \bar{B}\left[\omega_{k_j}, \frac{1}{j}\right] \qquad \text{for all} \quad p \in T_i.$$

Pick $p_i \in T_i$, $i = 1, 2, \ldots$, with $p_1 < p_2 < \cdots$. Then if $j < i$, we have x_{p_i}, $x_{p_j} \in \bar{B}(\omega_{k_j}, 1/j)$, so $d(x_{p_i}, x_{p_j}) \le 2/j \to 0$ as $j \to \infty$. Thus $\{x_{p_i}\}$ is a Cauchy sequence, hence converges (to a point of K_ε since K_ε is closed). Therefore $\{x_p\}$ has a subsequence converging to a point of K_ε, so K_ε is compact. ∎

We now apply the Daniell theory to obtain theorems on representation of positive linear functionals in a topological context.

4.3.9 ***Theorem.*** Let Ω be a compact Hausdorff space, and let E be a positive linear functional on $C(\Omega)$, with $E(1) = 1$. There is a unique probability measure P on $\mathscr{A}(\Omega)$ such that $E(f) = \int_\Omega f\, dP$ for all $f \in C(\Omega)$.

PROOF. Let $L = C(\Omega)$. If $f_n \in L, f_n \downarrow 0$, then $f_n \to 0$ uniformly (this is *Dini*'s *theorem*). For given $\delta > 0$, we have $\Omega = \bigcup_{n=1}^{\infty} \{f_n < \delta\}$; hence by compactness,

$$\Omega = \bigcup_{n=1}^{N} \{f_n < \delta\} \qquad \text{for some} \quad N$$

$$= \{f_N < \delta\} \qquad \text{by montonicity of} \quad \{f_n\}.$$

Thus $n \ge N$ implies $0 \le f_n(\omega) \le f_N(\omega) < \delta$ for all ω, proving uniform convergence.

Thus if $\delta > 0$ is given, eventually $0 \le f_n \le \delta$, so $0 \le E(f_n) \le E(\delta) = \delta$. Therefore $E(f_n) \downarrow 0$, hence E is a Daniell integral. The result follows from 4.2.9. ∎

A somewhat different result is obtained if we use the Daniell theory with hypothesis B.

4.3.10 ***Theorem.*** Let Ω be a compact Hausdorff space, and let E be a positive linear functional on $C(\Omega)$, with $E(1) = 1$. There is a unique probability measure P on $\mathscr{B}(\Omega)$ such that

(a) $E(f) = \int_\Omega f\,dP$ for all $f \in C(\Omega)$, and
(b) for all $A \in \mathscr{B}(\Omega)$,

$$P(A) = \inf\{P(V) \colon V \supset A, \quad V \text{ open}\}$$

or equivalently,

$$P(A) = \sup\{P(K) \colon K \subset A, \quad K \text{ compact}\}.$$

("Compact" may be replaced by "closed" since Ω is compact Hausdorff.)

PROOF. Let $L = C(\Omega)$. If $\{f_n, n \in D\}$ is a net in L and $f_n \downarrow 0$, then, as in 4.3.9, for any $\delta > 0$ we have $\Omega = \bigcup_{j \in F} \{f_j < \delta\}$ for some finite set $F \subset D$. If $N \in D$ and $N \geq j$ for all $j \in F$, then $\Omega = \{f_N < \delta\}$ by monotonicity of the net. Thus $n \geq N$ implies $0 \leq f_n \leq f_N < \delta$, and it follows as in 4.3.9 that $E(f_n) \downarrow 0$.

By 4.2.10 there is a probability measure P on $\sigma(L'') = \sigma(\mathscr{G})$ such that $E(f) = \int_\Omega f\,dP$ for all $f \in L$. But in fact $\mathscr{G}$ *is the class of open sets*, so that $\sigma(\mathscr{G}) = \mathscr{B}(\Omega)$. For if $f \in L''$, there is a net of nonnegative continuous functions $f_n \uparrow f$; hence for each real a, $\{f > a\} = \bigcup_n \{f_n > a\}$ is an open set. Thus if $G \in \mathscr{G}$, then $I_G \in L''$, so that $G = \{I_G > 0\}$ is open. Conversely if G is open and $\omega \in G$, there is a continuous function $f_\omega : \Omega \to [0, 1]$ such that $f_\omega(\omega) = 1$ and $f_\omega = 0$ on G^c. Thus $I_G = \sup_\omega f_\omega$, so that if for each finite set $F \subset G$ we define $g_F = \max\{f_\omega : \omega \in F\}$, and direct the sets F by inclusion, we obtain a monotone net of nonnegative continuous functions increasing to I_G. Therefore $I_G \in L''$, so that $G \in \mathscr{G}$.

Thus we have established the existence of a probability measure P on $\mathscr{B}(\Omega)$ satisfying part (a) of 4.3.10; part (b) follows since $P = \mu^*$ on $\sigma(\mathscr{G})$, and $\mathscr{G}$ is the class of open sets.

To prove uniqueness, let P' be another probability measure satisfying (a) and (b) of 4.3.10. If we can show that P' satisfies 4.2.10(b), it will follow from the uniqueness part of 4.2.10 that $P' = P$. Thus let $\{G_n\}$ be a net of open sets with $G_n \uparrow G$; since G is the union of the G_n, G is open. By hypothesis, given $\delta > 0$, there is a compact $K \subset G$ such that $P'(G) \leq P'(K) + \delta$. Now $G_n \cup K^c \uparrow G \cup K^c = \Omega$; hence by compactness and the monotonicity of $\{G_n\}$, $G_m \cup K^c = \Omega$ for some m, so that $K \subset G_m$. Consequently,

$$P'(G) \leq P'(K) + \delta \leq P'(G_m) + \delta \leq \lim_n P'(G_n) + \delta$$

and it follows that $P'(G_n) \uparrow P'(G)$. ∎

Property (b) of 4.3.10 is often referred to as the "regularity" of P. Since the word "regular" is used in so many different ways in the literature, let us state exactly what it will mean for us.

4.3.11 ***Definitions.*** If μ is a measure on $\mathscr{B}(\Omega)$, where Ω is a normal topological space, μ is said to be *regular* iff for each $A \in \mathscr{B}(\Omega)$,

$$\mu(A) = \inf\{\mu(V) \colon V \supset A, \quad V \text{ open}\}$$

and

$$\mu(A) = \sup\{\mu(C) \colon C \subset A, \quad C \text{ closed}\}.$$

Either one of these conditions implies the other if μ is finite, and if in addition, Ω is a compact Hausdorff space, we obtain property (b) of 4.3.10.

If $\mu = \mu^+ - \mu^-$ is a finite signed measure on $\mathscr{B}(\Omega)$, Ω normal, we say that μ is regular iff μ^+ and μ^- are regular (equivalently, iff the total variation $|\mu|$ is regular).

The following result connects 4.3.9 and 4.3.10.

4.3.12 ***Theorem.*** If P is a probability measure on $\mathscr{A}(\Omega)$, Ω compact Hausdorff, then P has a unique extension to a regular probability measure on $\mathscr{B}(\Omega)$.

PROOF. Let $E(f) = \int_\Omega f\,dP, f \in C(\Omega)$. Then E is a positive linear functional on $L = C(\Omega)$, and thus (see the proof of 4.3.10) if $\{f_n\}$ is a net in L decreasing to 0, then $E(f_n) \downarrow 0$. By 4.3.10 there is a unique regular probability measure P' on $\mathscr{B}(\Omega)$ such that $\int_\Omega f\,dP = \int_\Omega f\,dP'$ for all $f \in L$. But each f in L is measurable: $(\Omega, \mathscr{A}) \to (R, \mathscr{B}(R))$, hence by 1.5.5, $\int_\Omega f\,dP'$ is determined by the values of P' on Baire sets. Thus the condition that $\int_\Omega f\,dP = \int_\Omega f\,dP'$ for all $f \in L$ is equivalent to $P = P'$ on $\mathscr{A}(\Omega)$, by the uniqueness part of 4.3.9. ∎

In 4.3.9 and 4.3.10, the assumption $E(1) = 1$ is just a normalization, and if it is dropped, the results are the same, except that "unique probability measure" is replaced by "unique finite measure." Similarly, 4.3.12 applies equally well to finite measures.

Now let Ω be a compact Hausdorff space, and consider $L = C(\Omega)$ as a vector space over the reals; L is a Banach space with the sup norm. If E is a positive linear functional on L, we can show that E is continuous, and this will allow us to generalize 4.3.9 and 4.3.10 by giving representation theorems for continuous linear functionals on $C(\Omega)$. To prove continuity of E, note that if $f \in L$ and $\|f\| \le 1$, then $-1 \le f(\omega) \le 1$ for all ω; hence $-E(1) \le E(f) \le$

$E(1)$, that is, $|E(f)| \leq E(1)$. Therefore $\|E\| \leq E(1)$; in fact $\|E\| = E(1)$, as may be seen by considering the function that is identically 1.

The representation theorem we are about to prove will involve integration with respect to a finite signed measure $\mu = \mu^+ - \mu^-$; the integral is defined in the obvious way, namely,

$$\int_\Omega f\,d\mu = \int_\Omega f\,d\mu^+ - \int_\Omega f\,d\mu^-,$$

assuming the right side is well-defined.

4.3.13 ***Theorem.*** Let E be a continuous linear functional on $C(\Omega)$, Ω compact Hausdorff.

(a) There is a unique finite signed measure μ on $\mathscr{A}(\Omega)$ such that $E(f) = \int_\Omega f\,d\mu$ for all $f \in C(\Omega)$.

(b) There is a unique regular finite signed measure λ on $\mathscr{B}(\Omega)$ such that $E(f) = \int_\Omega f\,d\lambda$ for all $f \in C(\Omega)$.

Furthermore, any finite signed measure on $\mathscr{A}(\Omega)$ has a unique extension to a regular finite signed measure on $\mathscr{B}(\Omega)$; in particular λ is the unique extension of μ.

PROOF. The existence of the desired signed measures μ and λ will follow from 4.3.9 and 4.3.10 if we show that E is the difference of two positive linear functionals E^+ and E^-. If $f \geq 0$, $f \in C(\Omega)$, define

$$E^+(f) = \sup\{E(g) : 0 \leq g \leq f, \quad g \in C(\Omega)\}.$$

If $f \leq M$, then $E^+(f) \leq M\|E\| < \infty$ by continuity of E; hence E^+ is finite. If $0 \leq g_i \leq f_i$, $i = 1, 2$, with all functions in $C(\Omega)$, then

$$E(g_1) + E(g_2) = E(g_1 + g_2) \leq E^+(f_1 + f_2).$$

Take the sup over g_1 and g_2 to obtain $E^+(f_1) + E^+(f_2) \leq E^+(f_1 + f_2)$.

Now if $0 \leq g \leq f_1 + f_2$, $f_1, f_2 \geq 0$, define $g_1 = g \wedge f_1$, $g_2 = 0 \vee (g - f_1)$. Then $0 \leq g_1 \leq f_1$, $0 \leq g_2 \leq f_2$, and $g = g_1 + g_2$. Thus $E(g) = E(g_1) + E(g_2) \leq E^+(f_1) + E^+(f_2)$; hence $E^+(f_1 + f_2) \leq E^+(f_1) + E^+(f_2)$, and consequently

$$E^+(f_1 + f_2) = E^+(f_1) + E^+(f_2). \tag{1}$$

Clearly $E^+(af) = aE^+(f)$ if $a \geq 0$. Thus E^+ extends to a positive linear functional on $C(\Omega)$. Specifically, if $f = f^+ - f^-$, take $E^+(f) = E^+(f^+) - E^+(f^-)$. If f can also be represented as $g - h$, where g and h are nonnegative functions in $C(\Omega)$, then $E^+(f^+) - E^+(f^-) = E^+(g) - E^+(h)$ by (1), so the extension is well-defined. If $f \geq 0$, $f \in C(\Omega)$, define

$$E^-(f) = -\inf\{E(g): 0 \leq g \leq f, \quad g \in C(\Omega)\} = (-E)^+(f).$$

By the above argument, E^- extends to a positive linear functional on $C(\Omega)$. Now

$$\begin{aligned} E^-(f) &= -\inf\{E(f) - E(f-g): 0 \leq g \leq f, \quad g \in C(\Omega)\} \\ &= -E(f) + \sup\{E(h): 0 \leq h \leq f, \quad h \in C(\Omega)\} \\ &= -E(f) + E^+(f). \end{aligned}$$

Thus $E = E^+ - E^-$.

To prove that μ is unique, assume that τ is a finite signed measure on $\mathscr{A}(\Omega)$ such that $\int_\Omega f\,d\tau = 0$ for all $f \in C(\Omega)$. Then $\int_\Omega f\,d\tau^+ = \int_\Omega f\,d\tau^-$ for all $f \in C(\Omega)$, hence $\tau^+ = \tau^-$ (so $\tau = 0$) by 4.3.9. Uniqueness of λ is proved similarly, using 4.3.10.

Now if μ is any finite signed measure on $\mathscr{A}(\Omega)$, $E(f) = \int_\Omega f\,d\mu$ defines a continuous linear functional on $C(\Omega)$, so by what we have just proved there is a unique regular finite signed measure λ on $\mathscr{B}(\Omega)$ such that $\int_\Omega f\,d\mu = \int_\Omega f\,d\lambda$ for all $f \in C(\Omega)$. But as in 4.3.12, this condition is equivalent to $\mu = \lambda$ on $\mathscr{A}(\Omega)$, so μ has a unique extension to a regular finite signed measure λ on $\mathscr{B}(\Omega)$. ∎

Theorems 4.3.9, 4.3.10, and 4.3.13 are referred to as versions of the *Riesz representation theorem*. [This name is also given to the somewhat different result 3.3.4(a).] A result of this type for complex-valued functions and complex measures is given in Problem 6.

We are going to show that $\|E\| = |\mu|(\Omega) = |\lambda|(\Omega)$ in 4.3.13. To do this we need a result on the approximation of Borel measurable functions by continuous functions.

4.3.14 ***Theorem.*** Consider the measure space $(\Omega, \mathscr{F}, \mu)$, where Ω is a normal topological space, $\mathscr{F} = \mathscr{B}(\Omega)$, and μ is a regular measure on $\mathscr{F}$. If $0 < p < \infty$, $\varepsilon > 0$, and $f \in L^p(\Omega, \mathscr{F}, \mu)$, there is a continuous complex-valued function $g \in L^p(\Omega, \mathscr{F}, \mu)$ such that $\|f - g\|_p < \varepsilon$; furthermore, g can be chosen so that $\sup|g| \leq \sup|f|$. Thus the continuous functions are dense in L^p.

PROOF. This is done exactly as in 2.4.14, except that the application of Problem 12, Section 1.4, is now replaced by the hypothesis that μ is regular. ∎

4.3.15 ***Theorem.*** In the Riesz representation theorem 4.3.13, $\|E\| = |\mu|(\Omega) = |\lambda|(\Omega)$.

PROOF. If $f \in C(\Omega)$, $E(f) = \int_\Omega f\,d\lambda = \int_\Omega f\,d\lambda^+ - \int_\Omega f\,d\lambda^-$; hence

$$|E(f)| \le \int_\Omega |f|\;d\lambda^+ + \int_\Omega |f|\;d\lambda^- = \int_\Omega |f|\;d|\lambda| \le \|f\|\,|\lambda|(\Omega),$$

where $\|\ \|$ is the sup norm. Thus $\|E\| \le |\lambda|(\Omega)$.

Now let $A_1, \ldots, A_n$ be disjoint measurable subsets of Ω, and define $f = \sum_{j=1}^n x_j I_{A_j}$, where $x_j = 1$ if $\lambda(A_j) \ge 0$, and $x_j = -1$ if $\lambda(A_j) < 0$. By 4.3.14, if $\varepsilon > 0$, there is a continuous function g such that $\|g\| \le 1$ and $\int_\Omega |f - g|\,d|\lambda| < \varepsilon$; hence $E(g) = \int_\Omega g\,d\lambda \ge \int_\Omega f\,d\lambda - \varepsilon$. But

$$\int_\Omega f\;d\lambda = \sum_{j=1}^n |\lambda(A_j)|;$$

hence

$$\|E\| \ge E(g) \ge \sum_{j=1}^n |\lambda(A_j)| - \varepsilon.$$

But $\sum_{j=1}^n |\lambda(A_j)|$ may be taken arbitrarily close to $|\lambda|(\Omega)$ (see Problem 4, Section 2.1); hence $\|E\| \ge |\lambda|(\Omega)$. Since $|\lambda|(\Omega) = |\mu|(\Omega)$, the result follows. ∎

If Ω is compact Hausdorff and $M(\Omega)$ is the collection of finite signed measures on $\mathscr{A}(\Omega)$ (or equally well the regular finite signed measures on $\mathscr{B}(\Omega)$), Theorems 4.3.13 and 4.3.15 show that the map $E \to \mu$ (or $E \to \lambda$) is an isometric isomorphism of the conjugate space of $C(\Omega)$ and $M(\Omega)$, where the norm of an element $\mu \in M(\Omega)$ is taken as $|\mu|(\Omega)$.

We close this section with some further results on approximation by continuous functions.

4.3.16 ***Theorem.*** Let μ be a regular finite measure on $\mathscr{B}(\Omega)$, Ω normal. If f is a complex-valued Borel measurable function on Ω and $\delta > 0$, there is a continuous complex-valued function g on Ω such that

$$\mu\{\omega : f(\omega) \ne g(\omega)\} < \delta.$$

Furthermore, it is possible to choose g so that $\sup|g| \le \sup|f|$.

PROOF. First assume f is real-valued and $0 \le f < 1$. If $h_n(\omega) = (k-1)2^{-n}$ when $(k-1)2^{-n} \le f(\omega) < k2^{-n}$, $k = 1, 2, \ldots, n2^n$, $h_n(\omega) = n$ when $f(\omega) \ge n$, the h_n are nonnegative simple functions increasing to f. Let $f_n = h_n - h_{n-1}$,

$n = 1, 2, \ldots$ (with $h_0 = 0$), so that $f = \sum_{n=1}^{\infty} f_n$. Note that f_n has only two possible values, 0 and 2^{-n}. If $A_n = \{f_n \neq 0\}$, let C_n be a closed subset of A_n and V_n an open overset of A_n such that $\mu(V_n - C_n) < \delta 2^{-n}$. Since Ω is normal, there is a continuous $g_n: \Omega \to [0, 1]$ such that $g_n = 1$ on C_n and $g_n = 0$ off V_n. If $g = \sum_{n=1}^{\infty} 2^{-n} g_n$, then by the Weierstrass M-test, g is a continuous map of Ω into $[0, 1]$. We claim that if $\omega \notin \bigcup_{n=1}^{\infty} (V_n - C_n)$, a set of measure less than δ, then $f(\omega) = g(\omega)$. To see this, observe that for each n, $\omega \in C_n$ or $\omega \notin V_n$. If $\omega \in C_n \subset A_n$, then $2^{-n} g_n(\omega) = 2^{-n} = f_n(\omega)$, and if $\omega \notin V_n$, then $2^{-n} g_n(\omega) = 0 = f_n(\omega)$ since $\omega \notin A_n$.

This proves the existence of g when $0 \leq f < 1$; the extension to a complex-valued bounded f is immediate. If f is unbounded, write $f = f I_{\{|f| < n\}} + f I_{\{|f| \geq n\}} = f_1 + f_2$, where f_1 is bounded and $\mu\{f_2 \neq 0\} = \mu\{|f| \geq n\}$, which can be made less than $\delta/2$ for sufficiently large n. If g is continuous and $\mu\{f_1 \neq g\} < \delta/2$, then $\mu\{f \neq g\} < \delta$, as desired.

Finally, if $|f| \leq M < \infty$, and g approximates f as above, define $g_1(\omega) = g(\omega)$ if $|g(\omega)| \leq M$, $g_1(\omega) = Mg(\omega)/|g(\omega)|$ if $|g(\omega)| > M$. Then g_1 is continuous, $|g_1| \leq M$, and $f(\omega) = g(\omega)$ implies $|g(\omega)| \leq M$; hence $g_1(\omega) = g(\omega) = f(\omega)$. Therefore $\mu\{f \neq g_1\} \leq \mu\{f \neq g\} < \delta$, completing the proof. ∎

4.3.17 ***Corollaries.*** Assume the hypothesis of 4.3.16.

(a) There is a sequence of continuous complex-valued functions f_n on Ω converging to f a.e. $[\mu]$, with $|f_n| \leq \sup |f|$ for all n.

(b) Given $\varepsilon > 0$, there is a closed set $C \subset \Omega$ and a continuous complex-valued function g on Ω such that $\mu(C) \geq \mu(\Omega) - \varepsilon$ and $f = g$ on C, hence the restriction of f to C is continuous. If μ has the additional property that $\mu(A) = \sup\{\mu(K): K \subset A,\ K \text{ compact}\}$ for each $A \in \mathscr{B}(\Omega)$, then C may be taken as compact.

PROOF. (a) By 4.3.16, there is a continuous function f_n such that $|f_n| \leq M = \sup|f|$ and $\mu\{f_n \neq f\} < 2^{-n}$. If $A_n = \{f_n \neq f\}$ and $A = \lim \sup_n A_n$, then $\mu(A) = 0$ by the Borel–Cantelli lemma. But if $\omega \notin A$, then $f_n(\omega) = f(\omega)$ for sufficiently large n.

(b) By 4.3.16, there is a continuous g such that $\mu\{f \neq g\} < \varepsilon/2$. By regularity of μ, there is a closed set $C \subset \{f = g\}$ with $\mu(C) \geq \mu\{f = g\} - \varepsilon/2$. The set C has the desired properties. The proof under the assumption of approximation by compact subsets is the same, with C compact rather than closed. ∎

Corollary 4.3.17(b) is called *Lusin's theorem.*

Problems

1. Let F be a closed subset of the metric space Ω. Define $f_n(\omega) = e^{-nd(\omega, F)}$ where $d(\omega, F) = \inf\{d(\omega, y): y \in F\}$. Show that the f_n are continuous and $f_n \downarrow I_F$. Use this to give a direct proof (avoiding 4.3.2) that in a metric space, the Baire and Borel sets coincide.
2. Give an example of a measure space $(\Omega, \mathscr{F}, \mu)$, where Ω is a metric space and $\mathscr{F} = \mathscr{B}(\Omega)$, such that for some $A \in \mathscr{F}$, $\mu(A) \neq \sup\{\mu(K): K \subset A, K$ compact$\}$.
3. In 4.3.14, assume in addition that Ω is locally compact, and that $\mu(A) = \sup\{\mu(K): K$ compact subset of $A\}$ for all Borel sets A. Show that the continuous functions with compact support (that is, the continuous functions that vanish outside a compact subset) are dense in $L^p(\Omega, \mathscr{F}, \mu)$, $0 < p < \infty$. Also, as in 4.3.14, if $f \in L^p$ is approximated by the continuous function g with compact support, g may be chosen so that $\sup|g| \leq \sup|f|$.
4. (a) Let Ω be a normal topological space, and let H be the smallest class of real-valued functions on Ω that contains the continuous functions and is closed under pointwise limits of monotone sequences. Show that H is the class of Baire measurable functions, that is, H consists of all $f: (\Omega, \mathscr{A}) \to (R, \mathscr{B}(R))$ (use 4.1.4).
   (b) If H is as in part (a) and $\sigma(H)$ is the smallest σ-field $\mathscr{G}$ of subsets of Ω making all functions in H measurable (relative to $\mathscr{G}$ and $\mathscr{B}(R)$), show that $\sigma(H) = \mathscr{A}(\Omega)$; hence $\sigma(H)$ is the same as $\sigma(C(\Omega))$.
5. Let Ω be a normal topological space, and let K_0 be the class of all continuous real-valued functions on Ω. Having defined K_β for all ordinals β less than the ordinal α, define

$$K_\alpha = \bigcup\{K_\beta: \beta < \alpha\}',$$

   where, for any class D of real-valued functions, D' means the class of all real-valued functions that are pointwise limits of monotone sequences in D.

   Let

$$K = \bigcup\{K_\alpha: \alpha < \beta_1\},$$

   where β_1 is the first uncountable ordinal. Show that K is the class of Baire measurable functions.
6. Let E be a continuous linear functional on the space $C(\Omega)$ of all continuous complex-valued functions on the compact Hausdorff space Ω. Show that there is a unique complex measure μ on $\mathscr{A}(\Omega)$ such that $E(f) = \int_\Omega f\,d\mu$ for all $f \in C(\Omega)$, and a unique regular complex measure λ on $\mathscr{B}(\Omega)$ such that $E(f) = \int_\Omega f\,d\lambda$ for all $f \in C(\Omega)$. Furthermore, $\|E\| =$

$|\mu|(\Omega) = |\lambda|(\Omega)$, so that $C(\Omega)^*$ is isometrically isomorphic to the space of complex measures on $\mathscr{A}(\Omega)$, or equally well to the space of regular complex measures on $\mathscr{B}(\Omega)$. (If $\mu = \mu_1 + i\mu_2$ is a complex measure, $\int_\Omega f\,d\mu$ is defined as $\int_\Omega f\,d\mu_1 + i\int_\Omega f\,d\mu_2$, provided f is integrable with respect to μ_1 and μ_2. Also, μ is said to be regular iff μ_1 and μ_2 are regular; since $|\mu_1|, |\mu_2| \le |\mu| \le |\mu_1| + |\mu_2|$, this is equivalent to regularity of $|\mu|$. See Section 2.2, Problem 6, and Section 2.4, Problems 10 and 11 for the basic properties of complex measures.)

7. Let $M(\Omega, \mathscr{F})$ be the collection of finite signed measures on a σ-field $\mathscr{F}$ of subsets of Ω. Show that if we take $\|\mu\| = |\mu|(\Omega)$, $\mu \in M(\Omega, \mathscr{F})$, $M(\Omega, \mathscr{F})$ becomes a Banach space. (A similar result holds for the collection of complex measures.)

4.4 Measures on Uncountably Infinite Product Spaces

In Section 2.7 we considered probability measures on countably infinite product spaces. The result may be extended to uncountable products if certain topological assumptions are made about the individual factor spaces.

The product of uncountably many σ-fields is formed in essentially the same way as in the countable case.

4.4.1 ***Definitions and Comments.*** For t in the arbitrary index set T, let $(\Omega_t, \mathscr{F}_t)$ be a measurable space. Let $\prod_{t \in T} \Omega_t$ be the set of all functions $\omega = (\omega(t), t \in T)$ on T such that $\omega(t) \in \Omega_t$ for each $t \in T$. If $t_1, \ldots, t_n \in T$ and $B^n \subset \prod_{i=1}^n \Omega_{t_i}$, we define the set $B^n(t_1, \ldots, t_n)$ as $\{\omega \in \prod_{t \in T} \Omega_t\colon (\omega(t_1), \ldots, \omega(t_n)) \in B^n\}$. We call $B^n(t_1, \ldots, t_n)$ the *cylinder* with *base* B^n at $(t_1, \ldots, t_n)$; the cylinder is said to be *measurable* iff $B^n \in \prod_{i=1}^n \mathscr{F}_{t_i}$. If $B^n = B_1 \times \cdots \times B_n$, the cylinder is called a *rectangle*, a *measurable rectangle* iff $B_i \in \mathscr{F}_{t_i}$, $i = 1, \ldots, n$. Note that if all $\Omega_t = \Omega$, then $\prod_{t \in T} \Omega_t = \Omega^T$, the set of all functions from T to Ω.

For example, let $T = [0, 1]$, $\Omega_t = R$ for all $t \in T$, $B^2 = \{(u, v)\colon u > 3, 1 < v < 2\}$. Then

$$B^2(\tfrac{1}{2}, \tfrac{3}{4}) = \{x \in R^T\colon x(\tfrac{1}{2}) > 3, \quad 1 < x(\tfrac{3}{4}) < 2\}$$

[see Fig. 4.1, where $x_1 \in B^2(\frac{1}{2}, \frac{3}{4})$ and $x_2 \notin B^2(\frac{1}{2}, \frac{3}{4})$].

Exactly as in 2.7.1, the measurable cylinders form a field, as do the finite disjoint unions of measurable rectangles. The minimal σ-field over the measurable cylinders is denoted by $\prod_{t \in T} \mathscr{F}_t$, and called the *product* of the σ-fields $\mathscr{F}_t$. If $\Omega_t = S$ and $\mathscr{F}_t = \mathscr{S}$ for all t, $\prod_{t \in T} \mathscr{F}_t$ is denoted by $\mathscr{S}^T$. Again as in 2.7.1, $\prod_{t \in T} \mathscr{F}_t$ is also the minimal σ-field over the measurable rectangles.

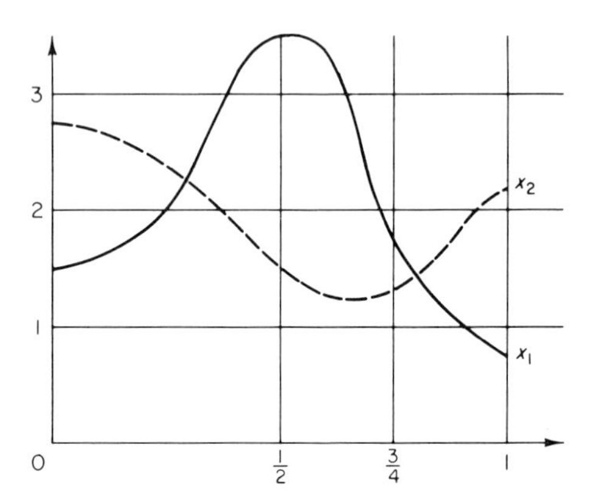

Figure 4.1.

We now consider the problem of constructing probability measures on $\prod_{t \in T} \mathscr{F}_t$. The approach will be as follows: Let $v = \{t_1, \dots, t_n\}$ be a finite subset of T, where $t_1 < t_2 < \cdots < t_n$. (If T is not a subset of R, some fixed total ordering is put on T.) Assume that for each such v we are given a probability measure P_v on $\prod_{i=1}^n \mathscr{F}_{t_i}$; $P_v(B)$ is to represent $P\{\omega \in \prod_{t \in T} \Omega_t$: $(\omega(t_1), \dots, \omega(t_n)) \in B\}$. We shall require that the P_v be "consistent"; to see what kind of consistency is needed, consider an example.

Suppose T is the set of positive integers and $\Omega_t = R$, $\mathscr{F}_t = \mathscr{B}(R)$ for all t. Suppose we know $P_{12345}(B^5) = P\{\omega: (\omega_1, \omega_2, \omega_3, \omega_4, \omega_5) \in B^5\}$ for all $B^5 \in \mathscr{B}(R^5)$. Then $P\{\omega: (\omega_2, \omega_3) \in B^2\} = P\{\omega: (\omega_1, \omega_2, \omega_3, \omega_4, \omega_5) \in R \times B^2 \times R^2\} = P_{12345}(R \times B^2 \times R^2)$, $B^2 \in \mathscr{B}(R^2)$. Thus once probabilities of sets involving the first five coordinates are specified, probabilities of sets involving (ω_2, ω_3) [as well as $(\omega_1, \omega_3, \omega_4)$, and so on], are determined. Thus the original specification of P_{23} must agree with the measure induced from P_{12345}. We are going to show that under appropriate topological assumptions, a consistent family of probability measures P_v determines a unique probability measure on $\prod_{t \in T} \mathscr{F}_t$.

Now to formalize: If $v = \{t_1, \dots, t_n\}$, $t_1 < \cdots < t_n$, the space $(\prod_{i=1}^n \Omega_{t_i}, \prod_{i=1}^n \mathscr{F}_{t_i})$ is denoted by $(\Omega_v, \mathscr{F}_v)$. If $u = \{t_{i1}, \dots, t_{ik}\}$ is a nonempty subset of v and $y = (y(t_1), \dots, y(t_n)) \in \Omega_v$, the k-tuple $(y(t_{i1}), \dots, y(t_{ik}))$ is denoted by y_u. Similarly if $\omega = (\omega(t), t \in T)$ belongs to $\prod_{t \in T} \Omega_t$, the notation ω_v will be used for $(\omega(t_1), \dots, \omega(t_n))$.

If P_v is a probability measure on $\mathscr{F}_v$, the *projection* of P_v on $\mathscr{F}_u$ is the probability measure $\pi_u(P_v)$ on $\mathscr{F}_u$ defined by

$$[\pi_u(P_v)](B) = P_v\{y \in \Omega_v: y_u \in B\}, \qquad B \in \mathscr{F}_u.$$

Similarly, if Q is a probability measure on $\prod_{t \in T} \mathscr{F}_t$, the projection of Q on $\mathscr{F}_v$ is defined by

$$[\pi_v(Q)](B) = Q\left\{\omega \in \prod_{t \in T} \Omega_t : \omega_v \in B\right\} = Q(B(v)), \qquad B \in \mathscr{F}_v.$$

We need one preliminary result.

4.4.2 ***Theorem.*** For each $n = 1, 2, \ldots$, suppose that $\mathscr{F}_n$ is the class of Borel sets of a separable metric space Ω_n. Let $\Omega = \prod_n \Omega_n$, with the product topology, and let $\mathscr{F} = \mathscr{B}(\Omega)$. [Note that Ω is metrizable, so that the Baire and Borel sets of Ω coincide; we may take

$$d(x, y) = \sum_n \frac{1}{2^n} \frac{d_n(x_n, y_n)}{1 + d_n(x_n, y_n)},$$

where d_n is the metric of Ω_n. Also, if each Ω_n is complete, so is Ω.]

Then $\mathscr{F}$ is the product σ-field $\prod_n \mathscr{F}_n$.

PROOF. The sets $\{\omega \in \Omega : \omega_1 \in A_1, \ldots, \omega_n \in A_n\}$, $n = 1, 2, \ldots$, where the A_i range over the countable base for Ω_i (recall that separability and second countability are equivalent in metric spaces), form a countable base for Ω. Since the sets are measurable rectangles, it follows that every open subset of Ω belongs to $\prod_n \mathscr{F}_n$; hence $\mathscr{F} \subset \prod_n \mathscr{F}_n$. On the other hand, for a fixed positive integer i let $\mathscr{C} = \{B \in \mathscr{B}(\Omega_i) : \{\omega \in \Omega : \omega_i \in B\} \in \mathscr{F}\}$. Then $\mathscr{C}$ is a σ-field containing the open sets of Ω_i, hence $\mathscr{C} = \mathscr{B}(\Omega_i)$. In other words, every measurable rectangle with one-dimensional base belongs to $\mathscr{F}$. Since an arbitrary measurable rectangle is a finite intersection of such sets, it follows that $\prod_n \mathscr{F}_n \subset \mathscr{F}$. ∎

We are now ready for the main result.

4.4.3 ***Kolmogorov Extension Theorem.*** For each t in the arbitrary index set T, let Ω_t be a complete, separable metric space, and $\mathscr{F}_t$ the Borel sets of Ω_t. Assume that for each finite nonempty subset v of T, we are given a probability measure P_v on $\mathscr{F}_v$. Assume the P_v are consistent, that is, $\pi_u(P_v) = P_u$ for each nonempty $u \subset v$.

Then there is a unique probability measure P on $\mathscr{F} = \prod_{t \in T} \mathscr{F}_t$ such that $\pi_v(P) = P_v$ for all v.

PROOF. We define the hoped-for measure on measurable cylinders by $P(B^n(v)) = P_v(B^n)$, $B^n \in \mathscr{F}_v$.

We must show that this definition makes sense since a given measurable cylinder can be represented in several ways. For example, if all $\Omega_t = R$ and $B^2 = (-\infty, 3) \times (4, 5)$, then

$$\begin{aligned} B^2(t_1, t_2) &= \{\omega\colon \omega(t_1) < 3, \quad 4 < \omega(t_2) < 5\} \\ &= \{\omega\colon \omega(t_1) < 3, \quad 4 < \omega(t_2) < 5, \quad \omega(t_3) \in R\} \\ &= B^3(t_1, t_2, t_3) \qquad \text{where} \quad B^3 = (-\infty, 3) \times (4, 5) \times R. \end{aligned}$$

It is sufficient to consider dual representation of the same measurable cylinder in the form $B^n(v) = B^k(u)$ where $k < n$ and $u \subset v$. But then

$$\begin{aligned} P_u(B^k) &= [\pi_u(P_v)](B^k) \qquad \text{by the consistency hypothesis} \\ &= P_v\{y \in \Omega_v\colon y_u \in B^k\} \qquad \text{by definition of projection.} \end{aligned}$$

But the assumption $B^n(v) = B^k(u)$ implies that if $y \in \Omega_v$, then $y \in B^n$ iff $y_u \in B^k$, hence $P_u(B^k) = P_v(B^n)$, as desired.

Thus, P is well-defined on measurable cylinders; the class $\mathscr{F}_0$ of measurable cylinders forms a field, and $\sigma(\mathscr{F}_0) = \mathscr{F}$.

Now if $A_1, \ldots, A_m$ are disjoint sets in $\mathscr{F}_0$, we may write (by introducing extra factors as in the above example) $A_i = B_i^n(v)$, $i = 1, \ldots, m$, where $v = \{t_1, \ldots, t_n\}$ is fixed and the B_i^n, $i = 1, \ldots, m$, are disjoint sets in $\mathscr{F}_v$. Thus

$$\begin{aligned} P\left(\bigcup_{i=1}^m A_i\right) &= P\left(\bigcup_{i=1}^m B_i^n(v)\right) \\ &= P_v\left(\bigcup_{i=1}^m B_i^n\right) \qquad \text{by definition of } P \\ &= \sum_{i=1}^m P_v(B_i^n) \qquad \text{since } P_v \text{ is a measure} \\ &= \sum_{i=1}^m P(A_i) \qquad \text{again by definition of } P. \end{aligned}$$

Therefore P is finitely additive on $\mathscr{F}_0$. To show that P is countably additive on $\mathscr{F}_0$, we must verify that P is continuous from above at $\varnothing$ and invoke 1.2.8(b). The Carathéodory extension theorem (1.3.10) then extends P to $\mathscr{F}$.

Let A_k, $k = 1, 2, \ldots$ be a sequence of measurable cylinders decreasing to $\varnothing$. If $P(A_k)$ does not approach 0, we have, for some $\varepsilon > 0$, $P(A_k) \geq \varepsilon > 0$ for all k. Suppose $A_k = B^{n_k}(v_k)$; by tacking on extra factors, we may assume that the numbers n_k and the sets v_k increase with k.

By 4.4.2, each Ω_{v_k} is a complete, separable metric space and $\mathscr{F}_{v_k} = \mathscr{B}(\Omega_{v_k})$. It follows from 4.3.8 that we can find a compact set $C^{n_k} \subset B^{n_k}$ such that $P_{v_k}(B^{n_k} - C^{n_k}) < \varepsilon/2^{k+1}$. Define $A_k' = C^{n_k}(v_k) \subset A_k$. Then $P(A_k - A_k') =$

$P_{v_k}(B^{n_k} - C^{n_k}) < \varepsilon/2^{k+1}$. In this way we approximate the given cylinders by cylinders with compact bases.

Now take

$$D_k = A_1' \cap A_2' \cap \cdots \cap A_k' \subset A_1 \cap A_2 \cap \cdots \cap A_k = A_k.$$

Then

$$P(A_k - D_k) = P\left(A_k \cap \bigcup_{i=1}^{k} A_i'^c\right) \leq \sum_{i=1}^{k} P(A_k \cap A_i'^c)$$

$$\leq \sum_{i=1}^{k} P(A_i - A_i') < \sum_{i=1}^{k} \varepsilon/2^{i+1} < \varepsilon/2.$$

Since $D_k \subset A_k' \subset A_k$, $P(A_k - D_k) = P(A_k) - P(D_k)$, consequently $P(D_k) > P(A_k) - \varepsilon/2$. In particular, D_k is not empty.

Now pick $x^k \in D_k$, $k = 1, 2, \ldots$. Say $A_1' = C^{n_1}(t_1, \ldots, t_{n_1}) = C^{n_1}(v_1)$ (note all $D_k \subset A_1'$). Consider the sequence

$$(x_{t_1}^1, \ldots, x_{t_{n_1}}^1), \quad (x_{t_1}^2, \ldots, x_{t_{n_1}}^2), \quad (x_{t_1}^3, \ldots, x_{t_{n1}}^3), \quad \ldots,$$

that is, $x_{v_1}^1, x_{v_1}^2, x_{v_1}^3, \ldots$.

Since the $x_{v_1}^n$ belong to C^{n_1}, a compact subset of Ω_{v_1}, we have a convergent subsequence $x_{v_1}^{r_{1n}}$ approaching some $x_{v_1} \in C^{n_1}$. If $A_2' = C^{n_2}(v_2)$ (so $D_k \subset A_2'$ for $k \geq 2$), consider the sequence $x_{v_2}^{r_{11}}, x_{v_2}^{r_{12}}, x_{v_2}^{r_{13}}, \ldots \in C^{n_2}$ (eventually), and extract a convergent subsequence $x_{v_2}^{r_{2n}} \to x_{v_2} \in C^{n_2}$.

Note that $(x_{v_2}^{r_{2n}})_{v_1} = x_{v_1}^{r_{2n}}$; as $n \to \infty$, the left side approaches $(x_{v_2})_{v_1}$, and since $\{r_{2n}\}$ is a subsequence of $\{r_{1n}\}$, the right side approaches x_{v_1}. Hence $(x_{v_2})_{v_1} = x_{v_1}$.

Continue in this fashion; at step i we have a subsequence

$$x_{v_i}^{r_{in}} \to x_{v_i} \in C^{n_i}, \qquad \text{and} \qquad (x_{v_i})_{v_j} = x_{v_j} \qquad \text{for} \quad j < i.$$

Pick any $\omega \in \prod_{t \in T} \Omega_t$ such that $\omega_{v_j} = x_{v_j}$ for all $j = 1, 2, \ldots$ (such a choice is possible since $(x_{v_i})_{v_j} = x_{v_j}$, $j < i$). Then $\omega_{v_j} \in C^{n_j}$ for each j; hence

$$\omega \in \bigcap_{j=1}^{\infty} A_j' \subset \bigcap_{j=1}^{\infty} A_j = \varnothing,$$

a contradiction. Thus P extends to a measure on $\mathscr{F}$, and by construction, $\pi_v(P) = P_v$ for all v.

Finally, if P and Q are two probability measures on $\mathscr{F}$ such that $\pi_v(P) = \pi_v(Q)$ for all finite $v \subset T$, then for any $B^n \in \mathscr{F}_v$,

$$P(B^n(v)) = [\pi_v(P)](B^n) = [\pi_v(Q)](B^n) = Q(B^n(v)).$$

Thus P and Q agree on measurable cylinders, and hence on $\mathscr{F}$ by the uniqueness part of the Carathéodory extension theorem. ∎

Problems

In Problems 1–7, $(\Omega_t, \mathscr{F}_t)$ is a measurable space for each $t \in T$, and $(\Omega, \mathscr{F}) = (\prod_{t \in T} \Omega_t, \prod_{t \in T} \mathscr{F}_t)$.

1. Let p_t be the projection map of Ω onto Ω_t, that is, $p_t(\omega) = \omega(t)$. If $(S, \mathscr{S})$ is a measurable space and $f: S \to \Omega$, show that f is measurable iff $p_t \circ f$ is measurable for all t.
2. If all $\Omega_t = R$, all $\mathscr{F}_t = \mathscr{B}(R)$, and T is a (nonempty) subset of R, how many sets are there in $\mathscr{F}$?
3. Show that if $B \in \mathscr{F}$, then membership in B is determined by a countable number of coordinates, that is, there is a countable set $T_0 \subset T$ and a set $B_0 \in \mathscr{F}_{T_0} = \prod_{t \in T_0} \mathscr{F}_t$ such that $\omega \in B$ iff $\omega_{T_0} \in B_0$, where $\omega_{T_0} = (\omega(t), t \in T_0)$.
4. If T is an open interval of reals and $\Omega_t = R$ (or $\overline{R}$), $\mathscr{F}_t = \mathscr{B}(R)$ (or $\mathscr{B}(\overline{R})$) for all t, use Problem 3 to show that the following sets do *not* belong to $\mathscr{F}$:
   (a) $\{\omega: \omega \text{ is continuous at } t_0\}$, where t_0 is a fixed element of T.
   (b) $\{\omega: \sup_{a \le t \le b} \omega(t) < c\}$, where $c \in R$ and $[a, b] \subset T$.
5. Assume each Ω_t is a compact metric space, with $\mathscr{F}_t$ the Baire (= Borel) subsets of Ω_t. Then by the Tychonoff theorem, Ω is compact in the topology of pointwise convergence. Show that $\mathscr{F}$ is the class of Baire sets of Ω, in other words, $\mathscr{A}(\prod_{t \in T} \Omega_t) = \prod_{t \in T} \mathscr{A}(\Omega_t)$, as follows:
   (a) If $A = \{\omega \in \Omega: \omega(t) \in F\}$, F a closed subset of Ω_t, show that $A = \{\omega \in \Omega: f(\omega) = 0\}$ for some $f \in C(\Omega)$. Conclude that $\mathscr{F} \subset \mathscr{A}(\Omega)$.
   (b) Use the Stone–Weierstrass theorem to show that the functions $f \in C(\Omega)$ depending on only one coordinate [that is, $f(\omega) = g_t(\omega(t))$ for some t, where $g_t \in C(\Omega_t)$] generate an algebra that is dense in $C(\Omega)$. Conclude that $\mathscr{A}(\Omega) \subset \mathscr{F}$.
6. Assume T is an open interval of reals, and $(\Omega_t, \mathscr{F}_t) = (\overline{R}, \mathscr{B}(\overline{R}))$ for all t; thus $\Omega = \overline{R}^T$, $\mathscr{F} = \mathscr{B}(\overline{R})^T$. Let $A = \{\omega \in \Omega: \omega \text{ is continuous at } t_0\}$, where t_0 is a fixed point of T.
   (a) Show that A is an $F_{\sigma\delta}$ (a countable intersection of F_σ sets); hence $A \in \mathscr{B}(\Omega)$.
   (b) Show that $A \notin \mathscr{A}(\Omega)$, so that we have an example of a compact Hausdorff space in which the Baire and Borel sets do not coincide.
7. (Alternative proof of the Kolmogorov extension theorem†) Assume the hypothesis of 4.4.3, with the stronger condition that each Ω_t is a compact metric space. Put the topology of pointwise convergence on Ω.
   (a) Let A be the set of functions in $C(\Omega)$ that depend on only a finite number of coordinates; that is, there is a finite set $v \subset T$ and a con-

† Nelson. E., *Ann. of Math.* **69**, 630 (1959).

tinuous $g: \Omega_v \to R$ such that $f(\omega) = g(\omega_v)$, $\omega \in \Omega$. Use the Stone–Weierstrass theorem to show that A is dense in $C(\Omega)$.

(b) If $f \in A$, define $E(f) = \int_\Omega g \, dP_v$. Show that E is well-defined and extends uniquely to a positive linear functional on $C(\Omega)$. Since $E(1) = 1$, the Riesz representation theorem 4.3.9 yields a unique probability measure P on $\mathscr{A}(\Omega)$ $(= \prod_{t \in T} \mathscr{F}_t$ by Problem 5) such that $E(f) = \int_\Omega f \, dP$ for all $f \in C(\Omega)$.

(c) Let v be a fixed finite subset of T, and let H be the collection of all functions $g: (\Omega_v, \mathscr{F}_v) \to (R, \mathscr{B}(R))$ such that if $f(\omega) = g(\omega_v)$, $\omega \in \Omega$, then $\int_\Omega f \, dP = \int_\Omega g \, dP_v$. Show that $I_A \in H$ for each open set $A \subset \Omega_v$, and then show that H contains all bounded Borel measurable functions on $(\Omega_v, \mathscr{F}_v)$. Conclude that $\pi_v(P) = P_v$. (Uniqueness of P is proved as in 4.4.3.)

8. The metric space Ω is said to be *Borel equivalent* to a subset of the metric space Ω' iff there is a one-to-one map $f: \Omega \to \Omega'$ such that $E = f(\Omega) \in \mathscr{B}(\Omega')$ and f and f^{-1} are Borel measurable. [Measurability of f^{-1} means that $f^{-1}: (E, \mathscr{B}(E)) \to (\Omega, \mathscr{B}(\Omega))$.]

(a) Let Ω be a complete separable metric space with metric d. [It may be assumed without loss of generality that $d(x, y) \le 1$ for all x, y, since the metric $d' = d/(1 + d)$ induces the same topology as d and is also complete.] Denote by $[0, 1]^\infty$ the space of all sequences of real numbers with components in $[0, 1]$, with the topology of pointwise convergence. (This space is metrizable; explicitly, we may take the metric as

$$d_0(x, y) = \sum_{n=1}^{\infty} \frac{1}{2^n} \frac{|x_n - y_n|}{1 + |x_n - y_n|}.)$$

If $D = \{\omega_1, \omega_2, \ldots\}$ is a countable dense subset of Ω, define $f: \Omega \to [0, 1]^\infty$ by $f(\omega) = \{d(\omega, \omega_n), n = 1, 2, \ldots\}$. Show that f is continuous and one-to-one, with a continuous inverse.

(b) Show that $f(\Omega)$ is a Borel subset of $[0, 1]^\infty$; thus Ω is Borel equivalent to a subset of $[0, 1]^\infty$.

(c) Let $S = \{0, 1\}$, and let $\mathscr{S}$ consist of all subsets of S. Show that there is a map $g: ([0, 1), \mathscr{B}[0, 1)) \to (S^\infty, \mathscr{S}^\infty)$ such that g is one-to-one, $g[0, 1) \in \mathscr{S}^\infty$, and both g and g^{-1} are measurable.

(d) Show that $([0, 1], \mathscr{B}[0, 1])$ and $(S^\infty, \mathscr{S}^\infty)$ are equivalent, that is, there is a map $h: [0, 1] \to S^\infty$ such that h is one-to-one onto, and h and h^{-1} are measurable.

(e) If $(\Omega_n, \mathscr{F}_n)$ is equivalent to $(S_n, \mathscr{S}_n)$, with associated map h_n, $n = 1, 2, \ldots$, show that $(\prod_n \Omega_n, \prod_n \mathscr{F}_n)$ is equivalent to $(\prod_n S_n, \prod_n \mathscr{S}_n)$. Thus by (d), $([0, 1]^\infty, (\mathscr{B}[0, 1])^\infty)$ is equivalent to $(S^\infty, \mathscr{S}^\infty)$.

Now by 4.4.2, $(\mathscr{B}[0, 1])^\infty$ is the minimal σ-field over the open sets of $[0, 1]^\infty$, that is, $(\mathscr{B}[0, 1])^\infty$ is the class of Borel sets $\mathscr{B}([0, 1]^\infty)$. It follows from these results that if Ω is a complete, separable metric space, Ω is Borel equivalent to a (Borel) subset of $[0, 1]$.

4.5 Weak Convergence of Measures

By the Riesz representation theorem, a continuous linear functional on $C(\Omega)$, where Ω is compact Hausdorff, may be identified with a regular finite signed measure on $\mathscr{B}(\Omega)$. Thus if $\{\mu_n\}$ is a sequence of such measures, weak* convergence of the sequence to the measure μ means that $\int_\Omega f\,d\mu_n \to \int_\Omega f\,d\mu$ for all $f \in C(\Omega)$. In this section we investigate this type of convergence in a somewhat different context; Ω will be a metric space, not necessarily compact, and all measures will be nonnegative. The results form the starting point for the study of the central limit theorem of probability.

4.5.1 ***Theorem.*** Let $\mu, \mu_1, \mu_2, \ldots$ be finite measures on the Borel sets of a metric space Ω. The following conditions are equivalent:

(a) $\int_\Omega f\,d\mu_n \to \int_\Omega f\,d\mu$ for all bounded continuous $f\colon \Omega \to R$.

(b) $\liminf_{n\to\infty} \int_\Omega f\,d\mu_n \geq \int_\Omega f\,d\mu$ for all bounded lower semicontinuous $f\colon \Omega \to R$.

(b′) $\limsup_{n\to\infty} \int_\Omega f\,d\mu_n \leq \int_\Omega f\,d\mu$ for all bounded upper semicontinuous $f\colon \Omega \to R$.

(c) $\int_\Omega f\,d\mu_n \to \int_\Omega f\,d\mu$ for all bounded $f\colon (\Omega, \mathscr{B}(\Omega)) \to (R, \mathscr{B}(R))$ such that f is continuous a.e. $[\mu]$.

(d) $\liminf_{n\to\infty} \mu_n(A) \geq \mu(A)$ for every open set $A \subset \Omega$, and $\mu_n(\Omega) \to \mu(\Omega)$.

(d′) $\limsup_{n\to\infty} \mu_n(A) \leq \mu(A)$ for every closed set $A \subset \Omega$, and $\mu_n(\Omega) \to \mu(\Omega)$.

(e) $\mu_n(A) \to \mu(A)$ for every $A \in \mathscr{B}(\Omega)$ such that $\mu(\partial A) = 0$ (∂A denotes the boundary of A).

PROOF. (a) *implies* (b): If $g \leq f$ and g is bounded continuous,

$$\liminf_{n\to\infty} \int_\Omega f\,d\mu_n \geq \liminf_{n\to\infty} \int_\Omega g\,d\mu_n = \int_\Omega g\,d\mu \qquad \text{by (a).}$$

But since f is lower semicontinuous (LSC), it is the limit of a sequence of continuous functions, and if $|f| \leq M$, all functions in the sequence can also be taken less than or equal to M in absolute value. (See the appendix on general topology, Section A6, for the basic properties of semicontinuous functions.) Thus if we take the sup over g in the above equation, we obtain (b).

(b) *is equivalent to* (b′): Note that f is LSC iff $-f$ is upper semicontinuous (USC).

(b) *implies* (c): Let $\underline{f}$ be the lower envelope of f (the sup of all LSC functions g such that $g \le f$) and $\bar{f}$ the upper envelope (the inf of all USC functions g such that $g \ge f$). Since $\underline{f}(x) = \liminf_{y \to x} f(y)$ and $\bar{f}(x) = \limsup_{y \to x} f(y)$, continuity of f at x implies $\underline{f}(x) = f(x) = \bar{f}(x)$. Furthermore, $\underline{f}$ is LSC and $\bar{f}$ is USC. Thus if f is bounded and continuous a.e. $[\mu]$,

$$
\begin{aligned}
\int_\Omega f\,d\mu = \int_\Omega \underline{f}\,d\mu &\le \liminf_{n\to\infty} \int_\Omega \underline{f}\,d\mu_n &&\text{by (b)}\\
&\le \liminf_{n\to\infty} \int_\Omega f\,d\mu_n &&\text{since } \underline{f} \le f\\
&\le \limsup_{n\to\infty} \int_\Omega f\,d\mu_n\\
&\le \limsup_{n\to\infty} \int_\Omega \bar{f}\,d\mu_n &&\text{since } f \le \bar{f}\\
&\le \int_\Omega \bar{f}\,d\mu &&\text{by (b}')\\
&= \int_\Omega f\,d\mu, &&\text{proving (c).}
\end{aligned}
$$

(c) *implies* (d): Clearly (c) implies (a), which in turn implies (b). If A is open, then I_A is LSC, so by (b), $\liminf_{n\to\infty} \mu_n(A) \ge \mu(A)$. Now $I_\Omega \equiv 1$, so $\mu_n(\Omega) \to \mu(\Omega)$ by (c).

(d) *is equivalent to* (d′): Take complements.

(d) *implies* (e): Let A^0 be the interior of A, $\bar{A}$ the closure of A. Then

$$
\begin{aligned}
\limsup_{n\to\infty} \mu_n(A) \le \limsup_{n\to\infty} \mu_n(\bar{A}) &\le \mu(\bar{A}) &&\text{by (d}')\\
&= \mu(A) &&\text{by hypothesis.}
\end{aligned}
$$

Also, using (d),

$$\liminf_{n\to\infty} \mu_n(A) \ge \liminf_{n\to\infty} \mu_n(A^0) \ge \mu(A^0) = \mu(A).$$

(e) *implies* (a): Let f be a bounded continuous function on Ω. If $|f| < M$, let $A = \{c \in R\colon \mu(f^{-1}\{c\}) \ne 0\}$; A is countable since the $f^{-1}\{c\}$ are disjoint and μ is finite. Construct a partition of $[-M, M]$, say $-M = t_0 < t_1 < \cdots < t_j = M$, with $t_i \notin A$, $i = 0, 1, \ldots, j$ (M may be increased if necessary). If $B_i = \{x\colon t_i \le f(x) < t_{i+1}\}$, $i = 0, 1, \ldots, j-1$, it follows from (e) that

$$\sum_{i=0}^{j-1} t_i\,\mu_n(B_i) \to \sum_{i=0}^{j-1} t_i\,\mu(B_i).$$

[Since $f^{-1}(t_i, t_{i+1})$ is open, $\partial f^{-1}[t_i, t_{i+1}) \subset f^{-1}\{t_i, t_{i+1}\}$, and $\mu f^{-1}\{t_i, t_{i+1}\} = 0$ since $t_i, t_{i+1} \notin A$.] Now

$$\begin{aligned}\left|\int_\Omega f\,d\mu_n - \int_\Omega f\,d\mu\right| \le \left|\int_\Omega f\,d\mu_n - \sum_{i=0}^{j-1} t_i\,\mu_n(B_i)\right| \\ + \left|\sum_{i=0}^{j-1} t_i\,\mu_n(B_i) - \sum_{i=0}^{j-1} t_i\,\mu(B_i)\right| \\ + \left|\sum_{i=0}^{j-1} t_i\,\mu(B_i) - \int_\Omega f\,d\mu\right|.\end{aligned}$$

The first term on the right may be written as

$$\left|\sum_{i=0}^{j-1} \int_{B_i} (f(x) - t_i)\,d\mu_n(x)\right|$$

and this is bounded by $\max_i(t_{i+1} - t_i)\mu_n(\Omega)$, which can be made arbitrarily small by choice of the partition since $\mu_n(\Omega) \to \mu(\Omega) < \infty$ by (e). The third term on the right is bounded by $\max_i(t_{i+1} - t_i)\mu(\Omega)$, which can also be made arbitrarily small. The second term approaches 0 as $n \to \infty$, proving (a). ∎

4.5.2 ***Comments.*** Another condition equivalent to those of 4.5.1 is that $\int_\Omega f\,d\mu_n \to \int_\Omega f\,d\mu$ for all bounded *uniformly* continuous $f: \Omega \to R$ (see Problem 1).

The proof of 4.5.1 works equally well if the sequence $\{\mu_n\}$ is replaced by a net.

The convergence described in 4.5.1 is sometimes called *weak* or *vague* convergence of measures. We shall write $\mu_n \xrightarrow{w} \mu$.

If the measures μ_n and μ are defined on $\mathscr{B}(R)$, there are corresponding distribution functions F_n and F on R. We may relate convergence of measures to convergence of distribution functions.

4.5.3 ***Definition.*** A *continuity point* of a distribution function F on R is a point $x \in R$ such that F is continuous at x, or $\pm\infty$ (thus by convention, ∞ and $-\infty$ are continuity points).

4.5.4 ***Theorem.*** Let $\mu, \mu_1, \mu_2, \ldots$ be finite measures on $\mathscr{B}(R)$, with corresponding distribution functions $F, F_1, F_2, \ldots$. The following are equivalent:

(a) $\mu_n \xrightarrow{w} \mu$.

(b) $F_n(a, b] \to F(a, b]$ at all continuity points a, b of F, where $F(a, b] = F(b) - F(a)$, $F(\infty) = \lim_{x\to\infty} F(x)$, $F(-\infty) = \lim_{x\to-\infty} F(x)$.

If all distribution functions are 0 at $-\infty$, condition (b) is equivalent to the statement that $F_n(x) \to F(x)$ at all points $x \in R$ at which F is continuous, and $F_n(\infty) \to F(\infty)$.

PROOF. (a) *implies* (b): If a and b are continuity points of F in R, then $(a, b]$ is a Borel set whose boundary has μ-measure 0. By 4.5.1(e), $\mu_n(a, b] \to \mu(a, b]$, that is, $F_n(a, b] \to F(a, b]$. If $a = -\infty$, the argument is the same, and if $b = \infty$, then (a, ∞) is a Borel set whose boundary has μ-measure 0, and the proof proceeds as before.

(b) *implies* (a): Let A be an open subset of R; write A as the disjoint union of open intervals $I_1, I_2, \ldots$. Then

$$\begin{aligned}\liminf_{n\to\infty} \mu_n(A) &= \liminf_{n\to\infty} \sum_k \mu_n(I_k) \\ &\geq \sum_k \liminf_{n\to\infty} \mu_n(I_k) \qquad \text{by Fatou's lemma.}\end{aligned}$$

Let $\varepsilon > 0$ be given. For each k, let I_k' be a right-semiclosed subinterval of I_k such that the endpoints of I_k' are continuity points of F, and $\mu(I_k') \geq \mu(I_k) - \varepsilon 2^{-k}$; the I_k' can be chosen since F has only countably many discontinuities. Then

$$\liminf_{n\to\infty} \mu_n(I_k) \geq \liminf_{n\to\infty} \mu_n(I_k') = \mu(I_k') \qquad \text{by (b).}$$

Thus

$$\liminf_{n\to\infty} \mu_n(A) \geq \sum_k \mu(I_k') \geq \sum_k \mu(I_k) - \varepsilon = \mu(A) - \varepsilon.$$

Since ε is arbitrary, we have $\mu_n \xrightarrow{w} \mu$ by 4.5.1(d). ∎

Condition (b) of 4.5.4 is sometimes called weak convergence of the sequence $\{F_n\}$ to F, written $F_n \xrightarrow{w} F$.

Problems

1. (a) If F is a closed subset of the metric space Ω, show that I_F is the limit of a decreasing sequence of uniformly continuous functions f_n, with $0 \leq f_n \leq 1$ for all n.
   (b) Show that in 4.5.1, $\mu_n \xrightarrow{w} \mu$ iff $\int_\Omega f\, d\mu_n \to \int_\Omega f\, d\mu$ for all bounded uniformly continuous $f\colon \Omega \to R$.
2. Show that in 4.5.1, $\mu_n \xrightarrow{w} \mu$ iff $\mu_n(A) \to \mu(A)$ for all *open* sets A such that $\mu(\partial A) = 0$.
3. Let Ω be a locally compact metric space. If $A \subset \Omega$, A is said to be *bounded* iff $A \subset K$ for some compact set K.

If $\mu, \mu_1, \mu_2, \ldots$ are measures on $\mathscr{B}(\Omega)$ that are finite on bounded Borel sets, we say that $\mu_n \xrightarrow{w} \mu$ iff $\int_\Omega f\,d\mu \to \int_\Omega f\,d\mu$ for all continuous functions $f: \Omega \to R$ with compact support. (The support of f is defined by $\operatorname{supp} f = \overline{\{x: f(x) \neq 0\}}$; note that f has compact support iff f vanishes outside a compact set.)

Show that the following conditions are equivalent:

(a) $\mu_n \xrightarrow{w} \mu$.
(b) $\mu_n(A) \to \mu(A)$ for all bounded Borel sets A with $\mu(\partial A) = 0$.
(c) $\int_\Omega f\,d\mu_n \to \int_\Omega f\,d\mu$ for all bounded $f: \Omega \to R$ such that f has compact support and is continuous a.e. $[\mu]$.

4.6 References

The approach to the Daniell integral and the Riesz representation theorem given in Sections 4.2 and 4.3 is based on Neveu (1965). If the hypothesis that L contains the constant functions is dropped in the development of the Daniell integral, the situation becomes more complicated. However, a representation theorem somewhat similar to 4.2.9 may be proved under the assumption that $f \in L$ implies $1 \wedge f \in L$. For a detailed proof of this result (known as *Stone's theorem*), see Royden (1968, Chapter 13).

There are other versions of the Riesz representation theorem appropriate in a locally compact Hausdorff space Ω. If E is a positive linear functional on the set $C_c(\Omega)$ of continuous real-valued functions on Ω with compact support, there is a unique regular measure μ on the Borel sets such that $E(f) = \int_\Omega f\,d\mu$ for each f in $C_c(\Omega)$. Here, regularity means that μ is finite on compact sets, $\mu(A) = \inf\{\mu(V): V \supset A,\ V \text{ open}\}$ for all Borel sets A, and $\mu(A) = \sup\{\mu(K): K \subset A,\ K \text{ compact}\}$ for all Borel sets A which are either open or of finite measure. Also, if E is a continuous linear functional on the Banach space $C_0(\Omega)$ of continuous real-valued functions on Ω that vanish at ∞ (f vanishes at ∞ iff, given $\varepsilon > 0$, there is a compact set K such that $|f| < \varepsilon$ off K), there is a unique regular finite signed measure on the Borel sets such that $E(f) = \int_\Omega f\,d\mu$ for all $f \in C_0(\Omega)$; furthermore, $\|E\| = |\mu|(\Omega)$. As in the compact case, the result may be extended to complex-valued functions and complex measures, as in Problem 6, Section 4.3. Proofs are given in Rudin (1966).

In probability theory, there is usually no reason to replace compact by locally compact spaces in order to take advantage of a more general version of the Riesz representation theorem. For if we have a random experiment whose outcomes are represented by points in a locally compact space, we simply compactify the space.

For an account of the theory of weak convergence of measures, with applications to probability, see Billingsley (1968) and Parthasarathy (1967).

Appendix on General Topology

A1 Introduction

The reader is assumed to be familiar with elementary set theory, including basic properties of ordinal and cardinal numbers [see, for example, Halmos (1960)]. Also, an undergraduate course in point-set topology is assumed; Simmons (1963) is a suitable text for such a course. In this appendix, we shall concentrate on aspects of general topology that are useful in functional analysis and probability. A good reference for collateral reading is Dugundji (1966).

Before proceeding, we mention one result, which, although usually covered in a first course in topology, deserves to be stated explicitly because of its fundamental role in the construction of topological vector spaces (see 3.5.1). Throughout the appendix, a *neighborhood* of a point x is an open set containing x, an *overneighborhood* of x is an overset of a neighborhood of x.

A1.1 ***Theorem.*** Let Ω be a set, and suppose that for each $x \in \Omega$, we are given a nonempty collection $\mathscr{V}(x)$ of subsets of Ω satisfying the following:

(a) $x \in$ each $U \in \mathscr{V}(x)$.
(b) If $U_1, U_2 \in \mathscr{V}(x)$, then $U_1 \cap U_2 \in \mathscr{V}(x)$.
(c) If $U \in \mathscr{V}(x)$ and $U \subset V$, then $V \in \mathscr{V}(x)$.

(d) If $U \in \mathscr{V}(x)$, there is a set $V \in \mathscr{V}(x)$ such that $V \subset U$ and $U \in \mathscr{V}(y)$ for each $y \in V$.

Then there is a unique topology on Ω such that for each x, $\mathscr{V}(x)$ is the system of overneighborhoods of x.

PROOF. If such a topology exists, a set U will be open iff U is an overneighborhood of each of its points, that is, $U \in \mathscr{V}(x)$ for each $x \in U$. Thus it suffices to show that $\mathscr{T} = \{U \subset \Omega\colon U \in \mathscr{V}(x) \text{ for each } x \in U\}$ is a topology, and for each x, the overneighborhood system $\mathscr{N}(x)$ coincides with $\mathscr{V}(x)$. If $U, V \in \mathscr{T}$, then by (b), $U \cap V \in \mathscr{V}(x)$ for each $x \in U \cap V$, so $U \cap V \in \mathscr{T}$. If $U_i \in \mathscr{T}$ for each $i \in I$, then (c) implies that $\bigcup\{U_i\colon i \in I\} \in \mathscr{T}$; (c) yields $\Omega \in \mathscr{T}$ as well. Since $\varnothing \in \mathscr{T}$ trivially, $\mathscr{T}$ is a topology.

If $U \in \mathscr{N}(x)$, there is a set $V \in \mathscr{T}$ with $x \in V \subset U$. But then $U \in \mathscr{V}(x)$ by (c). Conversely, if $U \in \mathscr{V}(x)$, let $W = \{x' \in U\colon U \in \mathscr{V}(x')\}$. If $x' \in W$, then by (d), there is a set $V \in \mathscr{V}(x')$ with $V \subset U$ and $U \in \mathscr{V}(y)$ for each $y \in V$. But then $V \subset W$, so by (c), $W \in \mathscr{V}(x')$; consequently, $W \in \mathscr{T}$ by definition of $\mathscr{T}$, and furthermore $x \in W$ by (a). Thus if $U \in \mathscr{V}(x)$, there is a set $W \in \mathscr{T}$ with $x \in W \subset U$; hence $U \in \mathscr{N}(x)$. ∎

For the remainder of the appendix, $\mathscr{U}(x)$ will always stand for the collection of neighborhoods of x.

A2 Convergence

If Ω is a metric space, the topology of Ω can be described entirely in terms of convergence of sequences. For example, a subset A of Ω is closed iff whenever $\{x_n, n = 1, 2, \ldots\}$ is a sequence of points in A and $x_n \to x$, we have $x \in A$. Also, $x \in \overline{A}$, the closure of A, iff there is a sequence of points in A converging to x. This result does not generalize to arbitrary topological spaces.

A2.1 ***Example.*** Let α be the first uncountable ordinal, and let Ω be the set of all ordinals less than or equal to α (recall that for ordinals, $a < b$ means $a \in b$). Put the order topology on Ω; this topology has as a base the sets $\Omega \cap (a, b) = \{x \in \Omega\colon a < x < b\}$ where a and b are arbitrary ordinals.

We show that α belongs to the closure of $\Omega - \{\alpha\}$, but no sequence in $\Omega - \{\alpha\}$ converges to α. If U is a neighborhood of α, then for some a, b, we have $\alpha \in \Omega \cap (a, b) \subset U$. Now $a < \alpha < b$; hence a is countable, and therefore so is $a + 1$. Thus $a + 1 \in U \cap (\Omega - \{\alpha\})$, proving that $\alpha \in \overline{\Omega - \{\alpha\}}$. But if $x_n \in \Omega - \{\alpha\}$, $n = 1, 2, \ldots$, then each x_n is countable; hence so is $c = \sup x_n$.

Since $c < \alpha$, $V = (c, \alpha]$ is a neighborhood of α and x_n is never in V, so that the sequence cannot possibly converge to α.

In order to describe an arbitrary topology in terms of convergence, we must consider objects more general than sequences. A sequence is a function on the positive integers; the desired generalization, called a "net" or "generalized sequence," is a function on a directed set.

A2.2 *Definitions.* A *directed set* is a set D on which there is defined a preordering (a reflexive and transitive relation), denoted by $\leq$, with the property that whenever $a, b \in D$, there is a $c \in D$ with $a \leq c$ and $b \leq c$. A *net* in a topological space Ω is a function from a directed set D into Ω. A net will be denoted by $\{x_n, n \in D\}$ or simply by $\{x_n\}$. The net $\{x_n\}$ is said to *converge* to the point x iff for every neighborhood U of x, the net is *eventually* in U, that is, there is an $n_0 \in D$ such that $x_n \in U$ for all $n \in D$ such that $n \geq n_0$.

The basic relations between convergence and topology in metric spaces can now be generalized to arbitrary topological spaces.

A2.3 *Theorem.* Let A be a subset of the topological space Ω.

(a) A point $x \in \Omega$ belongs to $\bar{A}$ iff there is a net $\{x_n\}$ in A such that $x_n \to x$.

(b) A is closed iff for every net $\{x_n\}$ in A such that $x_n \to x$, we have $x \in A$.

(c) A point $x \in \Omega$ is a cluster point of A (that is, every neighborhood of x contains a point of A other than x) iff there is a net in $A - \{x\}$ converging to x.

PROOF. (a) If $x_n \in A$ and $x_n \to x$, let $U \in \mathscr{U}(x)$. Then x_n is eventually in U, in particular $U \cap A \neq \varnothing$, and thus $x \in \bar{A}$. Conversely, if $x \in \bar{A}$, then for each $U \in \mathscr{U}(x)$, choose $x_U \in U \cap A$. Then $\{x_U\}$ becomes a net in A (with $U \leq V$ iff $U \supset V$) and $x_U \to x$.

(b) Suppose that A is closed. If $x_n \in A$, $x_n \to x$, then by (a), $x \in \bar{A}$; thus by hypothesis, $x \in A$. Conversely, if A is not closed, let $x \in \bar{A} - A$. By (a) we can find $x_n \in A$, $x_n \to x$. Since $x \notin A$, the result follows.

(c) If $x_n \in A - \{x\}$, $x_n \to x$, then if $U \in \mathscr{U}(x)$, x_n is eventually in U, hence $U \cap (A - \{x\}) \neq \varnothing$, and thus x is a cluster point of A. Conversely, if x is a cluster point of A, then for each $U \in \mathscr{U}(x)$, choose $x_U \in U \cap (A - \{x\})$; the x_U form a net in $A - \{x\}$ converging to x. ∎

A2.4 *Comments.* The reason that sequences are not adequate in A2.3 is that in choosing a point x_U in each neighborhood U of x, we are in general forced to make uncountably many choices. This would not be necessary if Ω were *first countable.* (A first countable space is one with a countable base of neighborhoods at each point; that, is, if $x \in \Omega$, there are countably many neighborhoods $V_1, V_2, \ldots$ of x such that for each neighborhood U of x, we have $V_n \subset U$ for some n; by replacing V_n by $\bigcap_{i=1}^{n} V_i$, we may assume without loss of generality that $V_{n+1} \subset V_n$ for all n.) In A2.3, the proof will go through if we simply choose $x_n \in V_n$ for $n = 1, 2, \ldots$. Thus *in a first countable space, sequences are adequate to describe the topology*; in other words, A2.3 holds with "net" replaced by "sequence."

A criterion for openness in terms of nets may also be given: The set V is open iff for every net $\{x_n\}$ in Ω such that $x_n \to x \in V$, we have $x_n \in V$ eventually. [The "only if" part follows from the definition of convergence; for the "if" part, apply A2.3(b) to V^c.]

A2.5 ***Definitions.*** Let $\{x_n, n \in D\}$ be a net, and suppose that we are given a directed set E and a map $k \to n_k$ of E into D. Then $\{x_{n_k}, k \in E\}$ is called a *subnet* of $\{x_n, n \in D\}$, provided that "as k becomes large, so does n_k"; that is, given $n_0 \in D$, there is a $k_0 \in E$ such that $k \geq k_0$ implies $n_k \geq n_0$. If $E = D =$ the positive integers, we obtain the usual notion of a *subsequence.*

If $\{x_n, n \in D\}$ is a net in the topological space Ω, the point $x \in \Omega$ is called an *accumulation point* of the net iff for each neighborhood U of x, x_n is *frequently* in U; in other words, given $n \in D$, there is an $m \in D$ with $m \geq n$ and $x_m \in U$.

A2.6 ***Theorem.*** Let $\{x_n, n \in D\}$ be a net in the topological space Ω. If $x \in \Omega$, x is an accumulation point of $\{x_n\}$ iff there is a subnet $\{x_{n_k}, k \in E\}$ converging to x.

PROOF. If $x_{n_k} \to x$, $U \in \mathscr{U}(x)$ and $n \in D$, then for some $k_0 \in E$, we have $x_{n_k} \in U$ for $k \geq k_0$. But by definition of subnet, there is a $k_1 \in E$ such that $k \geq k_1$ implies $n_k \geq n$. Thus if $k \geq k_0$ and $k \geq k_1$, we have $n_k \geq n$ and $x_{n_k} \in U$, proving that x is an accumulation point.

Conversely, let x be an accumulation point of $\{x_n, n \in D\}$. Let E be the collection of pairs (n, U), where $n \in D$, U is a neighborhood of x, and $x_n \in U$. Direct E by setting $(n, U) \leq (m, V)$ iff $n \leq m$ and $U \supset V$. If $k = (m, V) \in E$ let $n_k = m$. Given n and U, then if $k = (m, V) \geq (n, U)$, we have $n_k = m \geq n$, so that $\{x_{n_k}, k \in E\}$ is a subnet of $\{x_n, n \in D\}$. Now if U is a neighborhood of x, then $x_n \in U$ for some $n \in D$. If $k = (m, V) \geq (n, U)$, then $x_{n_k} = x_m \in V \subset U$; therefore $x_{n_k} \to x$. ∎

For some purposes, it is more convenient to specify convergence in a topological space by means of filters rather than nets. If $\{x_n, n \in D\}$ is a net in Ω, and $a \in D$, let $T_a = \{n \in D: n \geq a\}$, and let $x(T_a)$ be the set of all x_n, $n \geq a$. The $x(T_a)$, $a \in D$, are called the *tails* of the net. The collection $\mathscr{A}$ of tails is an example of a filterbase, which we now define.

A2.7 ***Definitions and Comments.*** Let $\mathscr{A}$ be a nonempty family of subsets of a set Ω. Then $\mathscr{A}$ is called a *filterbase* in Ω iff

(a) each $U \in \mathscr{A}$ is nonempty;
(b) if $U, V \in \mathscr{A}$, there is a $W \in \mathscr{A}$ with $W \subset U \cap V$.

If $\mathscr{F}$ is a nonempty family of subsets of Ω such that

(c) each $U \in \mathscr{F}$ is nonempty,
(d) if $U, V \in \mathscr{F}$, then $U \cap V \in \mathscr{F}$, and
(e) if $U \in \mathscr{F}$ and $U \subset V$, then $V \in \mathscr{F}$,

then $\mathscr{F}$ is called a *filter* in Ω. If $\mathscr{A}$ is a filterbase, then $\mathscr{F} = \{U \subset \Omega: U \supset V$ for some $V \in \mathscr{A}\}$ is a filter, called the *filter generated by* $\mathscr{A}$. If $\mathscr{A}$ is the collection of neighborhoods of a given point x in a topological space, $\mathscr{A}$ is a filterbase, and the filter generated by $\mathscr{A}$ is the system of overneighborhoods of x.

A filterbase $\mathscr{A}$ in a topological space Ω is said to *converge* to the point x (notation $\mathscr{A} \to x$) iff for each $U \in \mathscr{U}(x)$ there is a set $A \in \mathscr{A}$ such that $A \subset U$. A filter $\mathscr{F}$ in Ω is said to converge to x iff each $U \in \mathscr{U}(x)$ belongs to $\mathscr{F}$. Thus a filterbase $\mathscr{A}$ converges to x iff the filter generated by $\mathscr{A}$ converges to x.

If $\{x_n, n \in D\}$ is a net, then $x_n \to x$ iff for each $U \in \mathscr{U}(x)$ we have $x(T_a) \subset U$ for some $a \in D$, that is, $x_n \to x$ iff the associated filterbase converges to x.

Convergence in a topological space may be described using filterbases instead of nets. The analog of Theorem A2.3 is the following:

A2.8 ***Theorem.*** Let B be a subset of the topological space Ω.

(a) A point $x \in \Omega$ belongs to $\bar{B}$ iff there is a filterbase $\mathscr{A}$ in B such that $\mathscr{A} \to x$.

(b) B is closed iff for every filterbase $\mathscr{A}$ in B such that $\mathscr{A} \to x$, we have $x \in B$.

(c) A point $x \in \Omega$ is a cluster point of B iff there is a filterbase in $B - \{x\}$ converging to x.

PROOF. (a) If $\mathscr{A} \to x$ and $U \in \mathscr{U}(x)$, then $A \subset U$ for some $A \in \mathscr{A}$, in particular, $U \cap B \neq \varnothing$; thus $x \in \bar{B}$. Conversely, if $x \in \bar{B}$, then $U \cap B \neq \varnothing$ for each $U \in \mathscr{U}(x)$. Let $\mathscr{A}$ be the collection of sets $U \cap B$, $U \in \mathscr{U}(x)$. Then $\mathscr{A}$ is a filterbase in B and $\mathscr{A} \to x$.

(b) If B is closed, $\mathscr{A}$ is a filterbase in B, and $\mathscr{A} \to x$, then $x \in \bar{B}$ by (a), hence $x \in B$ by hypothesis. Conversely, if B is not closed and $x \in \bar{B} - B$, by (a) there is a filterbase $\mathscr{A}$ in B with $\mathscr{A} \to x$. Since $x \notin B$, the result follows.

(c) If there is such a filterbase $\mathscr{A}$ and $U \in \mathscr{U}(x)$, then $U \supset A$ for some $A \in \mathscr{A}$; in particular, $U \cap (B - \{x\}) \neq \varnothing$, so x is a cluster point of B. Conversely, if x is a cluster point of B, let $\mathscr{A}$ consist of all sets $U \cap (B - \{x\})$, $U \in \mathscr{U}(x)$. Then $\mathscr{A}$ is a filterbase in $B - \{x\}$ and $\mathscr{A} \to x$. ▌

If Ω is first countable, the filterbases in A2.8 may be formed using a countable system of neighborhoods of x, so that in a first countable space, the topology may be described by filterbases containing countably many sets.

A2.9 ***Definitions.*** The filterbase $\mathscr{B}$ is said to be *subordinate* to the filterbase $\mathscr{A}$ iff for each $A \in \mathscr{A}$ there is a $B \in \mathscr{B}$ with $B \subset A$; this means that the filter generated by $\mathscr{A}$ is included in the filter generated by $\mathscr{B}$.

If $\{x_{n_k}, k \in E\}$ is a subnet of $\{x_n, n \in D\}$, the filterbase determined by the subnet is subordinate to the filterbase determined by the original net. For if $n_0 \in D$, there is a $k_0 \in E$ such that $k \geq k_0$ implies $n_k \geq n_0$. Therefore

$$\{x_{n_k} \colon k \in E, \quad k \geq k_0\} \subset \{x_n \colon n \in D, \quad n \geq n_0\}.$$

If $\mathscr{A}$ is a filterbase in the topological space Ω, the point $x \in \Omega$ is called an *accumulation point* of $\mathscr{A}$ iff $U \cap A \neq \varnothing$ for all $U \in \mathscr{U}(x)$ and all $A \in \mathscr{A}$, in other words, $x \in \bar{A}$ for all $A \in \mathscr{A}$. We may now prove the analog of Theorem A2.6.

A2.10 ***Theorem.*** Let $\mathscr{A}$ be a filterbase in the topological space Ω. If $x \in \Omega$, x is an accumulation point of $\mathscr{A}$ iff there is a filterbase $\mathscr{B}$ subordinate to $\mathscr{A}$ with $\mathscr{B} \to x$; in other words, some overfilter of $\mathscr{A}$ converges to x.

PROOF. If $\mathscr{B}$ is subordinate to $\mathscr{A}$ and $\mathscr{B} \to x$, let $U \in \mathscr{U}(x)$, $A \in \mathscr{A}$. Then $U \supset B$ and $A \supset B_1$ for some $B, B_1 \in \mathscr{B}$; hence $U \cap A \supset B \cap B_1$, which is nonempty since $\mathscr{B}$ is a filterbase. Therefore $x \in \bar{A}$. Conversely, if x is an accumulation point of $\mathscr{A}$, let $\mathscr{B}$ consist of all sets $U \cap A$, $U \in \mathscr{U}(x)$, $A \in \mathscr{A}$. Then $\mathscr{A} \subset \mathscr{B}$ (take $U = \Omega$), hence $\mathscr{B}$ is subordinate to $\mathscr{A}$; since $\mathscr{B} \to x$, the result follows. ▌

A2.11 ***Definition.*** An *ultrafilter* is a maximal filter, that is, a filter included in no properly larger filter. (By Zorn's lemma, every filter is included in an ultrafilter.)

A2.12 ***Theorem.*** Let $\mathscr{F}$ be a filter in the set Ω.

(a) $\mathscr{F}$ is an ultrafilter iff for each $A \subset \Omega$ we have $A \in \mathscr{F}$ or $A^c \in \mathscr{F}$.

(b) If $\mathscr{F}$ is an ultrafilter and $p: \Omega \to \Omega'$, the filter $\mathscr{G}$ generated by the filterbase $p(\mathscr{F}) = \{p(F): F \in \mathscr{F}\}$ is an ultrafilter in Ω'.

(c) If Ω is a topological space and $\mathscr{F}$ is an ultrafilter in Ω, $\mathscr{F}$ converges to each of its accumulation points.

PROOF. (a) If $\mathscr{F}$ is an ultrafilter and $A \notin \mathscr{F}$, necessarily $A \cap B = \varnothing$ for some $B \in \mathscr{F}$. For if not, let $\mathscr{A}$ consist of all sets $A \cap B$, $B \in \mathscr{F}$; then $\mathscr{A}$ is a filterbase generating a filter larger than $\mathscr{F}$. But $A \cap B = \varnothing$ implies $B \subset A^c$; hence $A^c \in \mathscr{F}$. Conversely, if the condition is satisfied, let $\mathscr{F}$ be included in the filter $\mathscr{G}$. If $A \in \mathscr{G}$ and $A \notin \mathscr{F}$, then $A^c \in \mathscr{F} \subset \mathscr{G}$, a contradiction since $A \cap A^c = \varnothing$.

(b) Let $A \subset \Omega'$; by (a), either $p^{-1}(A) \in \mathscr{F}$ or $p^{-1}(A^c) \in \mathscr{F}$. If $p^{-1}(A) \in \mathscr{F}$, then $A \supset pp^{-1}(A) \in p(\mathscr{F})$; hence $A \in \mathscr{G}$. Similarly, if $p^{-1}(A^c) \in \mathscr{F}$, then $A^c \in \mathscr{G}$. By (a), $\mathscr{G}$ is an ultrafilter.

(c) Let x be an accumulation point of $\mathscr{F}$. If $U \in \mathscr{U}(x)$ and $U \notin \mathscr{F}$, then $U^c \in \mathscr{F}$ by (a). But $U \cap U^c = \varnothing$, contradicting the fact that x is an accumulation point of $\mathscr{F}$. ∎

We have associated with each net $\{x_n, n \in D\}$ the filterbase $\{x(T_a): a \in D\}$ of tails of the net, and have seen that convergence of the net is equivalent to convergence of the filterbase. We now prove a converse result.

A2.13 ***Theorem.*** If $\mathscr{A}$ is a filterbase in the set Ω, there is a net in Ω such that the collection of tails of the net coincides with $\mathscr{A}$.

PROOF. Let D be all ordered pairs (a, A) where $a \in A$ and $A \in \mathscr{A}$; define $(a, A) \le (b, B)$ iff $B \subset A$. If (a, A) and (b, B) belong to D, choose $C \in \mathscr{A}$ with $C \subset A \cap B$; for any $c \in C$ we have $(c, C) \ge (a, A)$ and $(c, C) \ge (b, B)$, hence D is directed. If we set $x_{(a,A)} = a$ we obtain a net in Ω with $x(T_{(a,A)}) = A$. ∎

We conclude this section with a characterization of continuity.

A2.14 ***Theorem.*** Let $f: \Omega \to \Omega'$, where Ω and Ω' are topological spaces. The following are equivalent:

(a) The function f is continuous on Ω; that is, $f^{-1}(V)$ is open in Ω whenever V is open in Ω'.

(b) For every net $\{x_n\}$ in Ω converging to the point $x \in \Omega$, the net $\{f(x_n)\}$ converges to $f(x)$.

(c) For every filterbase $\mathscr{A}$ in Ω converging to the point $x \in \Omega$, the filterbase $f(\mathscr{A})$ converges to $f(x)$.

PROOF. Let $\{x_n\}$ be a net and $\mathscr{A}$ a filterbase such that the tails of the net coincide with the elements of the filterbase. If, say, $x(T_a) = A \in \mathscr{A}$, then $f(A) = \{f(x_n): n \in D, n \geq a\}$. Thus the tails of the net $\{f(x_n)\}$ coincide with the elements of $f(\mathscr{A})$. It follows that (b) and (c) are equivalent.

If f is continuous and $x_n \to x$, let V be a neighborhood of $f(x)$. Then $f^{-1}(V)$ is a neighborhood of x; hence x_n is eventually in $f^{-1}(V)$, so that $f(x_n)$ is eventually in V. Thus (a) implies (b). Conversely, if (b) holds and C is closed in Ω', let $\{x_n\}$ be a net in $f^{-1}(C)$ converging to x. Then $f(x_n) \to f(x)$ by (b), and since C is closed we have $f(x) \in C$ by A2.3(b). Thus $x \in f^{-1}(C)$, hence $f^{-1}(C)$ is closed, proving continuity of f. ▮

A3 Product and Quotient Topologies

In the Euclidean plane R^2, a base for the topology may be formed from sets $U \times V$, where U and V are open subsets of R; in fact U and V can be taken to be open intervals, so that $U \times V$ is an open rectangle. If

$$\{(x_n, y_n), \quad n = 1, 2, \ldots\}$$

is a sequence in R^2, then $(x_n, y_n) \to (x, y)$ iff $x_n \to x$ and $y_n \to y$, that is, convergence in R^2 is "pointwise" or "coordinatewise" convergence.

In general, given an arbitrary collection of topological spaces Ω_i, $i \in I$, let Ω be the cartesian product $\prod_{i \in I} \Omega_i$, which is the collection of all families $(x_i, i \in I)$; that is, all functions on I such that $x_i \in \Omega_i$ for each i. We shall place a topology on Ω such that convergence in the topology coincides with pointwise convergence.

A3.1 ***Definition.*** The *product topology* (also called the *topology of pointwise convergence*) on $\Omega = \prod_{i \in I} \Omega_i$ has as a base all sets of the form

$$\{x \in \Omega: x_{i_k} \in U_{i_k}, \quad k = 1, \ldots, n\}$$

where the U_{i_k} are open in Ω_{i_k} and n is an arbitrary positive integer. (Since the intersection of two sets of this type is a set of this type, the sets do in fact form a base.)

If p_i is the projection of Ω onto Ω_i, the product topology is the weakest topology making each p_i continuous; in other words, the product topology is included in any topology that makes each p_i continuous.

The product topology has the following properties:

A3.2 ***Theorem.*** Let $\Omega = \prod_{i \in I} \Omega_i$, with the product topology.

(a) If $\{x^{(n)}, n \in D\}$ is a net in Ω and $x \in \Omega$, then $x^{(n)} \to x$ iff $x_i^{(n)} \to x_i$ for each i.

(b) A map f from a topological space Ω_0 into Ω is continuous iff $p_i \circ f$ is continuous for each i.

(c) If $f_i\colon \Omega_0 \to \Omega_i$, $i \in I$, and we define $f\colon \Omega_0 \to \Omega$ by $f(x) = (f_i(x), i \in I)$, then f is continuous iff each f_i is continuous.

(d) The projections p_i are open maps of Ω onto Ω_i.

PROOF. (a) If $x^{(n)} \to x$, then $x_i^{(n)} = p_i(x^{(n)}) \to p_i(x) = x_i$ by continuity of the p_i. Conversely, assume $x_i^{(n)} \to x_i$ for each i. Let

$$V = \{y \in \Omega\colon y_{i_k} \in U_{i_k}, \quad k = 1, \ldots, r\},$$

be a basic neighborhood of x. Since $x_{i_k} \in U_{i_k}$, there is an $n_k \in D$ with $x_{i_k}^{(n)} \in U_{i_k}$ for $n \geq n_k$. Therefore, if $n \in D$ and $n \geq n_k$ for all $k = 1, \ldots, r$, we have $x \in V$, so that $x^{(n)} \to x$.

(b) The "only if" part follows by continuity of the p_i. Conversely, assume each $p_i \circ f$ continuous. If $x^{(n)} \to x$, then $p_i(f(x^{(n)})) \to p_i(f(x))$ by hypothesis; hence $f(x^{(n)}) \to f(x)$ by (a).

(c) We have $f_i = p_i \circ f$, so (b) applies.

(d) The result follows from the observation that

$$p_i\{x \in \Omega\colon x_{i_k} \in U_{i_k}, \quad k = 1, \ldots, n\} = \begin{cases} \Omega_i & \text{if } \; i \neq \text{any } i_k, \\ U_{i_k} & \text{if } \; i = i_k \text{ for some } k. \end{cases} \quad ∎$$

Note that if Ω is the collection of all functions from a topological space S to a topological space T, then $\Omega = \prod_{i \in I} \Omega_i$, where $I = S$ and $\Omega_i = T$ for all i. If $\{f_n\}$ is a net of functions from S to T, and $f\colon S \to T$, then $f_n \to f$ in the product topology iff $f_n(s) \to f(s)$ for all $s \in S$.

We now consider quotient spaces.

A3.3 ***Definition.*** Let Ω_0 be a topological space, and p a map of Ω_0 onto a set Ω. The *identification topology* on Ω is the strongest topology making p continuous, that is, the open subsets of Ω are the sets U such that $p^{-1}(U)$ is open in Ω_0. When Ω has the identification topology, it is called an *identification space*, and p is called an *identification map*.

A quotient topology may be regarded as a particular identification topology. Let R be an equivalence relation on the topological space Ω_0, Ω_0/R the set of equivalence classes, and $p: \Omega_0 \to \Omega_0/R$ the canonical projection: $p(x) = [x]$, the equivalence class containing x. The *quotient space* of Ω_0 by R is Ω_0/R with the identification topology determined by p.

In fact, any identification space may be regarded as a quotient space. To see this, we need two preliminary results.

A3.4 ***Lemma.*** If Ω has the identification topology determined by $p: \Omega_0 \to \Omega$, then $g: \Omega \to \Omega_1$ is continuous iff $g \circ p: \Omega_0 \to \Omega_1$ is continuous.

PROOF. The "only if" part follows from the continuity of p. If $g \circ p$ is continuous and V is open in Ω_1, then $(g \circ p)^{-1}(V) = p^{-1}(g^{-1}(V))$ is open in Ω_0. By definition of the identification topology, $g^{-1}(V)$ is open in Ω. ∎

A3.5 ***Lemma.*** Let $p: \Omega_0 \to \Omega$ be an identification, and let $h: \Omega_0 \to \Omega_1$ be continuous. Assume that $h \circ p^{-1}$ is single-valued, in other words, $p(x) = p(y)$ implies $h(x) = h(y)$. Then $h \circ p^{-1}$ is continuous.

PROOF. Since $h = (h \circ p^{-1}) \circ p$, the result follows from A3.4. (Note that p^{-1} is defined on all of Ω since p is onto.) ∎

A3.6 ***Theorem.*** Let $f: \Omega_0 \to \Omega$ be an identification. Define an equivalence relation R on Ω_0 by calling x and y equivalent iff $f(x) = f(y)$. Let p be the canonical projection of Ω_0 onto Ω_0/R. Then Ω_0/R, with the quotient topology, is homeomorphic to Ω.

PROOF. We have $f(x) = f(y)$ iff $p(x) = p(y)$, so by A3.5, $f \circ p^{-1}$ and $p \circ f^{-1}$ are both continuous. Since these functions are inverses of each other, they define a homeomorphism of Ω and Ω_0/R. ∎

The following result gives conditions under which a given topology arises from an identification.

A3.7 ***Theorem.*** Let p be a map of Ω_0 onto Ω. If p is continuous and either open or closed, it is an identification, that is, the identification topology on Ω determined by p coincides with the original topology on Ω.

PROOF. Since p is continuous and the identification topology is the largest one making p continuous, the original topology is included in the identification topology. If p is an open map and U is an open subset of Ω in the identification topology, $p^{-1}(U)$ is open in Ω_0, hence $p(p^{-1}(U)) = U$ is open in the original topology. If p is a closed map, the same argument applies, with "open" replaced by "closed." ∎

A4 Separation Properties and Other Ways of Classifying Topological Spaces

Topological spaces may be classified as to how well disjoint sets may be separated, as follows. (The results of this section are generally discussed in a first course in topology, and will not be proved.)

A4.1 ***Definitions.*** A topological space Ω is said to be a T_0 *space* iff given any two distinct points x and y, *at least one* point has a neighborhood not containing the other; Ω is a T_1 *space* iff *each* point has a neighborhood not containing the other; Ω is a T_2 (*or Hausdorff*) *space* iff x and y have disjoint neighborhoods. This is equivalent to uniqueness of limits of nets (or filterbases). Also, Ω is said to be a T_3 (*or regular*) *space* iff Ω is T_2 and for each closed set C and point $x \notin C$, there are disjoint open sets U and V with $x \in U$ and $C \subset V$; Ω is said to be a T_4 (*or normal*) *space* iff Ω is T_2 and for each pair of disjoint closed sets A and B there are disjoint open sets U and V with $A \subset U$ and $B \subset V$.

It follows from the definitions that T_i implies T_{i-1}, $i = 4, 3, 2, 1$. Also, the T_1 property is equivalent to the statement that $\{x\}$ is closed for each x. The space Ω is regular iff it is Hausdorff and for each open set U and point $x \in U$, there is a $V \in \mathscr{U}(x)$ with $\overline{V} \subset U$. The space Ω is normal iff it is Hausdorff and for each closed set A and open set $U \supset A$, there is an open set V with $A \subset V \subset \overline{V} \subset U$.

A metric space is T_4, for if A and B are disjoint closed sets, we may take $U = \{x: d(x, A) - d(x, B) < 0\}$, $V = \{x: d(x, A) - d(x, B) > 0\}$, where $d(x, A) = \inf\{d(x, y): y \in A\}$.

A4.2 ***Urysohn's Lemma.*** Let Ω be a Hausdorff space. Then Ω is normal iff for each pair of disjoint closed sets A and B, there is a continuous function $f: \Omega \to [0, 1]$ with $f = 0$ on A and $f = 1$ on B.

A4.3 ***Tietze Extension Theorem.*** Let Ω be a Hausdorff space. Then Ω is normal iff for every closed set $A \subset \Omega$ and every continuous real-valued function f defined on A, f has an extension to a continuous real-valued function F on Ω. Furthermore, if $|f| < c$ (respectively, $|f| \leq c$) on A, then $|F|$ can be taken less than c (respectively, less than or equal to c) on Ω.

A4.4 ***Theorem.*** Let A be a closed subset of the normal space Ω. There is a continuous $f: \Omega \to [0, 1]$ such that $A = f^{-1}\{0\}$ iff A is a G_δ, that is, a countable intersection of open sets.

A4.5 ***Definitions and Comments.*** A topological space Ω is *second countable* iff there is a countable base for the topology, *first countable* iff there is a countable base at each point (see A2.4). Second countability implies first countability but not conversely. Any metric space is first countable.

If Ω is second countable, it is necessarily *separable*, that is, there is a countable dense subset of Ω. Furthermore, if Ω is second countable, it is *Lindelöf*, that is, for every family of open sets V_i, $i \in I$, such that $\bigcup_i V_i = \Omega$, there is a countable subfamily whose union is Ω (for short, every open covering of Ω has a countable subcovering).

In a metric space, the separable, second countable, and Lindelöf properties are equivalent as follows. Second countability always implies separability and Lindelöf. If Ω is separable with a countable dense set $\{x_1, x_2, \ldots\}$, then the balls $B(x_i, r) = \{y \in \Omega: d(y, x_i) < r\}$, $i = 1, 2, \ldots$, r rational, form a countable base. If Ω is Lindelöf, the cover by balls $B(x, 1/n)$, $x \in \Omega$, has a countable subcover $\{B(x_{ni}, 1/n), i = 1, 2, \ldots\}$, and the sets $B(x_{ni}, 1/n)$, $i, n = 1, 2, \ldots$, form a countable base.

This result implies that any space that is separable but not second countable (or not Lindelöf), or Lindelöf but not second countable (or not separable) cannot be metrizable, that is, there is no metric whose topology coincides with the original one.

A4.6 ***Definitions and Comments.*** A topological space Ω is said to be *completely regular* iff Ω is Hausdorff and for each $x \in \Omega$ and closed set $C \subset \Omega$ with $x \notin C$, there is a continuous $f: \Omega \to [0, 1]$ such that $f(x) = 1$ and $f = 0$ on C.

If A is a subset of the normal space Ω, then A, with the relative topology, is completely regular (this follows quickly from Urysohn's lemma). Also, complete regularity implies regularity. Thus complete regularity is in between regularity and normality; for this reason, completely regular spaces are sometimes called $T_{3\frac{1}{2}}$ spaces.

Now if Ω_0 is a Hausdorff space and $\mathscr{F}$ is the family of continuous maps $f: \Omega_0 \to [0, 1]$, let $\Omega = \prod \{I_f : f \in \mathscr{F}\}$ where each $I_f = [0, 1]$. Let $e: \Omega_0 \to \Omega$ be the evaluation map, that is, $e(x) = (f(x), f \in \mathscr{F})$.

With the product topology on Ω, e is continuous, and e is one-to-one iff $\mathscr{F}$ distinguishes points, in other words, given $x, y \in \Omega_0$, $x \neq y$, there is an $f \in \mathscr{F}$ such that $f(x) \neq f(y)$. Finally, if $\mathscr{F}$ distinguishes points from closed sets, that is, if Ω_0 is completely regular, then e is an open map of Ω_0 onto $e(\Omega_0) \subset \Omega$. [If $\{x_n\}$ is a net and $e(x_n) \to e(x)$ but $x_n \nrightarrow x$, there is a neighborhood U of x such that x_n is not eventually in U; that is, given m, there is an $n \geq m$ with $x_n \notin U$. Choose $f \in \mathscr{F}$ with $f(x) = 1$ and $f = 0$ on $\Omega_0 - U$. Then for each m, we have $f(x_n) = 0$ for some $n \geq m$, so that $f(x_n) \nrightarrow f(x)$. But then $e(x_n) \nrightarrow e(x)$, a contradiction.]

It follows that if Ω_0 is completely regular, it is homeomorphic to a subset of a normal space. [Since Ω is a product of Hausdorff spaces, it is Hausdorff; the Tychonoff theorem, to be proved later, shows that Ω is compact, and hence normal (see A5.3(d) and A5.4).]

Since $e(\Omega_0)$ is determined completely by $\mathscr{F}$, we may say that the continuous functions are adequate to describe the topology of Ω_0.

A5 Compactness

The notion of compactness appears in virtually all areas of mathematics. The original compactness result was the Heine–Borel theorem: If $[0, 1] \subset \bigcup_i V_i$, where the V_i are open subsets of R, then in fact $[0, 1]$ is covered by finitely many V_i, that is, $[0, 1] \subset \bigcup_{k=1}^n V_{i_k}$ for some $V_{i_1}, \ldots, V_{i_n}$. In general, we have the following definition:

A5.1 ***Definition.*** The topological space Ω is *compact* iff every open covering of Ω has a finite subcovering.

There are several ways of expressing this idea.

A5.2 ***Theorem.*** If Ω is a topological space, the following are equivalent:

(a) Ω is compact.

(b) Each family of closed sets $C_i \subset \Omega$ with the finite intersection property (all finite intersections of the C_i are nonempty) has nonempty intersection. Equivalently, for every family of closed subsets of Ω with empty intersection, there is a finite subfamily with empty intersection.

(c) Every net in Ω has an accumulation point in Ω; in other words (by A2.6), every net in Ω has a subnet converging to a point of Ω.

(d) Every filterbase in Ω has an accumulation point in Ω, that is (by A2.10), every filterbase in Ω has a convergent filterbase subordinate to it, or equally well, every filter in Ω has an overfilter converging to a point of Ω.

(e) Every ultrafilter in Ω converges to a point of Ω.

PROOF. Parts (a) and (b) are equivalent by the duality between open and closed sets. If $\{x_n\}$ is a net and $\mathscr{A}$ is a filterbase whose elements coincide with the tails of the net, then the accumulation points of the net and of the filterbase coincide, so that (c) and (d) are equivalent. Now (d) implies (e) by A2.12(c), and (e) implies (d) since every filter is included in an ultrafilter. To prove that (b) implies (d), observe that if $\mathscr{A}$ is a filterbase, the sets $A \in \mathscr{A}$, hence the sets $\bar{A}$, $A \in \mathscr{A}$, have the finite intersection property, so by (b), there is a point $x \in \bigcap\{\bar{A} : A \in \mathscr{A}\}$. Finally, we prove that (d) implies (b). If the closed sets C_i have the finite intersection property, the finite intersections of the C_i form a filterbase, which by (d) has an accumulation point x. But then x belongs to all C_i. ∎

It is important to note that if $A \subset \Omega$, the statement that every covering of A by sets open in Ω has a finite subcovering is equivalent to the statement that every covering of A by sets open in A (in the relative topology inherited from Ω) has a finite subcovering. Thus when we talk about a compact subset of a topological space, there is no ambiguity.

The following results follow quickly from the definition of compactness.

A5.3 ***Theorem.*** (a) If Ω is compact and f is continuous on Ω, then $f(\Omega)$ is compact.

(b) A closed subset C of a compact space Ω is compact.

(c) If A and B are disjoint compact subsets of the Hausdorff space Ω, there are disjoint open sets U and V such that $A \subset U$ and $B \subset V$. In particular (take $A = \{x\}$), a compact subset of a Hausdorff space is closed.

(d) A compact Hausdorff space is normal.

(e) If A is a compact subset of the regular space Ω, and A is a subset of the open set U, there is an open set V with $A \subset V \subset \bar{V} \subset U$.

PROOF. (a) This is immediate from the definition of compactness.

(b) If C is covered by sets U open in Ω, the sets U together with $\Omega - C$ cover Ω. By compactness there is a finite subcover.

(c) If $x \notin B$ and $y \in B$, there are disjoint neighborhoods $U_y(x)$ and $V(y)$ of x and y. The $V(y)$ cover B; hence there is a finite subcover $V(y_i)$, $i = 1, \dots, n$. Then $U' = \bigcap_{i=1}^n U_{y_i}(x)$ and $V' = \bigcup_{i=1}^n V(y_i)$ are disjoint open sets with $x \in U'$ and $B \subset V'$. If we repeat the process for each $x \in A$, we obtain disjoint open sets $U(x)$ and $V(x)$ as above. The $U(x)$ cover A; hence there is a finite subcover $U(x_i)$, $i = 1, \dots, m$. Take $U = \bigcup_{i=1}^m U(x_i)$, $V = \bigcap_{i=1}^m V(x_i)$.

(d) If A and B are disjoint closed sets, they are compact by (b); the result then follows from (c).

(e) If $x \in A$, regularity yields an open set $V(x)$ with $x \in V(x)$ and $\overline{V(x)} \subset U$. The $V(x)$ cover A; so for some $x_1, \ldots, x_n$, we have

$$A \subset \bigcup_{i=1}^{n} V(x_i) \subset \bigcup_{i=1}^{n} \overline{V(x_i)} = \overline{\bigcup_{i=1}^{n} V(x_i)} \subset U. \quad \blacksquare$$

The following is possibly the most important compactness result.

A5.4 ***Tychonoff Theorem.*** If Ω_i is compact for each $i \in I$, then $\Omega = \prod_{i \in I} \Omega_i$ is compact in the product topology.

PROOF. Let $\mathscr{F}$ be an ultrafilter in Ω. If p_i is the projection of Ω onto Ω_i, then by A2.12(b), $p_i(\mathscr{F})$ is a filterbase that generates an ultrafilter in Ω_i. By hypothesis, $p_i(\mathscr{F})$ converges to some $x_i \in \Omega_i$, and it follows that $\mathscr{F} \to x = (x_i, i \in I)$. [To see this, observe that if $\mathscr{A}$ is a filterbase in Ω and $\{x^n\}$ is a net whose tails are the elements of $\mathscr{A}$, then the tails of $\{p_i(x^n)\}$ are the elements of $p_i(\mathscr{A})$. Since $x^n \to x$ iff $p_i(x^n) \to p_i(x)$ for all i, by A3.2(a), it follows that $\mathscr{A} \to x$ iff $p_i(\mathscr{A}) \to p_i(x)$ for all i.] The Tychonoff theorem now follows from A5.2(e). $\blacksquare$

The following result is often used to infer that the inverse of a particular one-to-one continuous map is continuous.

A5.5 ***Theorem.*** Let $f: \Omega \to \Omega_1$ where Ω is compact, Ω_1 is Hausdorff, and f is continuous. Then f is a closed map; consequently if f is one-to-one onto, it is a homeomorphism.

PROOF. By A5.3(a), (b) and (c). $\blacksquare$

A5.6 ***Corollary.*** Let $p: \Omega_0 \to \Omega$ be an identification, and let $h: \Omega_0 \to \Omega_1$ be continuous; assume $h \circ p^{-1}$ is single valued, and hence continuous by A3.5. Assume also that Ω_0 is compact (hence so is Ω because p is onto) and Ω_1 is Hausdorff. If $h \circ p^{-1}$ is one-to-one onto, it is a homeomorphism.

PROOF. Apply A5.5 with $f = h \circ p^{-1}$. $\blacksquare$

Corollary A5.6 is frequently applied in constructing quotient spaces. For example, if one pair of opposite edges of a rectangle are identified, we obtain a cylinder. Formally, let $I^2 = \{(x, y): 0 \le x \le 1, 0 \le y \le 1\}$. Define an equivalence relation R on I^2 by specifying that $(0, y)$ be equivalent to $(1, y)$, $0 \le y \le 1$, with the other equivalence classes consisting of single points.

Let $h(x, y) = (e^{i2\pi x}, y)$, $(x, y) \in I^2$; h maps I^2 onto a cylinder C. If p is the canonical projection of I^2 onto I^2/R, then A5.6 implies that $h \circ p^{-1}$ is a homeomorphism of I^2/R and C.

In some situations, for example in metric spaces, there are alternative ways of expressing the idea of compactness.

A5.7 ***Definition.*** A topological space Ω is said to be *countably compact* iff every countable open covering of Ω has a finite subcover.

A5.8 ***Theorem.*** For any topological space Ω, the following properties (a)–(d) are equivalent, and each implies (e). If Ω is T_1, then all five properties are equivalent.

(a) Ω is countably compact.

(b) Each sequence of closed subsets of Ω with the finite intersection property has nonempty intersection.

(c) Every sequence in Ω has an accumulation point.

(d) Every countable filterbase in Ω has an accumulation point.

(e) Every infinite subset of Ω has a cluster point.

PROOF. The equivalence of (a), (b), and (d) is proved exactly as in A5.2, and (d) implies (c) because the tails of a sequence form a countable filterbase. To prove that (c) implies (d), let $\mathscr{A} = \{A_1, A_2, \ldots\}$ be a countable filterbase, and choose $x_n \in \bigcap_{i=1}^n A_i$, $n = 1, 2, \ldots$. If x is an accumulation point of $\{x_n\}$ and $U \in \mathscr{U}(x)$, then for each n there is an $m \ge n$ such that $x_m \in U$, hence $U \cap \bigcap_{i=1}^m A_i \ne \varnothing$. It follows that $U \cap A_n \ne \varnothing$ for all n, and consequently x is an accumulation point of $\mathscr{A}$. To prove that (c) implies (e), pick a sequence $\{x_n\}$ of distinct points from the infinite set A and observe that if x is an accumulation point of the sequence, then x is a cluster point of A. Finally, we show that (e) implies (a) if Ω is T_1. Say $U_1, U_2, \ldots$ form a countable open covering of Ω with no finite subcover. Choose $x_1 \notin U_1$; having chosen distinct $x_1, \ldots, x_k$ with $x_j \notin \bigcup_{i=1}^j U_i$, $j = 1, \ldots, k$, then $x_1, \ldots, x_k$ all belong to some finite union $\bigcup_{i=1}^n U_i$, $n \ge k+1$; choose $x_{k+1} \notin \bigcup_{i=1}^n U_i$ (hence $x_{k+1} \notin \bigcup_{i=1}^{k+1} U_i$ and $x_1, \ldots, x_{k+1}$ are distinct). In this way we form an infinite set $A = \{x_1, x_2, \ldots\}$ with no cluster point. For if x is such a point, x belongs to U_n for some n. Since Ω is T_1, there is a set $U \in \mathscr{U}(x)$ such that $U \subset U_n$ and $x_i \notin U$, $i = 1, 2, \ldots, n-1$ (unless $x_i = x$). Since $x_k \notin U_n$ for $k \ge n$, U contains no point of A distinct from x. ▮

A5.9 ***Definitions and Comments.*** The topological space Ω is said to be *sequentially compact* iff every sequence in Ω has a convergent subsequence. By A5.8(c), sequential compactness implies countable compactness. In a first countable space, countable and sequential compactness are equivalent. For if x is an accumulation point of the sequence $\{x_n\}$ and $V_1, V_2, \ldots$ (with $V_{n+1} \subset V_n$ for all n) form a countable base at x, then for each k we may find $n_k \geq k$ such that $x_{n_k} \in V_k$. Thus we have a subsequence converging to x.

A5.10 ***Theorem.*** In a second countable space or a metric space, compactness, countable compactness, and sequential compactness are equivalent.

PROOF. A second countable space is Lindelöf (see A4.5), so compactness and countable compactness are equivalent. It is first countable, so countable compactness and sequential compactness are equivalent; this result holds in a metric space also, because a metric space is first countable.

Now a sequentially compact metric space Ω is *totally bounded*, that is, for each $\varepsilon > 0$, Ω can be covered by finitely many balls of radius ε. (If not, inductively pick $x_1, x_2, \ldots$ with $x_{n+1} \notin \bigcup_{i=1}^{n} B(x_i, \varepsilon)$; then $\{x_n\}$ can have no convergent subsequence.) Thus for each positive integer n, Ω can be covered by finitely many balls $B(x_{ni}, 1/n)$, $i = 1, 2, \ldots, k_n$. If $\{U_j, j \in J\}$ is an arbitrary open covering of Ω, for each ball $B(x_{ni}, 1/n)$ we choose, if possible, a set U_{ni} of the covering such that $U_{ni} \supset B(x_{ni}, 1/n)$. If $x \in \Omega$, then x belongs to a ball $B(x, \varepsilon)$ included in some U_j; hence $x \in B(x_{ni}, 1/n) \subset B(x, \varepsilon) \subset U_j$ for some n and i; therefore $x \in U_{ni}$. Thus the U_{ni} form a countable subcover, and Ω, which is countably compact, must in fact be compact. ∎

Note that a compact metric space is Lindelöf, hence (see A4.5) is second countable and separable.

A5.11 ***Definition.*** A Hausdorff space is said to be *locally compact* iff each $x \in \Omega$ has a *relatively compact* neighborhood, that is, a neighborhood whose closure is compact. (Its follows that a compact Hausdorff space is locally compact.)

A5.12 ***Theorem.*** The following are equivalent, for a Hausdorff space Ω:

(a) Ω is locally compact.

(b) For each $x \in \Omega$ and $U \in \mathscr{U}(x)$, there is a relatively compact open set V with $x \in V \subset \overline{V} \subset U$. (Thus a locally compact space is regular; furthermore, the relatively compact open sets form a base for the topology.)

(c) If K is compact, U is open, and $K \subset U$, there is a relatively compact open set V with $K \subset V \subset \overline{V} \subset U$.

PROOF. It is immediate that (c) implies (a), and (b) implies (c) is proved by applying (b) to each point of K and using compactness. To prove that (a) implies (b), let x belong to the open set U. By (a), there is a neighborhood V_1 of x such that $K = \overline{V}_1$ is compact. Now K is compact Hausdorff, and hence regular, and $x \in U \cap V_1$, which is open in Ω, and hence open in K. Thus (see A4.1) there is a set W open in Ω such that $x \in W \cap K$ and the closure of $W \cap K$ in K, namely, $\overline{W} \cap K$, is a subset of $U \cap V_1$.

Now $x \in W \cap V_1$ and $\overline{W \cap V_1} \subset \overline{W} \cap K \subset U$, so $V = W \cap V_1$ is the desired relatively compact neighborhood. ∎

The following properties of locally compact spaces are often useful:

A5.13 ***Theorem.*** (a) Let Ω be a locally compact Hausdorff space. If $K \subset U \subset \Omega$, with K compact and U open, there is a continuous $f: \Omega \to [0, 1]$ such that $f = 0$ on K and $f = 1$ on $\Omega - U$. In particular, a locally compact Hausdorff space is completely regular.

(b) Let Ω be locally compact Hausdorff, or, more generally, completely regular. If A and B are disjoint subsets of Ω with A compact and B closed, there is a continuous $f: \Omega \to [0, 1]$ such that $f = 0$ on A and $f = 1$ on B.

(c) Let Ω be locally compact Hausdorff, and let $A \subset U \subset \Omega$, with A compact and U open. Then there are sets B and V with $A \subset V \subset B \subset U$, where V is open and σ-compact (a countable union of compact sets) and B is compact and is also a G_δ (a countable intersection of open sets). Consequently (take $A = \{x\}$) the σ-compact open sets form a base for the topology.

PROOF. (a) Let $K \subset V \subset \overline{V} \subset U$, with V open, $\overline{V}$ compact [see A5.12(c)]. $\overline{V}$ is normal, so there is a continuous $g: \overline{V} \to [0, 1]$ with $g = 0$ on K, $g = 1$ on $\overline{V} - V$. Define $f = g$ on $\overline{V}$, $f = 1$ on $\Omega - V$. (On $\overline{V} - V$, $g = 1$ so f is well-defined.) Now f is continuous on Ω (look at preimages of closed sets), so f is the desired function.

(b) By complete regularity, for each $x \in A$, there is a continuous $f_x: \Omega \to [0, 1]$ with $f_x(x) = 0$, $f_x = 1$ on B. By compactness,

$$A \subset \bigcup_{i=1}^{n} \{x : f_{x_i}(x) < \tfrac{1}{2}\}$$

for some $x_1, \ldots, x_n$. Let $g = \prod_{i=1}^{n} f_{x_i}$; then $g = 1$ on B and $0 \le g < \frac{1}{2}$ on A, so $f = \max(2g - 1, 0)$ is the desired function.

(c) By A.5.12(c), we may assume without loss of generality that U is included in a compact set K. By (b) there is a continuous $f: \Omega \to [0, 1]$ with $f = 0$ on A, $f = 1$ off U. Let $V = \{x: f(x) < \frac{1}{2}\}$, $B = \{x: f(x) \leq \frac{1}{2}\}$. Then V is open, B is closed, and $A \subset V \subset B \subset U \subset K$; hence B is compact. Now B is a G_δ because

$$B = \bigcap_{n=1}^{\infty} \left\{x: f(x) < \frac{1}{2} + \frac{1}{n}\right\}, \quad \text{and} \quad V = \bigcup_{n=1}^{\infty} \left\{x: f(x) \leq \frac{1}{2} - \frac{1}{n}\right\} \subset K,$$

so V is σ-compact. (In fact V is a countable union of compact G_δ's.) ▌

A5.14 ***Corollary.*** Let Ω be completely regular, A a compact G_δ subset of Ω. There is a continuous $f: \Omega \to [0, 1]$ such that $f^{-1}\{0\} = A$.

PROOF. If $A = \bigcap_{n=1}^{\infty} U_n$, the U_n open, by A5.13(b) there are continuous $f_n: \Omega \to [0, 1]$ with $f_n = 0$ on A and $f_n = 1$ off U_n. Let $f = \sum_{n=1}^{\infty} f_n 2^{-n}$. ▌

A5.15 ***Theorem.*** If Ω is Hausdorff, the following are equivalent:

(a) Ω is locally compact Lindelöf.
(b) Ω can be expressed as $\bigcup_{n=1}^{\infty} U_n$ where $\overline{U}_n$ is a compact subset of U_{n+1} for each n.
(c) Ω is locally compact and σ-compact.

PROOF. (a) *implies* (b): The relatively compact open sets form a base by A5.12(b), in particular they cover Ω. Extract a countable subcover $\{V_1, V_2, \ldots\}$ using Lindelöf, and let $U_1 = V_1$, and for $n > 1$, $U_n = V_n \cup W_n$ where W_n is a relatively compact open set and $W_n \supset \bigcup_{i=1}^{n-1} \overline{U}_i$.

(b) *implies* (c): We have $\Omega = \bigcup_{n=1}^{\infty} \overline{U}_n$, proving σ-compactness. If $x \in \Omega$, then $x \in U_n$ for some n, with $\overline{U}_n$ compact, proving local compactness.

(c) *implies* (a): If $\Omega = \bigcup_{n=1}^{\infty} K_n$, K_n compact, and $\{U_i\}$ is an open covering of Ω, extract a finite subcovering of each K_n, and put the sets together to form a countable subcovering of Ω. ▌

The *Urysohn metrization theorem*, which we shall not prove, states that for a second countable space Ω, metrizability and regularity are equivalent. This result yields the following corollary to A5.15.

A5.16 ***Theorem.*** If Ω is locally compact Hausdorff, then Ω is second countable iff Ω is metrizable and σ-compact.

Proof. If Ω is second countable, it is metrizable by A5.12(b) and the Urysohn metrization theorem. Also, Ω is Lindelöf (see A4.5), hence is σ-compact by A5.15. Conversely, if Ω is metrizable and σ-compact, then Ω is Lindelöf by A5.15, hence second countable (see A4.5). ∎

Finally, we consider the important *one-point compactification.*

A5.17 ***Theorem.*** Let Ω be locally compact Hausdorff, and let $\Omega^* = \Omega \cup \{\infty\}$, where ∞ stands for any element not belonging to Ω. Put the following topology on Ω^*: U is open in Ω^* iff U is open in Ω or U is the complement (in Ω^*) of a compact subset of Ω. Then:

(a) If $U \subset \Omega$, U is open in Ω iff U is open in Ω^*; thus Ω^* induces the original topology on Ω.

(b) Ω^* is compact Hausdorff.

PROOF. Part (a) follows from the definition of the topology. To prove (b), let $\{U_i\}$ be an open covering of Ω^*. Then ∞ belongs to some U_i and the remaining U_j cover the compact set $\Omega^* - U_i$, so the cover reduces to a finite subcover. Since Ω is Hausdorff, distinct points of Ω have disjoint neighborhoods. If $x \in \Omega$, let V be an open subset of Ω with $x \in V$ and $\overline{V} \subset \Omega$, where $\overline{V}$, the closure of V in Ω, is compact [see A 5.12(b)]. Then V and $\Omega^* - \overline{V}$ are disjoint neighborhoods of x and ∞. ∎

A6 Semicontinuous Functions

If $f_1, f_2, \ldots$ are continuous maps from the topological space Ω to the extended reals $\overline{R}$, and $f_n(x)$ increases to a limit $f(x)$ for each x, f need not be continuous; however, f is lower semicontinuous. Functions of this type play an important role in many aspects of analysis and probability.

A6.1 ***Definition.*** Let Ω be a topological space. The function $f: \Omega \to \overline{R}$ is said to be *lower semicontinuous* (LSC) on Ω iff $\{x \in \Omega: f(x) > a\}$ is open in Ω for each $a \in \overline{R}$, *upper semicontinuous* (USC) on Ω iff $\{x \in \Omega: f(x) < a\}$ is open in Ω for each $a \in \overline{R}$. Thus f is LSC iff $-f$ is USC. Note that f is continuous iff it is both LSC and USC.

We have the following criterion for semicontinuity.

A6.2 ***Theorem.*** The function f is LSC on Ω iff, for each net $\{x_n\}$ converging to a point $x \in \Omega$, we have $\liminf_n f(x_n) \geq f(x)$, where $\liminf_n f(x_n)$ means

$\sup_n \inf_{k \geq n} f(x_k)$. Hence f is USC iff $\limsup_n f(x_n) \leq f(x)$ whenever $x_n \to x$. (In a first countable space, "net" may be replaced by "sequence.")

PROOF. Let f be LSC. If $x_n \to x$ and $b < f(x)$, then $x \in f^{-1}(b, \infty]$, an open subset of Ω, hence eventually $x_n \in f^{-1}(b, \infty]$, that is, $f(x_n) > b$ eventually. Thus $\liminf_n f(x_n) \geq f(x)$. Conversely if $x_n \to x$ implies

$$\liminf_n f(x_n) \geq f(x),$$

we show that $V = \{x : f(x) > a\}$ is open. Let $x_n \to x$, where $f(x) > a$. Then $\liminf_n f(x_n) > a$, hence $f(x_n) > a$ eventually, that is, $x_n \in V$ eventually. Thus (see A2.4) V is open. ∎

We now prove a few properties of semicontinuous functions.

A6.3 ***Theorem.*** Let f be LSC on the compact space Ω. Then f attains its infimum. (Hence if f is USC on the compact space Ω, f attains its supremum.)

PROOF. If $b = \inf f$, there is a sequence of points $x_n \in \Omega$ with $f(x_n) \to b$. By compactness, we have a subnet x_{n_k} converging to some $x \in \Omega$. Since f is LSC, $\liminf_k f(x_{n_k}) \geq f(x)$. But $f(x_{n_k}) \to b$, so that $f(x) \leq b$; consequently $f(x) = b$. ∎

A6.4 ***Theorem.*** If f_i is LSC on Ω for each $i \in I$, then $\sup_i f_i$ is LSC; if I is finite, then $\min_i f_i$ is LSC. (Hence if f_i is USC for each i, then $\inf_i f_i$ is USC, and if I is finite, then $\max_i f_i$ is USC.)

PROOF. Let $f = \sup_i f_i$; then $\{x : f(x) > a\} = \bigcup_{i \in I} \{x : f_i(x) > a\}$; hence $\{x : f(x) > a\}$ is open. If $g = \min(f_1, f_2, \ldots, f_n)$, then

$$\{x : g(x) > a\} = \bigcap_{i=1}^{n} \{x : f_i(x) > a\}$$

is open. ∎

A6.5 ***Theorem.*** Let $f : \Omega \to \overline{R}$, Ω any topological space, f arbitrary. Define

$$\underline{f}(x) = \liminf_{y \to x} f(y), \qquad x \in \Omega;$$

that is,

$$\underline{f}(x) = \sup_V \inf_{y \in V} f(y),$$

where V ranges over all neighborhoods of x. [If Ω is a metric space, then $\underline{f}(x) = \sup_{n=1,2,\ldots} \inf_{d(x,y)<1/n} f(y)$.]

Then $\underline{f}$ is LSC on Ω and $\underline{f} \le f$; furthermore if g is LSC on Ω and $g \le f$, then $g \le \underline{f}$. Thus $\underline{f}$, called the *lower envelope* of f, is the sup of all LSC functions that are less than or equal to f (there is always at least one such function, namely the function constant at $-\infty$).

Similarly, if $\bar{f}(x) = \limsup_{y \to x} f(y) = \inf_V \sup_{y \in V} f(y)$, then $\bar{f}$, the *upper envelope* of f, is USC and $\bar{f} \ge f$; in fact $\bar{f}$ is the inf of all USC functions that are greater than or equal to f.

PROOF. It suffices to consider $\underline{f}$. Let $\{x_n\}$ be a net in Ω with $x_n \to x$ and $\liminf_n \underline{f}(x_n) < b < \underline{f}(x)$. If V is a neighborhood of x, we can choose n such that $x_n \in V$ and $\underline{f}(x_n) < b$. Since V is also a neighborhood of x_n, we have

$$b > \underline{f}(x_n) \ge \inf_{y \in V} f(y),$$

so

$$\underline{f}(x) = \sup_V \inf_{y \in V} f(y) \le b < \underline{f}(x),$$

a contradiction. By A6.2, $\underline{f}$ is LSC, and $\underline{f} \le f$ by definition of $\underline{f}$. Finally if g is LSC, $g \le f$, then $\underline{f}(x) = \liminf_{y \to x} f(y) \ge \liminf_{y \to x} g(y) \ge g(x)$ since g is LSC. [If $\sup_V \inf_{y \in V} g(y) < b < g(x)$, then for each V pick $x_V \in V$ with $g(x_V) < b$. If $V_1 \le V_2$ means that $V_2 \subset V_1$, the x_V form a net converging to x, while $\liminf_V g(x_V) \le b < g(x)$, contradicting A6.2.] ∎

It can be shown that if Ω is completely regular, every LSC function on Ω is the sup of a family of continuous functions. If Ω is a metric space, the family can be assumed countable, as we now prove.

A6.6 ***Theorem.*** Let Ω be a metric space, f a LSC function on Ω. There is a sequence of continuous functions $f_n \colon \Omega \to \bar{R}$ such that $f_n \uparrow f$. (Thus if f is USC, there is a sequence of continuous functions $f_n \downarrow f$.) If $|f| \le M < \infty$, the f_n may be chosen so that $|f_n| \le M$ for all n.

PROOF [following Hausdorff (1962)]. First assume $f \ge 0$ and finite-valued. If d is the metric on Ω, define $g(x) = \inf\{f(z) + td(x, z) \colon z \in \Omega\}$, where $t > 0$ is fixed; then $0 \le g \le f$ since $g(x) \le f(x) + td(x, x) = f(x)$.

If $x, y \in \Omega$, then $f(z) + td(x, z) \le f(z) + td(y, z) + td(x, y)$. Take the inf over z to obtain $g(x) \le g(y) + td(x, y)$. By symmetry,

$$|g(x) - g(y)| \le td(x, y),$$

hence g is continuous on Ω.

Now set $t = n$; in other words let $f_n(x) = \inf\{f(z) + nd(x, z): z \in \Omega\}$. Then $0 \le f_n \uparrow h \le f$. But given $\varepsilon > 0$, for each n we can choose $z_n \in \Omega$ such that

$$f_n(x) + \varepsilon > f(z_n) + nd(x, z_n) \ge nd(x, z_n).$$

But $f_n(x) + \varepsilon \le f(x) + \varepsilon$, and it follows that $d(x, z_n) \to 0$. Since f is LSC, $\liminf_{n \to \infty} f(z_n) \ge f(x)$; thus $f(z_n) > f(x) - \varepsilon$ eventually. But now

$$\begin{aligned} f_n(x) &> f(z_n) - \varepsilon + nd(x, z_n) \ge f(z_n) - \varepsilon \\ &> f(x) - 2\varepsilon \qquad \text{for large enough} \quad n. \end{aligned}$$

It follows that $0 \le f_n \uparrow f$. If $|f| \le M < \infty$, then $f + M$ is LSC and nonnegative; if $0 \le g_n \uparrow f + M$, then $f_n = g_n - M \uparrow f$ and $|f_n| \le M$.

In general, let $h(x) = \frac{1}{2}\pi + \arctan x$, $x \in \bar{R}$; then h is an order-preserving homeomorphism of $\bar{R}$ and $[0, \pi]$. If f is LSC, then $h \circ f$ is finite-valued, LSC, and nonnegative, so that we can find continuous functions g_n such that $g_n \uparrow h \circ f$. Let $f_n = h^{-1} \circ g_n$; then $f_n \uparrow f$. ▮

A7 The Stone–Weierstrass Theorem

In this section, we consider the family $C(\Omega)$ of continuous *real-valued* functions on a compact Hausdorff space Ω. The set $C(\Omega)$ becomes a Banach space under the sup norm $\|f\| = \sup\{|f(x)|: x \in \Omega\}$. If $A \subset C(\Omega)$, A is called a *subalgebra* of $C(\Omega)$ iff for all $f_1, f_2 \in A$, $a, b \in R$, we have $af_1 + bf_2 \in A$ and $f_1 f_2 \in A$. Thus A is a vector space under addition and scalar multiplication, and a ring under ordinary multiplication. The main theorem of this section gives conditions under which A is dense in $C(\Omega)$. We assume throughout that Ω has at least two points; the Stone–Weierstrass theorem will be seen to be trivial if Ω has only a single point.

A7.1 ***Lemma.*** The function given by $f(x) = |x|$, $-1 \le x \le 1$, can be uniformly approximated by polynomials having no constant term.

PROOF. If $0 \le y \le a \le 1$, then $0 \le y \le y + \frac{1}{2}(a^2 - y^2) \le a$. (For $y^2 - 2y + 2a - a^2$ decreases when $0 \le y \le 1$, and its value when $y = a$ is 0.) If we define $y_0 = 0$, $y_{n+1} = y_n + \frac{1}{2}(a^2 - y_n^2)$, the above argument shows that $0 \le y_n \le y_{n+1} \le a$ for all n. Let $n \to \infty$ to obtain $y_n \to y$, where $y = y + \frac{1}{2}(a^2 - y^2)$; hence $y = a$.

Thus, changing the notation, define

$$p_0(x) = 0, p_{n+1}(x) = p_n(x) + \tfrac{1}{2}(x^2 - p_n^2(x)), \ -1 \le x \le 1.$$

By induction, p_n is a polynomial having no constant term, and $0 \le p_n(x) \le p_{n+1}(x) \to (x^2)^{1/2} = |x|$. Since the domain $[-1, 1]$ is compact and p_n converges monotonically to a continuous limit p, the convergence is uniform by Dini's theorem (see 4.3.9 for the details). ▌

A7.2 ***Lemma.*** Let A be a subset of $C(\Omega)$ that is closed under the lattice operations; that is, if $f, g \in A$, then $\max(f, g) \in A$ and $\min(f, g) \in A$.

If $f \in C(\Omega)$ and f can be approximated at each pair of points by functions in A [if $x \neq y$, there is a sequence of functions $f_n \in A$ with $f_n(x) \to f(x)$ and $f_n(y) \to f(y)$], then $f \in \bar{A}$.

PROOF. Given $\delta > 0$, $x, y \in \Omega$, there is a function $f_{xy} \in A$ with $|f - f_{xy}| < \delta$ at the points x and y. Let $U(x, y) = \{z \in \Omega;\ f_{xy}(z) < f(z) + \delta\}$, $V(x, y) = \{z \in \Omega: f_{xy}(z) > f(z) - \delta\}$. Since $x \in U(x, y)$, the $U(x, y)$ cover Ω for each fixed y; by compactness, $\Omega = \bigcup_{i=1}^{n} U(x_i, y)$ for some $x_1, \ldots, x_n$.

Let $f_y = \min\{f_{x_i y}: 1 \le i \le n\} \in A$ by hypothesis, and let

$$V(y) = \bigcap_{i=1}^{n} V(x_i, y).$$

If $x \in \Omega$, then $x \in U(x_i, y)$ for some i, so $f_y(x) \le f_{x_i y}(x) < f(x) + \delta$; if $x \in V(y)$, then $f_{x_i y}(x) > f(x) - \delta$ for all i, hence $f_y(x) > f(x) - \delta$.

But the $V(y)$ cover Ω [note $y \in V(y)$] so by compactness, $\Omega = \bigcup_{i=1}^{m} V(y_i)$ for some $y_1, \ldots, y_m$. Define $f_\delta = \max\{f_{y_i}: 1 \le i \le m\} \in A$ by hypothesis.

If $x \in \Omega$, then $f_\delta(x) < f(x) + \delta$; since $x \in V(y_i)$ for some i, we have $f_\delta(x) > f(x) - \delta$. Thus $\|f - f_\delta\| \le \delta$ and, since δ is arbitrary, it follows that $f \in \bar{A}$. ▌

A7.3 ***Lemma.*** Let A be a subalgebra of $C(\Omega)$, and assume that A is a closed subset of $C(\Omega)$. Then $f \in A$ implies $|f| \in A$; also, $f, g \in A$ implies $\max(f, g) \in A$ and $\min(f, g) \in A$.

PROOF. If $g \in A$ and p is a polynomial with no constant term, then $p \circ g \in A$. If $f \in A$ and $\|f\| = M$, set $g = f/M$. Then $|g(x)| \le 1$ for all x, so by A7.1, $\|p \circ g - |g|\,\|$ can be made arbitrarily small by appropriate choice of p; consequently $|g|$, hence $|f|$, belongs to A.

Now $f + g = \max(f, g) + \min(f, g)$, $|f - g| = \max(f, g) - \min(f, g)$. Addition and then subtraction of these equations completes the proof. ▌

A7.4 ***Lemma.*** Let A be a subalgebra of $C(\Omega)$ that *separates points*; that is, if $x \neq y$, there is an $f \in A$ with $f(x) \neq f(y)$. Assume that for each $x \in \Omega$, there is an $f \in A$ with $f(x) \neq 0$.

If $x, y \in \Omega$, $x \neq y$, and r and s are arbitrary real numbers, there is an $f \in A$ with $f(x) = r, f(y) = s$.

PROOF. If this is not the case, then for all $f_1, f_2 \in A$, the equations

$$af_1(x) + bf_2(x) = r \qquad \text{and} \qquad af_1(y) + bf_2(y) = s$$

have no solution for a and b. Thus the determinant $f_1(x)f_2(y) - f_1(y)f_2(x)$ is 0. In particular, if $f_1 = f$, $f_2 = f^2$, we obtain $f(x)f^2(y) = f(y)f^2(x)$, or $f(x)f(y)(f(y) - f(x)) = 0$ for all $f \in A$. Thus if $g \in A$ and $g(x) \neq g(y)$, we have $g(x)g(y) = 0$. If, say, $g(x) = 0$, then (taking $f_1 = g$ in the above determinant) $g(x)f_2(y) = g(y)f_2(x)$ for all $f_2 \in A$. Since $g(x) = 0$ we must have $g(y) \neq 0$; hence $f_2(x) = 0$ for all $f_2 \in A$, contradicting the hypothesis. The argument is symmetrical if $g(y) = 0$. ∎

A7.5 ***Stone–Weierstrass Theorem.*** Let A be a subalgebra of $C(\Omega)$ that separates points.

(a) If for each $x \in \Omega$, there is an $f \in A$ with $f(x) \neq 0$ (in particular, if the constant functions belong to A), then $\bar{A} = C(\Omega)$; thus A is (uniformly) dense in $C(\Omega)$.

(b) If all $f \in A$ vanish at a particular $x \in \Omega$ (there can be at most one such point since A separates points), then $\bar{A} = \{g \in C(\Omega): g(x) = 0\}$.

PROOF. (a) Since A is an algebra, so is $\bar{A}$; by A7.3, $\bar{A}$ is closed under the lattice operations. Let $g \in C(\Omega)$; if $x, y \in \Omega$, $x \neq y$, then by A7.4 there is an $f \in A \subset \bar{A}$ such that $f(x) = r = g(x)$ and $f(y) = s = g(y)$. Thus g can be approximated at each pair of points by functions in $\bar{A}$. By A7.2, $g \in \bar{A}$; in other words, $\bar{A} = C(\Omega)$.

(b) If $f(x) = 0$ for all $f \in C(\Omega)$, let B be the algebra generated by A and the constant functions; thus $B = \{f + c : f \in A, c \text{ a constant function}\}$. Then B satisfies the hypothesis of (a), so $\bar{B} = C(\Omega)$. Now if $\delta > 0$, $g \in C(\Omega)$, and $g(x) = 0$, then for some $f \in A$ and some c we have $\|g - (f + c)\| < \delta/2$. But $g(x) = f(x) = 0$; hence $|c| < \delta/2$, and therefore $\|g - f\| < \delta$. Thus $g \in \bar{A}$, so that $\{g \in C(\Omega);\ g(x) = 0\} \subset \bar{A}$. But the reverse inclusion holds by hypothesis, completing the proof. ∎

A7.6 ***Corollary.*** The Stone–Weierstrass theorem holds equally well with $C(\Omega)$ taken as the collection of all continuous *complex-valued* functions on Ω, if we add the hypothesis that whenever $f \in A$, the complex conjugate $\bar{f}$ also belongs to A.

PROOF. (a) Let $B \subset A$ be the set of functions in A that assume only real values. If $x \neq y$, there is an $f = f_1 + if_2 \in A$ with $f(x) \neq f(y)$. Thus either $f_1(x) \neq f_1(y)$ or $f_2(x) \neq f_2(y)$. Now $f_1 = (f + \bar{f})/2$ and $f_2 = (f - \bar{f})/2i$ belong to A by hypothesis; hence $f_1, f_2 \in B$ since f_1 and f_2 are real-valued. Thus B separates points. Furthermore, if $x \in \Omega$, then $f(x) \neq 0$ for some $f \in A$; but $|f|^2 = f\bar{f} \in B$, so B satisfies the hypothesis of A7.5(a). If $f = f_1 + if_2 \in C(\Omega)$, then f_1 and f_2 can be uniformly approximated by functions in B; hence f can be uniformly approximated by functions in A.

(b) This follows from (a) exactly as in A7.5. ▌

If Ω is the unit disk in the complex plane and A is the collection of all analytic functions on Ω, then A is a closed subalgebra of $C(\Omega)$ that separates points and contains the constant functions; but $\bar{A} \neq C(\Omega)$. Thus the hypothesis $f \in A$ implies $\bar{f} \in A$ cannot be dropped in A7.6.

A8 Topologies on Function Spaces

In this section we examine spaces consisting of functions from an arbitrary set Ω to another set Ω_1. Some structure will be imposed on Ω_1; it can be a metric space or, more generally, a gauge space, which we now define.

A8.1 ***Definitions and Comments.*** A *gauge space* is a space Ω_1 whose topology is determined by a family $\mathscr{D} = \mathscr{D}(\Omega_1)$ of pseudometrics; thus a subbase for the topology is formed by the sets $B_d(x, \delta) = \{y \in \Omega_1 : d(x, y) < \delta\}$, $x \in \Omega_1$, $\delta > 0$, $d \in \mathscr{D}$. [A pseudometric has all the properties of a metric except that $d(x, y)$ can be 0 for $x \neq y$.] Convergence of x_n to x in Ω_1 means $d(x_n, x) \to 0$ for all $d \in \mathscr{D}$.

Since $B_{d_1}(x, \delta) \cap B_{d_2}(x, \delta) = B_d(x, \delta)$, where d is the pseudometric $\max(d_1, d_2)$, it follows that the sets $B_{d^+}(x, \delta)$, $d^+ = \max(d_{i_1}, \ldots, d_{i_n})$, $n = 1, 2, \ldots$, the $d_{i_k} \in \mathscr{D}$, form a base for the topology. Denote by $\mathscr{D}^+ = \mathscr{D}^+(\Omega_1)$ the collection of all pseudometrics d^+.

The gauge space Ω_1 is Hausdorff iff whenever $x \neq y$, there is a $d \in \mathscr{D}$ with $d(x, y) \neq 0$; in this case, $\mathscr{D}$ is said to be *separating*.

The Hausdorff gauge spaces coincide with the completely regular spaces. To see this, let x belong to the open set U in the gauge space Ω_1. If $d \in \mathscr{D}^+$ and $B_d(x, \delta) \subset U$, let $f(y) = \min(1, d(x, y)/\delta)$. Then f is continuous, $0 \leq f \leq 1$, $f(x) = 0$, and $f = 1$ off U. Conversely, if Ω_1 is completely regular, it can be embedded in a product of closed unit intervals (see A4.6). Now a product of gauge spaces is a gauge space [take $d(x, y) = d_i(x_i, y_i)$, where the d_i are pseudometrics for the ith coordinate space], and a subspace of a gauge space is a gauge space, and the result follows.

If $\mathscr{D}$ consists of countably many pseudometrics $d_1, d_2, \ldots$, let

$$d = \sum_{n=1}^{\infty} \min(d_n, 2^{-n});$$

the single pseudometric d incudes the topology of Ω_1; hence Ω_1 is pseudometrizable (metrizable if Ω_1 is Hausdorff).

We now discuss function space topologies.

A8.2 ***Definitions and Comments.*** Let $F(\Omega, \Omega_1)$ be the collection of all functions from the set Ω to the gauge space Ω_1. If $\mathscr{A}$ is a family of subsets of Ω, the topology $\mathscr{T}_{\mathscr{A}}$ of *uniform convergence on members of* $\mathscr{A}$ has as subbase the sets

$$\{g \in F(\Omega, \Omega_1) : \sup_{x \in A} d(g(x), \ f(x)) < \delta\},$$

$$f \in F(\Omega, \Omega_1), \qquad A \in \mathscr{A}, \qquad \delta > 0, \qquad d \in \mathscr{D}.$$

Convergence relative to $\mathscr{T}_{\mathscr{A}}$ means uniform convergence on each $A \in \mathscr{A}$.

For example, if $\mathscr{A} = \{\Omega\}$, we obtain the topology of uniform convergence on Ω, also called the *uniform topology*; if Ω is a topological space and $\mathscr{A}$ consists of all compact subsets, we obtain the topology of uniform convergence on compact sets. If $\mathscr{A}$ is the collection of all singletons $\{x\}$, $x \in \Omega$, we have the topology of pointwise convergence.

Adaptations of standard proofs in metric spaces show the following:

A8.3 ***Theorem.*** If $C(\Omega, \Omega_1)$ is the collection of all continuous maps from the topological space Ω to the gauge space Ω_1, then $C(\Omega, \Omega_1)$ is closed in $F(\Omega, \Omega_1)$ relative to the topology of uniform convergence on Ω. In other words, a uniform limit of continuous functions is continuous.

PROOF. If $f_n \in C(\Omega, \Omega_1)$, $f_n \to f$ uniformly on Ω, $x_0 \in \Omega$, and $d \in \mathscr{D}$, then

$$d(f(x), f(x_0)) \le d(f(x), f_n(x)) + d(f_n(x), f_n(x_0)) + d(f_n(x_0), f(x_0)).$$

The result follows just as in the metric space proof. ∎

A8.4 ***Theorem.*** If f is a continuous mapping from the compact gauge space Ω to the gauge space Ω_1, then f is *uniformly continuous*; that is, if $d \in \mathscr{D}(\Omega_1)$ and $\varepsilon > 0$, there is a $d^+ \in \mathscr{D}^+(\Omega)$ and a $\delta > 0$ such that if $x_1, x_2 \in \Omega$ and $d^+(x_1, x_2) < \delta$, then $d(f(x_1), f(x_2)) < \varepsilon$.

PROOF. If f is not uniformly continuous, there is an $\varepsilon > 0$ and $e \in \mathscr{D}(\Omega_1)$ such that for all $\delta > 0$, $d^+ \in \mathscr{D}^+(\Omega)$, there are points $x, y \in \Omega$ with $d^+(x, y) < \delta$ but $e(f(x), f(y)) \geq \varepsilon$. If we choose such (x, y) for each $\delta > 0$ and $d \in \mathscr{D}(\Omega)$, we obtain a net (x_n, y_n) with $d(x_n, y_n) \to 0$ but $e(f(x_n), f(y_n)) \geq \varepsilon$ for all n. (Take $(\delta_1, d) \geq (\delta_2, d')$ iff $\delta_1 \leq \delta_2$.) By compactness we find a subnet with $(x_{n_k}, y_{n_k}) \to (x_0, y_0)$ for some $x_0, y_0 \in \Omega$; thus $f(x_{n_k}) \to f(x_0)$, $f(y_{n_k}) \to f(y_0)$ by continuity. But $d(x_{n_k}, y_{n_k}) \to 0$, so $d(x_0, y_0) = 0$ and consequently, $e(f(x_0), f(y_0)) = 0$ by continuity of f, a contradiction. ∎

We now prove a basic compactness theorem in function spaces.

A8.5 ***Arzela–Ascoli Theorem.*** Let Ω be a compact topological space, Ω_1 a Hausdorff gauge space, and $G \subset C(\Omega, \Omega_1)$, with the uniform topology. Then G is compact iff the following three conditions are satisfied:

(a) G is closed,
(b) $\{g(x) : g \in G\}$ is a relatively compact subset of Ω_1 for each $x \in \Omega$, and
(c) G is *equicontinuous* at each point of Ω; that is, if $\varepsilon > 0$, $d \in \mathscr{D}(\Omega_1)$, $x_0 \in \Omega$, there is a neighborhood U of x_0 such that if $x \in U$, then

$$d(g(x), g(x_0)) < \varepsilon$$

for all $g \in G$.

PROOF. We first note two facts about equicontinuity.

(1) If $M \subset F(\Omega, \Omega_1)$, where Ω is a topological space and Ω_1 is a gauge space, and M is equicontinuous at x_0, the closure of M in the topology of pointwise convergence is also equicontinuous at x_0.

(2) If M is equicontinuous at all $x \in \Omega$, then on M, the topology of pointwise convergence coincides with the topology of uniform convergence on compact subsets.

To prove (1), let $f_n \in M$, $f_n \to f$ pointwise; if $d \in \mathscr{D}(\Omega_1)$, we have

$$d(f(x), f(x_0)) \leq d(f(x), f_n(x)) + d(f_n(x), f_n(x_0)) + d(f_n(x_0), f(x_0)).$$

If $\delta > 0$, the third term on the right will eventually be less than $\delta/3$ by the pointwise convergence, and the second term will be less than $\delta/3$ for x in some neighborhood U of x_0, by equicontinuity. If $x \in U$, the first term is eventually less than $\delta/3$ by pointwise convergence, and the result follows.

To prove (2), let $f_n \in M$, $f_n \to f$ pointwise, and let K be a compact subset of Ω; fix $\delta > 0$ and $d \in \mathscr{D}(\Omega_1)$. If $x \in K$, equicontinuity yields a neighborhood

$U(x)$ such that $y \in U(x)$ implies $d(f_n(y), f_n(x)) < \delta/3$ for all n. By compactness, $K \subset \bigcup_{i=1}^{r} U(x_i)$ for some $x_1, \ldots, x_r$. Then

$$d(f(x), f_n(x)) \le d(f(x), f(x_i)) + d(f(x_i), f_n(x_i)) + d(f_n(x_i), f_n(x)).$$

If $x \in K$, then $x \in U(x_i)$ for some i; thus the third term on the right is less than $\delta/3$ for all n, so that the first term is less than or equal to $\delta/3$. The second term is eventually less than $\delta/3$ by pointwise convergence, and it follows that $f_n \to f$ uniformly on K.

Now assume (a)–(c) hold. Since $G \subset \prod_{x \in \Omega} \overline{\{g(x) : g \in G\}}$, which is pointwise compact by (b) and the Tychonoff theorem A5.4, the pointwise closure G_0 of G is pointwise compact. Thus if $\{g_n\}$ is a net in G, there is a subnet converging pointwise to some $g \in G_0$. By (c) and (1), G_0 is equicontinuous at each point of Ω; hence by (2), the subnet converges uniformly to g. But g is continuous by A8.3; hence $g \in G$ by (a).

Conversely, assume G compact. Since Ω_1 is Hausdorff, so is $C(\Omega, \Omega_1)$ [as well as $F(\Omega, \Omega_1)$], hence G is closed, proving (a). The map $g \to g(x)$ of G into Ω_1 is continuous, and (b) follows from A5.3(a). Finally, if G is not equicontinuous at x, there is an $\varepsilon > 0$ and a $d \in \mathscr{D}(\Omega_1)$ such that for each neighborhood U of x there is an $x_U \in U$ and $g_U \in G$ with $d(g_U(x), g_U(x_U)) \ge \varepsilon$. If $U \ge V$ means $U \subset V$, the g_U form a net in G, so there is a subnet converging uniformly to a limit $g \in G$. But $x_U \to x$; hence $g_U(x_U) \to g(x)$, a contradiction.

[The last step follows from the fact that the map $(x, g) \to g(x)$ of $\Omega \times G$ into Ω_1 is continuous. To see this, let $x_n \to x$, and $g_n \to g$ uniformly on Ω; if $d \in \mathscr{D}(\Omega_1)$, then

$$d(g_n(x_n), g(x)) \le d(g_n(x_n), g(x_n)) + d(g(x_n), g(x)).$$

The first term approaches 0 by uniform convergence, and the second term approaches 0 by continuity of g.] ∎

If G satisfies (b) and (c) of A8.5, but not necessarily (a), the closure of G in the uniform topology satisfies all three hypotheses, and hence is compact. In the special case when Ω_1 is the set of complex numbers, $C(\Omega, \Omega_1) = C(\Omega)$ is a Banach space, and in particular, compactness and sequential compactness are equivalent. We thus obtain the most familiar form of the Arzela–Ascoli theorem: If $f_1, f_2, \ldots$ is a sequence of continuous complex-valued functions on the compact space Ω, and if the f_n are pointwise bounded and equicontinuous, there is a uniformly convergent subsequence. In fact the f_n must be uniformly bounded by the continuity of the map $(x, g) \to g(x)$ referred to above.

A9 Complete Metric Spaces and Category Theorems

A9.1 ***Definitions and Comments.*** A metric space (or the associated metric d) is said to be *complete* iff each Cauchy sequence (that is, each sequence such that $d(x_n, x_m) \to 0$ as $n, m \to \infty$) converges to a point in the space.

Any compact metric space is complete since a Cauchy sequence with a convergent subsequence converges.

It is important to recognize that *completeness is not a topological property.* In other words, two metrics, only one of which is complete, may be equivalent, that is, induce the same topology. As an example, the chordal metric d' on the complex plane [$d'(z_1, z_2)$ is the Euclidean distance between the stereographic projections of z_1 and z_2 on the Riemann sphere] is not complete, but the equivalent Euclidean metric d is complete. Note also that the sequence $z_n = n$ is d'-Cauchy but not d-Cauchy, so the notion of a Cauchy sequence is not topological.

The subset A of the topological space Ω is said to be *nowhere dense* iff the interior of $\bar{A}$ is empty, in other words, iff the complement of $\bar{A}$ is dense. A set $B \subset \Omega$ is said to be of *category* 1 *in* Ω iff B can be expressed as a countable union of nowhere dense subsets of Ω; otherwise B is of *category* 2 *in* Ω.

The following result is the best known category theorem.

A9.2 ***Baire Category Theorem.*** Let (Ω, d) be a complete metric space. If A_n is closed in Ω for each $n = 1, 2, \ldots,$ and $\bigcup_{n=1}^{\infty} A_n = \Omega$, then the interior $A_n{}^0$ is nonempty for some n. Therefore Ω is of category 2 in itself.

PROOF. Assume $A_n{}^0 = \varnothing$ for all n. Then $A_1 \neq \Omega$, and since $\Omega - A_1$ is open, there is a ball $B(x_1, \delta_1) \subset \Omega - A_1$ with $0 < \delta_1 < \frac{1}{2}$. Now $B(x_1, \frac{1}{2}\delta_1) \not\subset A_2$ since $A_2{}^0 = \varnothing$; therefore $B(x_1, \frac{1}{2}\delta_1) - A_2$ is a nonempty open set, hence includes a ball $B(x_2, \delta_2)$ with $0 < \delta_2 < \frac{1}{4}$. Inductively, we find $B(x_n, \delta_n)$, $0 < \delta_n < 2^{-n}$, such that $B(x_n, \delta_n) \subset B(x_{n-1}, \frac{1}{2}\delta_{n-1}) - A_n$ [in particular, $B(x_n, \delta_n) \cap A_n = \varnothing$]. Now if $n < m$,

$$d(x_n, x_m) \leq \sum_{i=n}^{m-1} d(x_i, x_{i+1}) < \sum_{i=n}^{m-1} 2^{-i} \to 0 \qquad \text{as} \quad n, m \to \infty,$$

so by completeness, x_n converges to some $x \in \Omega$. Since $x_k \in B(x_n, \frac{1}{2}\delta_n)$ for $k > n$, we have $x \in B(x_n, \delta_n)$ for all n, hence x cannot belong to any A_n, a contradiction. ∎

In A9.2, if for each $n = 1, 2, \ldots, U_n$ is an open dense subset of Ω, then $\bigcap_{n=1}^{\infty} U_n \neq \varnothing$, for if the intersection is empty, then Ω is the union of the nowhere dense sets $\Omega - U_n$. We prove a stronger result:

A9.3 ***Theorem.*** If U_n is an open dense subset of the complete metric space Ω $(n = 1, 2, \ldots)$, then $\bigcap_{n=1}^{\infty} U_n$ is dense.

PROOF. Consider any ball B. Then $\bar{B}$ is a closed subset of the complete metric space Ω, hence $\bar{B}$ is complete. Now $U_n \cap B$ is dense in B since B is open; hence $U_n \cap B$ (and therefore $U_n \cap \bar{B}$) is dense in $\bar{B}$. By the remark after A9.2, $(\bigcap_{n=1}^{\infty} U_n) \cap \bar{B} \neq \varnothing$. Thus $\bigcap_{n=1}^{\infty} U_n$ meets every closed ball, hence every open ball, hence every nonempty open set. ▮

Other important spaces satisfy the conclusion of A9.3:

A9.4 ***Theorem.*** If U_n is an open dense subset of the locally compact Hausdorff space Ω $(n = 1, 2, \ldots)$, then $\bigcap_{n=1}^{\infty} U_n$ is dense.

PROOF. Let U be a nonempty open subset of Ω; we must show that $U \cap \bigcap_{n=1}^{\infty} U_n \neq \varnothing$. Since U_1 is dense, $U \cap U_1 \neq \varnothing$, and by local compactness (see A5.12), there is a nonempty, relatively compact, open set V_1 with $\bar{V}_1 \subset U \cap U_1$.

Now $V_1 \cap U_2 \neq \varnothing$ so we may choose a nonempty, relatively compact, open set V_2 with $\bar{V}_2 \subset V_1 \cap U_2$. Inductively we select nonempty, relatively compact, open sets V_n with $\bar{V}_n \subset V_{n-1} \cap U_n$.

Since the $\bar{V}_n$ decrease with n, they have the finite intersection property, and since all $\bar{V}_n$ are subsets of the compact set $\bar{V}_1$, $\bigcap_{n=1}^{\infty} \bar{V}_n \neq \varnothing$. If x belongs to this intersection, then $x \in U \cap \bigcap_{n=1}^{\infty} U_n$. ▮

A9.5 ***Definition.*** The topological space Ω is said to be a *Baire space* iff the intersection of countably many open dense subsets of Ω is dense in Ω. By A9.3 and A9.4, complete metric spaces and locally compact Hausdorff spaces are Baire spaces.

A9.6 ***Theorem.*** (a) Ω is a Baire space iff for each A of category 1 in Ω, $A^0 = \varnothing$; that is, $\Omega - A$ is dense.

(b) Let Ω be a Baire space. If $\Omega = \bigcup_{n=1}^{\infty} A_n$ where the A_n are closed, then $A_n{}^0 \neq \varnothing$ for some n. Thus Ω is of category 2 in itself.

PROOF. (a) Assume Ω is a Baire space, and let $A = \bigcup_{n=1}^{\infty} A_n$, where the A_n are nowhere dense. Then $\Omega - A = \bigcap_{n=1}^{\infty} (\Omega - A_n) \supset \bigcap_{n=1}^{\infty} (\Omega - \bar{A}_n)$, which is dense by hypothesis.

Conversely, if every set of category 1 in Ω has empty interior, let U_n be dense in Ω, $n = 1, 2, \ldots$. Then $\Omega - U_n$ is (closed and) nowhere dense, so

$\bigcup_{n=1}^{\infty} (\Omega - U_n)$ is of category 1 in Ω. By hypothesis, the complement of this set, namely $\bigcap_{n=1}^{\infty} U_n$, is dense.

(b) If $A_n{}^0 = \varnothing$ for all n, then Ω is of category 1 in itself, contradicting (a). ▮

There are spaces that cannot be expressed as a countable union of nowhere dense sets, but are not Baire spaces. (As an example, consider the free union of a Baire space and a non-Baire space.)

A9.7 ***Definition.*** The topological space Ω is said to be *topologically complete* iff there is a complete metric d on Ω that induces the given topology.

The following sequence of results is aimed at characterizing topologically complete spaces.

A9.8 ***Theorem.*** Let (Ω, d) be a complete metric space. If U is an open subset of Ω, U is topologically complete.

PROOF. If $x, y \in U$, let $d'(x, y) = d(x, y) + |f(x) - f(y)|$, where $f(x) = [d(x, \Omega - U)]^{-1}$. Then d' is a metric on U, and d-convergence is equivalent to d'-convergence by continuity of f. Thus d' induces the original topology. If $\{x_n\}$ is d'-Cauchy, it is d-Cauchy; hence it converges (relative to d) to some $x \in \Omega$. But $x \in U$; for if not, then $f(x_n) \to f(x) = \infty$, contradicting $\{x_n\}$ d'-Cauchy. Since d and d' are equivalent on U, x_n converges to x relative to d'. ▮

A9.9 ***Theorem.*** Let (Ω, d) be a complete metric space, and U_n, $n = 1, 2, \ldots,$ open subsets of Ω. If $A = \bigcap_{n=1}^{\infty} U_n$, that is, if A is a G_δ, then A, with the relative topology, is topologically complete.

PROOF. Let d_n be a complete metric for U_n with d_n equivalent to d [d_n exists by A9.8; we may assume that $d_n \le 1$, for if not replace d_n by $d_n/(1 + d_n)$.] Define $g: A \to B = \prod_{i=1}^{\infty} U_i$ by $g(x) = (x, x, \ldots)$, and define a metric d_0 on B by $d_0(x, y) = \sum_{i=1}^{\infty} 2^{-i} d_i(x_i, y_i)$. Then d_0 induces the product topology on B, and a sequence $\{x^{(n)}\}$ in B is d_0-Cauchy iff $\{x_i^{(n)}\}$ is d_i-Cauchy for each i. It follows that d_0 is complete. (Thus we have shown that a countable product of topologically complete spaces is topologically complete.)

Now g is a homeomorphism of A and $g(A) \subset B$, and in fact $g(A)$ is closed in B. For if $(x_n, x_n, \ldots) \to y \in B$, then for each i we have $d_i(x_n, y_i) \to 0$ as $n \to \infty$, so that $x_n \to y_i$. But then y_i is the same for all i, say $y_i = y_0$. Since

$y_i \in U_i$ for each i we have $y_0 \in \bigcap_{i=1}^{\infty} U_i = A$, hence $y = (y_0, y_0, \ldots) \in g(A)$, proving $g(A)$ closed. Since B is complete, so is $g(A)$, and since g is a homeomorphism, A is topologically complete. ▮

It follows from A9.9 that the irrational numbers are topologically complete. In fact, a complete metric for the set of irrationals in (0, 1) is given by $d(.a_1 a_2 \ldots, .b_1 b_2, \ldots) = 1/n$, where n is the first integer for which $a_n \neq b_n$ (in the decimal expansion).

A9.10 ***Theorem.*** Let A be a topologically complete subspace of the Hausdorff space Ω. If A is dense in Ω, then A is a G_δ.

PROOF. Let $U_n = \{x \in \Omega$: for some neighborhood V of x, the diameter of $V \cap A$ is less than $1/n\}$. (The diameter is calculated using a particular complete metric d for A.) If $x \in U_n$, then $x \in V \subset U_n$, so U_n is open in Ω. We are going to show that $A = \bigcap_{n=1}^{\infty} U_n$.

Let $x \in A$, and let $V_0 = \{y \in A : d(x, y) < 1/4n\}$. Then V_0 is open in A; hence $V_0 = V \cap A$, where V is open in Ω. Now $x \in V_0 \subset V$, and if $y, z \in V_0$, we have $d(y, z) \leq d(y, x) + d(x, z) < 1/2n$, so that V_0 has diameter less than or equal to $1/2n < 1/n$. It follows that $x \in \bigcap_{n=1}^{\infty} U_n$.

Conversely, assume $x \in \bigcap_{n=1}^{\infty} U_n$. Since A is dense, there is a net $x_i \in A$ with $x_i \to x$. For each n, let V be a neighborhood of x such that $V \cap A$ has diameter less than $1/n$. Eventually $x_i \in V$, and it follows that for some index i_n we have $d(x_i, x_j) < 1/n$ for $i, j \geq i_n$. (In other words, the net is Cauchy.) It may be assumed that $i_n \leq i_{n+1}$ for all n; then the x_{i_n} form a Cauchy sequence which converges to a limit $y \in A$ by completeness. Since Ω is Hausdorff, we have $y = x$; hence $x \in A$. ▮

We now prove the main theorem on topological completeness.

A9.11 ***Theorem.*** Let Ω be a metrizable topological space. Then Ω is topologically complete iff it is an absolute G_δ; in other words, Ω is a G_δ in every metric space in which it is topologically embedded.

PROOF. If Ω is topologically complete and is embedded in the metric space Ω_1, then Ω is dense in $\overline{\Omega}$; so by A9.10, Ω is a G_δ in $\overline{\Omega}$. Thus Ω can be expressed as $\bigcap_{n=1}^{\infty} W_n$, where $W_n = U_n \cap \overline{\Omega}$, U_n open in Ω_1. But $\overline{\Omega}$ is a closed subset of the metric space Ω_1, so $\overline{\Omega}$ is a G_δ in Ω_1 $(\overline{\Omega} = \bigcap_{n=1}^{\infty} \{x \in \Omega_1 : d(x, \overline{\Omega}) < 1/n\})$. It follows that Ω is a G_δ in Ω_1.

Conversely, let Ω be an absolute G_δ. Embed Ω in its completion (we assume as known the standard process of completing a metric space by forming equivalence classes of Cauchy sequences). By A9.9, Ω is topologically complete. ▌

We now establish topological completeness for a wide class of spaces.

A9.12 ***Theorem.*** A locally compact metric space is topologically complete.

PROOF. If Ω is such a space and $\overline{\Omega}$ is its completion, then Ω is open in $\overline{\Omega}$. For if $x \in \Omega$, there is, by local compactness, an open (in $\overline{\Omega}$) set V with $x \in V \cap \Omega$ and $\overline{V} \cap \Omega$ compact. In fact $V \subset \Omega$, proving Ω open. For if $y \in V$ and $y \notin \Omega$, then for each $U \in \mathscr{U}(y)$, $U \cap V \cap \Omega \neq \varnothing$ since Ω is dense; consequently, $U \cap \overline{V} \cap \Omega \neq \varnothing$. The sets $U \cap \overline{V} \cap \Omega$, $U \in \mathscr{U}(y)$, form a filterbase $\mathscr{B}$ in $\overline{V} \cap \Omega$ converging to y, and since $\overline{\Omega}$ is Hausdorff, y is the only possible accumulation point of $\mathscr{B}$. But $y \notin \overline{V} \cap \Omega$, contradicting compactness of $\overline{V} \cap \Omega$ [see A5.2(d)]. By A9.8, Ω is topologically complete. ▌

A10 Uniform Spaces

We now give an alternative way of describing gauge spaces.

A10.1 ***Definitions and Comments.*** Let V be a relation on the set Ω, that is, a subset of $\Omega \times \Omega$. Then V is called a *connector* iff the diagonal

$$D = \{(x, x) : x \in \Omega\}$$

is a subset of V. A nonempty collection $\mathscr{H}$ of connectors is called a *uniformity* iff

(a) for all $V_1, V_2 \in \mathscr{H}$, there is a $W \in \mathscr{H}$ with $W \subset V_1 \cap V_2$ (in other words $\mathscr{H}$ is a filterbase), and

(b) for each $V \in \mathscr{H}$, there is a $W \in \mathscr{H}$ with $WW^{-1} \subset V$.

[The relation WW^{-1} is the composition of the two relations W and W^{-1}; that is, $(x, z) \in WW^{-1}$ iff for some $y \in \Omega$ we have $(x, y) \in W^{-1}$ and $(y, z) \in W$; $(x, y) \in W^{-1}$ means $(y, x) \in W$.]

If in addition, $\mathscr{H}$ is a filter, that is, if $V \in \mathscr{H}$ and $V \subset W$, then $W \in \mathscr{H}$, then $\mathscr{H}$ is called a *uniform structure*. In particular, if $\mathscr{H}$ is a uniformity, the filter generated by $\mathscr{H}$ is a uniform structure.

As an example, let $\mathscr{D}$ be a family of pseudometrics on Ω, and let $\mathscr{H}$ consist of all sets of the form $V = \{(x, y): d^+(x, y) < \delta\}$, $\delta > 0$, $d^+ \in \mathscr{D}^+$; then $\mathscr{H}$ is a uniformity on Ω. [To see that condition (b) is satisfied, let

$$W = \{(x, y): d^+(x, y) < \delta/2\};$$

then $WW^{-1} \subset V$ by the triangle inequality.]

By analogy with the pseudometric case, if V belongs to the uniformity $\mathscr{H}$, we say that x and y are *V-close* iff $(x, y) \in V$.

Two uniformities $\mathscr{H}_1$ and $\mathscr{H}_2$ are said to be *equivalent* iff they generate the same uniform structure; in other words, given $V_2 \in \mathscr{H}_2$, there is a $V_1 \in \mathscr{H}_1$ with $V_1 \subset V_2$ and given $W_1 \in \mathscr{H}_1$, there is a $W_2 \in \mathscr{H}_2$ with $W_2 \subset W_1$. Two uniform structures that are equivalent must be equal; hence *there is exactly one uniform structure in each equivalence class of uniformities.*

If $\mathscr{H}$ is a uniform structure and $V, W \in \mathscr{H}$, $WW^{-1} \subset V$, then if D is the diagonal, we have

$$W^{-1} = DW^{-1} \subset WW^{-1} \subset V,$$

hence $W \subset V^{-1}$, so that $V^{-1} \in \mathscr{H}$. But then

$$V \cap V^{-1} \in \mathscr{H} \text{ (and } V \cap V^{-1} \subset V).$$

Now $U = V \cap V^{-1}$ is *symmetric*, that is, $U = U^{-1}$. Thus if $\mathscr{H}$ is a uniform structure, the symmetric sets in $\mathscr{H}$ form a uniformity that generates $\mathscr{H}$, so that *every uniformity is equivalent to a uniformity containing only symmetric sets.*

We are going to show later that every uniform structure is generated by a uniformity corresponding to a family of pseudometrics. As a preliminary, we establish the following result:

A10.2 ***Metrization Lemma.*** Let U_n, $n = 0, 1, 2, \ldots$, be a sequence of subsets of $\Omega \times \Omega$ such that each U_n is a connector, $U_0 = \Omega \times \Omega$, and U_{n+1}^3 $(= U_{n+1}U_{n+1}U_{n+1}) \subset U_n$ for all n. Then there is a function d from $\Omega \times \Omega$ to the nonnegative reals, such that d satisfies the triangle inequality and $U_n \subset \{(x, y): d(x, y) < 2^{-n}\} \subset U_{n-1}$ for all $n = 1, 2, \ldots$. If each U_n is symmetric, there is a pseudometric satisfying this condition.

PROOF. Define $f(x, y) = 0$ if $(x, y) \in U_n$ for all n; $f(x, y) = \frac{1}{2}$ if $(x, y) \in U_0 - U_1$; $f(x, y) = \frac{1}{4}$ if $(x, y) \in U_1 - U_2$, and in general, $f(x, y) = 2^{-n}$ if $(x, y) \in U_{n-1} - U_n$. Since $U_{n+1} \subset U_{n+1}^3 \subset U_n$, f is well-defined on $\Omega \times \Omega$; furthermore,

$$(x, y) \in U_n \quad \text{iff} \quad f(x, y) < 2^{-n}. \tag{1}$$

If $x, y \in \Omega$, let $d(x, y) = \inf \sum_{i=0}^{n} f(x_i, x_{i+1})$, where the inf is taken over all finite sequences $x_0, x_1, \ldots, x_{n+1}$ with $x_0 = x$, $x_{n+1} = y$.

Since a chain from x to y followed by a chain from y to z yields a chain from x to z, d satisfies the triangle inequality, and since $d(x, y) \le f(x, y)$, $U_n \subset \{(x, y)\colon d(x, y) < 2^{-n}\}$ by (1). If the U_n are symmetric, then so is d; hence d is a pseudometric.

Now we claim that

$$f(x_0, x_{n+1}) \le 2 \sum_{i=0}^{n} f(x_i, x_{i+1}) \qquad \text{for any} \quad x_0, \ldots, x_{n+1}. \tag{2}$$

This is clear for $n = 0$, so assume the inequality for all integers up to $n - 1$. Let $a = \sum_{i=0}^{n} f(x_i, x_{i+1})$. If $a = 0$, then $(x_i, x_{i+1}) \in \bigcap_{n=0}^{\infty} U_n$, and since $U_i^3 \subset U_{i-1}$ we have $(x_0, x_{n+1}) \in U_i$ for all i, and there is nothing further to prove. Thus assume $a > 0$.

Let k be the largest integer such that $\sum_{i=0}^{k-1} f(x_i, x_{i+1}) \le a/2$ [if $f(x_0, x_1) > a/2$, take $k = 0$]. Then $\sum_{i=0}^{k} f(x_i, x_{i+1}) > a/2$, so $\sum_{i=k+1}^{n} f(x_i, x_{i+1}) \le a/2$. By induction hypothesis, $f(x_0, x_k) \le 2 \sum_{i=0}^{k-1} f(x_i, x_{i+1}) \le 2a/2 = a$ and $(x_{k+1}, x_{n+1}) \le 2 \sum_{i=k+1}^{n} f(x_i, x_{i+1}) \le 2a/2 = a$. Also,

$$f(x_k, x_{k+1}) \le \sum_{i=0}^{n} f(x_i, x_{i+1}) = a.$$

Let m be the smallest nonnegative integer such that $2^{-m} \le a$. If $m \le 2$, then (2) holds automatically since $f \le \frac{1}{2}$; thus assume $m > 2$. Then $2^{-m} \le a < 2^{-(m-1)}$, so $f(x_0, x_k) \le a < 2^{-(m-1)}$, and therefore by (1), (x_0, x_k) [and similarly, (x_k, x_{k+1}), (x_{k+1}, x_{n+1})] belongs to U_{m-1}. Thus $(x_0, x_{n+1}) \in U_{m-1}^3 \subset U_{m-2}$, which implies that $f(x_0, x_{n+1}) < 2^{-(m-2)}$; that is,

$$f(x_0, x_{n+1}) \le 2^{-(m-1)} \le 2a,$$

proving (2).

Finally, if $d(x, y) < 2^{-n}$, then for any chain from x to y, (2) yields (with $x_0 = x$, $x_{n+1} = y$) $f(x, y) \le 2d(x, y) < 2^{-(n-1)}$, so $(x, y) \in U_{n-1}$. ∎

A10.3 ***Metrization Theorem.*** If $\mathscr{H}$ is a uniformity, the following conditions are equivalent:

(a) $\mathscr{H}$ is *pseudometrizable*; that is, there is an equivalent uniformity generated by a single pseudometric. (See the example in A10.1 for the construction of the uniformity corresponding to a family of pseudometrics.)

(b) $\mathscr{H}$ has a *countable base*; that is, there are countably many sets V_n in the uniform structure generated by $\mathscr{H}$ such that if $U \in \mathscr{H}$, then $V_n \subset U$ for some n.

PROOF. If $\mathscr{H}$ is pseudometrizable with pseudometric d, the sets

$$\{(x, y): d(x, y) < r\},$$

r rational, form a countable base. Conversely, assume $\mathscr{H}$ has a countable base $V_0, V_1, V_2, \ldots$. We may assume that the V_n are symmetric (if not, consider $V_n \cap V_n^{-1}$), $V_n \supset V_{n+1}$ for all n (if not, consider $\bigcap_{i=0}^n V_i$), and $V_0 = \Omega \times \Omega$. Take $U_0 = V_0$, U_1 a symmetric set in the uniform structure $\mathscr{F}$ generated by $\mathscr{H}$ such that $U_1^3 \subset V_1$ $(\subset U_0)$; let U_2 be a symmetric set in $\mathscr{F}$ such that $U_2^3 \subset V_2 \cap U_1$. In general, let U_n be a symmetric set in $\mathscr{F}$ such that $U_n^3 \subset V_n \cap U_{n-1}$. (To see that the U_n exist, let W_n be a symmetric set in $\mathscr{F}$ with $W_n^2 \subset V_n \cap U_{n-1}$. If U_n is a symmetric set in $\mathscr{F}$ such that $U_n^2 \subset W_n$, then $U_n^3 = U_n^3 D \subset U_n^4 \subset W_n^2 \subset V_n \cap U_{n-1}$. Also, since $U_n \subset U_n^3 \subset V_n$, the U_n form a base for $\mathscr{H}$.)

By A10.2, there is a pseudometric d such that $U_n \subset \{(x, y): d(x, y) < 2^{-n}\} \subset U_{n-1}$, $n = 1, 2, \ldots$. It follows that $\mathscr{H}$ is equivalent to the uniformity of d. ∎

We now construct a topology arising from a uniformity.

A10.4 ***Definitions and Comments.*** Let $\mathscr{H}$ be a uniformity. The *topology induced* by $\mathscr{H}$ is defined by specifying the system $\mathscr{V}(x)$ of overneighborhoods of each point x. If $V \in \mathscr{H}$, let $V(x) = \{y: (x, y) \in V\}$. We take $\mathscr{V}(x)$ as the class of all oversets of the sets $V(x)$, $V \in \mathscr{H}$.

It is immediate that the first three axioms for an overneighborhood system are satisfied (see A1.1). We verify the last axiom [A1.1(d)]. If $V \in \mathscr{H}$, let U be a symmetric set in the uniform structure generated by $\mathscr{H}$ such that $U^2 \subset V$. If $W \in \mathscr{H}$ and $W \subset U$, we claim that $W(y) \subset V(x)$ for each $y \in W(x)$. For if $z \in W(y)$, then $(y, z) \in W$; but also $(x, y) \in W$, and hence $(x, z) \in W^2 \subset U^2 \subset V$, so $z \in V(x)$, as asserted. Thus $V(x) \in \mathscr{V}(y)$ for all $y \in W(x) \subset V(x)$, as required in A1.1(d).

It follows that there is a unique topology having (for each x) the sets $V(x)$, $V \in \mathscr{H}$, as a base for the overneighborhoods of x. Furthermore, any uniformity equivalent to $\mathscr{H}$ will induce the same topology. A topological space whose topology is determined in this way is called a *uniform space*.

In particular, let $\mathscr{D}$ be a family of pseudometrics on Ω, and let $\mathscr{H}$ consist of all sets $V = \{(x, y): d^+(x, y) < \delta\}$, $\delta > 0$, $d^+ \in \mathscr{D}^+$; then

$$V(x) = \{y: d^+(x, y) < \delta\}.$$

Thus the topology induced by $\mathscr{H}$ coincides with the gauge space topology determined by $\mathscr{D}$ (see A8.1). In other words, *every gauge space is a uniform space*. Conversely, *every uniform space is a gauge space*, as we now prove.

A10.5 ***Theorem.*** Let $\mathscr{F}$ be a uniform structure, and let $\mathscr{H}$ be the uniformity generated by the pseudometrics d such that the sets $V_{d\delta} = \{(x, y): d(x, y) < \delta\}$ belong to $\mathscr{F}$ for all $\delta > 0$. Then $\mathscr{F}$ and $\mathscr{H}$ are equivalent, so that the topology induced by $\mathscr{F}$ is identical to the gauge space topology determined by the pseudometrics d.

PROOF. It follows from the definition of $\mathscr{H}$ that $\mathscr{H} \subset \mathscr{F}$. Now if U is a symmetric set in $\mathscr{F}$, set $U_0 = \Omega \times \Omega$, $U_1 = U$, U_2 a symmetric set in $\mathscr{F}$ such that $U_2{}^3 \subset U_1$, and in general, let U_n be a symmetric set in $\mathscr{F}$ such that $U_n{}^3 \subset U_{n-1}$. By A10.2, there is a pseudometric d on Ω with

$$U_n \subset V_{d, 2^{-n}} \subset U_{n-1}$$

for all n. Since $\mathscr{F}$ is a uniform structure, $V_{d, 2^{-n}} \in \mathscr{F}$ for all n, hence $V_{d\delta} \in \mathscr{F}$ for all $\delta > 0$. But $U_2 \subset V_{d, 1/4} \subset U_1 = U$, proving the equivalence of $\mathscr{F}$ and $\mathscr{H}$. ▌

Since uniform spaces and gauge spaces coincide, the concept of uniform continuity, defined previously (see A8.4) for mappings of one gauge space to another, may now be translated into the language of uniform spaces. If $f: \Omega \to \Omega_1$, where Ω and Ω_1 are uniform spaces with uniformities $\mathscr{U}$ and $\mathscr{V}$, f is uniformly continuous iff given $V \in \mathscr{V}$, there is a $U \in \mathscr{U}$ such that $(x, y) \in U$ implies $(f(x), f(y)) \in V$.

We now consider separation properties in uniform spaces.

A10.6 ***Theorem.*** Let Ω be a uniform space, with uniformity $\mathscr{H}$. The following are equivalent:

(a) $\bigcap\{V: V \in \mathscr{H}\}$ is the diagonal D.
(b) Ω is T_2.
(c) Ω is T_1.
(d) Ω is T_0.

PROOF. If (a) holds and $x \neq y$, then $(x, y) \notin V$ for some $V \in \mathscr{H}$. If W is a symmetric set in the uniform structure generated by $\mathscr{H}$, and $W^2 \subset V$, then $W(x)$ and $W(y)$ are disjoint overneighborhoods of x and y, proving (b). If (d) holds and $x \neq y$, there is a set $V \in \mathscr{H}$ such that $y \notin V(x)$ [or a set $W \in \mathscr{H}$ with $x \notin W(y)$]. But then $(x, y) \notin V$ for some $V \in \mathscr{H}$, proving (a). ▌

Finally, we discuss topological groups and topological vector spaces as uniform spaces.

A *topological group* is a group on which there is defined a topology which makes the group operations continuous ($x_n \to x$, $y_n \to y$ implies $x_n y_n \to xy$;

$x_n \to x$ implies $x_n^{-1} \to x^{-1}$). Familiar examples are the integers, with ordinary addition and the discrete topology; the unit circle $\{z\colon |z| = 1\}$ in the complex plane, with multiplication of complex numbers and the Euclidean topology; all nonsingular $n \times n$ matrices of complex numbers, with matrix multiplication and the Euclidean topology (on R^{n^2}). If $\Omega = \prod_i \Omega_i$, where the Ω_i are topological groups, then Ω is a topological group with the product topology if multiplication is defined by $xy = (x_i y_i,\ i \in I)$.

A10.7 ***Theorem.*** Let Ω be a topological group, and let $\mathscr{H}$ consist of all sets $V_N = \{(x, y) \in \Omega \times \Omega\colon yx^{-1} \in N\}$, where N ranges over all overneighborhoods of the identity element e in Ω. Then $\mathscr{H}$ is a uniformity, and the topology induced by $\mathscr{H}$ coincides with the original topology. In particular, if Ω is Hausdorff, it is completely regular.

PROOF. The diagonal is a subset of each V_N, and $V_N \cap V_M = V_{N \cap M}$, so only the last condition [A10.1(b)] for a uniformity need be checked. Let $f(x, y) = xy^{-1}$, a continuous map of $\Omega \times \Omega$ into Ω (by definition of a topological group). If N is an overneighborhood of e, there is an overneighborhood M of e such that $f(M \times M) \subset N$.

We claim that $V_M V_M^{-1} \subset V_N$. For if $(x, y) \in V_M^{-1}$ and $(y, z) \in V_M$, then $(y, x) \in V_M$; hence $xy^{-1} \in M$ and $zy^{-1} \in M$. But $zx^{-1} = (zy^{-1})(yx^{-1}) = (zy^{-1})(xy^{-1})^{-1} \in f(M \times M) \subset N$. Consequently, $(x, z) \in V_N$, so $\mathscr{H}$ is a uniformity.

Let U be an overneighborhood of the point x in the original topology. If $N = Ux^{-1} = \{yx^{-1}\colon y \in U\}$, then $V_N(x) = \{y\colon (x, y) \in V_N\} = \{y\colon yx^{-1} \in N\} = Nx = U$; therefore U is an overneighborhood of x in the topology induced by $\mathscr{H}$. Conversely, let U be an overneighborhood of x in the uniform space topology; then $V_N(x) \subset U$ for some overneighborhood N of e. But $V_N(x) = Nx$, and the map $y \to yx$, carrying N onto Nx, is a homeomorphism of Ω with itself. Thus Nx, hence U, is an overneighborhood of x in the original topology. ∎

A10.8 ***Theorem.*** Let Ω be a topological group, with uniformity $\mathscr{H}$ as defined in A10.7. The following are equivalent:

(a) $\mathscr{H}$ is pseudometrizable (see A10.3).

(b) Ω is pseudometrizable; that is, there is a pseudometric that induces the given topology.

(c) Ω has a countable base of neighborhoods at the identity e (hence at every x, since the map $N \to Nx$ sets up a one-to-one correspondence between neighborhoods of e and neighborhoods of x).

PROOF. We obtain (a) implies (b) by the definition of the topology induced by $\mathscr{H}$ (see A10.4). Since every pseudometric space is first countable, it follows that (b) implies (c). Finally, if (c) holds and the sets $N_1, N_2, \ldots$ form a countable base at e, then by definition of $\mathscr{H}$, the sets $V_{N_1}, V_{N_2}, \ldots$ form a countable base for $\mathscr{H}$; hence, by A10.3, $\mathscr{H}$ is pseudometrizable. ▮

A *topological vector space* is a vector space with a topology that makes addition and scalar multiplication continuous (see 3.5). In particular, a topological vector space is an abelian topological group under addition. The uniformity $\mathscr{H}$ consists of sets of the form $V_N = \{(x, y): y - x \in N\}$, where N is an overneighborhood of 0. Theorem A10.8 shows that a topological vector space is pseudometrizable iff there is a countable base at 0.

The following fact is needed in the proof of the open mapping theorem (see 3.5.9):

A10.9 ***Theorem.*** If L is a pseudometrizable topological vector space, there is an invariant pseudometric $[d(x, y) = d(x + z, y + z)$ for all $z]$ that induces the topology of L.

PROOF. The pseudometric may be constructed by the method given in A10.2 and A10.3, and furthermore, the symmetric sets needed in A10.3 may be taken as sets in the uniformity $\mathscr{H}$ itself rather than in the uniform structure generated by $\mathscr{H}$. For if N is an overneighborhood of 0, so is $-N = \{-x: x \in N\}$ since the map $x \to -x$ is a homeomorphism. Thus $M = N \cap (-N)$ is an overneighborhood of 0. But then V_M is symmetric and $V_M \subset V_N$.

Now by definition of V_M, we have $(x, y) \in V_M$ iff $(x + z, y + z) \in V_M$ for all z, and it follows from the proof of A10.2 that the pseudometric d is invariant. ▮

Bibliography

Apostol, T.M., "Mathematical Analysis." Addison-Wesley, Reading, Massachusetts, 1957.

Bachman, G., and Narici, L., "Functional Analysis." Academic Press, New York, 1966.

Billingsley, P., "Convergence of Probability Measures." Wiley, New York, 1968.

Dubins, L., and Savage, L., "How to Gamble If You Must." McGraw-Hill, New York, 1965.

Dugundji, J., "Topology." Allyn and Bacon, Boston, 1966.

Dunford, N., and Schwartz, J. T., "Linear Operators." Wiley (Interscience), New York, Part 1, 1958; Part 2, 1963; Part 3, 1970.

Halmos, P. R., "Measure Theory." Van Nostrand, Princeton, New Jersey, 1950.

Halmos, P. R., "Introduction to Hilbert Space." Chelsea, New York, 1951.

Halmos, P. R., "Naive Set Theory." Van Nostrand, Princeton, New Jersey, 1960.

Hausdorff, F., "Set Theory." Chelsea, New York, 1962.

Kelley, J. L., and Namioka, I., "Linear Topological Spaces." Van Nostrand, Princeton, New Jersey, 1963.

Liusternik, L., and Sobolev, V., "Elements of Functional Analysis." Ungar, New York, 1961.

Loève, M., "Probability Theory." Van Nostrand, Princeton, New Jersey, 1955; 2nd ed., 1960; 3rd ed., 1963.

Neveu, J., "Mathematical Foundations of the Calculus of Probability." Holden-Day, San Francisco, 1965.

Parthasarathy, K., "Probability Measures on Metric Spaces." Academic Press, New York, 1967.

Royden, H. L. "Real Analysis." Macmillan, New York, 1963; 2nd ed., 1968.

Rudin, W., "Real and Complex Analysis." McGraw-Hill, New York, 1966.

Schaefer, H., "Topological Vector Spaces." Macmillan, New York, 1966.

Simmons, G., "Introduction to Topology and Modern Analysis." McGraw-Hill, New York, 1963.

Taylor, A. E., "Introduction to Functional Analysis." Wiley, New York, 1958.

Titchmarsh, E. C., "The Theory of Functions." Oxford Univ. Press, London and New York, 1939.

Yosida, K., "Functional Analysis." Springer-Verlag, Berlin and New York, 1968.

Solutions to Problems

Chapter 1

Section 1.1

2. We have $\limsup_n A_n = (-1, 1]$, $\liminf_n A_n = \{0\}$.
3. Using $\limsup_n A_n = \{\omega : \omega \in A_n$ for infinitely many $n\}$, $\liminf_n A_n = \{\omega : \omega \in A_n$ for all but finitely many $n\}$, we obtain

$$\liminf_n A_n = \{(x, y): x^2 + y^2 < 1\},$$

$$\limsup_n A_n = \{(x, y): x^2 + y^2 \leq 1\} - \{(0, 1), (0, -1)\}.$$

4. If $x = \limsup_{n \to \infty} x_n$, then $\limsup_n A_n$ is either $(-\infty, x)$ or $(-\infty, x]$. For if $y \in A_n$ for infinitely many n, then $x_n > y$ for infinitely many n; hence $x \geq y$. Thus $\limsup_n A_n \subset (-\infty, x]$. But if $y < x$, then $x_n > y$ for infinitely many n, so $y \in \limsup_n A_n$. Thus $(-\infty, x) \subset \limsup_n A_n$, and the result follows. The same result is valid for lim inf; the above analysis applies, with "eventually" replacing "for infinitely many n."

Section 1.2

4. (a) If $-\infty \le a < b < c < \infty$, then $\mu(a, c] = \mu(a, b] + \mu(b, c]$, and $\mu(a, \infty) = \mu(a, b] + \mu(b, \infty)$; finite additivity follows quickly. If $A_n = (-\infty, n]$, then $A_n \uparrow R$, but $\mu(A_n) = n \not\to \mu(R) = 0$. Thus μ is not continuous from below, hence not countably additive.
   (b) Finiteness of μ follows from the definition; since $\mu(-\infty, n] \to \infty$, μ is unbounded.
5. We have $\mu(\bigcup_{i=1}^{\infty} A_i) \ge \mu(\bigcup_{i=1}^{n} A_i) = \sum_{i=1}^{n} \mu(A_i)$ for all n; let $n \to \infty$ to obtain the desired result.
8. The minimal σ-field $\mathscr{F}$ (which is also the minimal field) consists of the collection $\mathscr{G}$ of all (finite) unions of sets of the form $B_1 \cap B_2 \cap \cdots \cap B_n$, where B_i is either A_i or A_i^c. For any σ-field containing $A_1, \ldots, A_n$ must contain all sets in $\mathscr{G}$; hence $\mathscr{G} \subset \mathscr{F}$. But $\mathscr{G}$ is a σ-field; hence $\mathscr{F} \subset \mathscr{G}$. Since there are 2^n disjoint sets of the form $B_1 \cap \cdots \cap B_n$, and each such set may or may not be included in a typical set in $\mathscr{F}$, $\mathscr{F}$ has at most 2^{2^n} members. The upper bound is attained if all sets $B_1 \cap \cdots \cap B_n$ are nonempty.
9. (*a*) As in Problem 8, any field over $\mathscr{C}$ must contain all sets in $\mathscr{G}$; hence $\mathscr{G} \subset \mathscr{F}$. But $\mathscr{G}$ is a field; hence $\mathscr{F} \subset \mathscr{G}$. For if $A_i = \bigcap_{j=1}^{r} B_{ij}$, then $(\bigcup_{i=1}^{n} A_i)^c = \bigcap_{i=1}^{n} \bigcup_{j=1}^{r} B_{ij}^c$, which belongs to $\mathscr{G}$ because of the distributive law $A \cap (B \cup C) = (A \cap B) \cup (A \cap C)$.
   (b) Note that the complement of a finite intersection $\bigcap_{j=1}^{r} B_{ij}$ belongs to $\mathscr{D}$; for example, if $B_1, B_2 \in \mathscr{C}$, then

$$
\begin{aligned}
&(B_1 \cap B_2^c)^c \\
&\quad = B_1^c \cup B_2 \\
&\quad = (B_1^c \cap B_2) \cup (B_1^c \cap B_2^c) \cup (B_2 \cap B_1) \cup (B_2 \cap B_1^c) \in \mathscr{D}.
\end{aligned}
$$

   Now $\mathscr{D}$ is closed under finite intersection by the distributive law, and it follows from this and the above remark that $\mathscr{D}$ is closed under complementation and is therefore a field. Just as in the proof that $\mathscr{F} = \mathscr{G}$, we find that $\mathscr{F} = \mathscr{D}$.
   (c) This is immediate from (a) and (b).
11. (a) Let $A_n \in \mathscr{S}$, $n = 1, 2, \ldots$. Then A_n belongs to some $\mathscr{C}_{\alpha_n}$, and we may assume $\alpha_1 \le \alpha_2 \le \cdots$, so $\mathscr{C}_{\alpha_1} \subset \mathscr{C}_{\alpha_2} \subset \cdots$. Let $\alpha = \sup_n \alpha_n < \beta_1$. Then all $\mathscr{C}_{\alpha_n} \subset \mathscr{C}_{\alpha}$, hence all $A_n \in \mathscr{C}_{\alpha}$. Thus $\bigcup_n A_n \in \mathscr{C}_{\alpha+1} \subset \mathscr{S}$, so $\bigcup_n A_n \in \mathscr{S}$. If $A \in \mathscr{S}$, then A belongs to some $\mathscr{C}_{\alpha}$; hence $A^c \in \mathscr{C}_{\alpha+1} \subset \mathscr{S}$.
    (b) We have card $\mathscr{C}_{\alpha} \le c$ for all α. For this is true for $\alpha = 0$, by hypothesis. If it is true for all $\beta < \alpha$, then $\bigcup_{\beta<\alpha} \mathscr{C}_{\beta}$ has cardinality at

most (card α)$c = c$. Now if $\mathscr{D}$ has cardinality c, then $\mathscr{D}'$ has cardinality at most $c^{\aleph_0} = (2^{\aleph_0})^{\aleph_0} = 2^{\aleph_0} = c$. Thus card $\mathscr{C}_\alpha \leq c$. It follows that $\bigcup_{\alpha<\beta_1} \mathscr{C}_\alpha$ has cardinality at most c.

Section 1.3

3. (a) Since $\lambda(\varnothing) = 0$ we have $\Omega \in \mathscr{M}$, and $\mathscr{M}$ is clearly closed under complementation. If $E, F \in \mathscr{M}$ and $A \subset \Omega$, then

$$\begin{aligned}\lambda[A \cap (E \cup F)] &= \lambda[A \cap (E \cup F) \cap E] \\ &\quad + \lambda[A \cap (E \cup F) \cap E^c] \qquad \text{since} \quad E \in \mathscr{M} \\ &= \lambda(A \cap E) + \lambda(A \cap F \cap E^c).\end{aligned}$$

Thus

$$\begin{aligned}&\lambda[A \cap (E \cup F)] + \lambda[A \cap (E \cup F)^c] \\ &\quad = \lambda(A \cap E) + \lambda(A \cap E^c \cap F) + \lambda(A \cap E^c \cap F^c) \\ &\quad = \lambda(A \cap E) + \lambda(A \cap E^c) \qquad \text{since} \quad F \in \mathscr{M} \\ &\quad = \lambda(A) \qquad \text{since} \quad E \in \mathscr{M}.\end{aligned}$$

This proves that $\mathscr{M}$ is a field. Also, if E and F are disjoint we have

$$\begin{aligned}\lambda[A \cap (E \cup F)] &= \lambda[A \cap (E \cup F) \cap E] + \lambda[A \cap (E \cup F) \cap E^c] \\ &= \lambda(A \cap E) + \lambda(A \cap F \cap E^c) \\ &= \lambda(A \cap E) + \lambda(A \cap F) \qquad \text{since} \quad E \cap F = \varnothing.\end{aligned}$$

Now if the E_n are disjoint sets in $\mathscr{M}$ and $F_n = \bigcup_{i=1}^n E_i \uparrow E$, then

$$\begin{aligned}\lambda(A) &= \lambda(A \cap F_n) + \lambda(A \cap F_n{}^c) \\ &\qquad \text{since } F_n \text{ belongs to the field } \mathscr{M} \\ &\geq \lambda(A \cap F_n) + \lambda(A \cap E^c) \\ &\qquad \text{since } E^c \subset F_n{}^c \text{ and } \lambda \text{ is monotone} \\ &= \sum_{i=1}^n \lambda(A \cap E_i) + \lambda(A \cap E^c) \\ &\qquad \text{by what we have proved above.}\end{aligned}$$

Since n is arbitrary,

$$\begin{aligned}\lambda(A) &\geq \sum_{n=1}^\infty \lambda(A \cap E_n) + \lambda(A \cap E^c) \\ &\geq \lambda(A \cap E) + \lambda(A \cap E^c) \qquad \text{by countable subadditivity of } \lambda.\end{aligned}$$

Thus $E \in \mathscr{M}$, proving that $\mathscr{M}$ is a σ-field.

Now $\lambda(A \cap E) + \lambda(A \cap E^c) \geq \lambda(A)$ by subadditivity, hence

$$\lambda(A) = \sum_{n=1}^{\infty} \lambda(A \cap E_n) + \lambda(A \cap E^c).$$

Replace A by $A \cap E$ to obtain $\lambda(A \cap E) = \sum_{n=1}^{\infty} \lambda(A \cap E_n)$, as desired.

(b) All properties are immediate except for countable subadditivity. If $A = \bigcup_{n=1}^{\infty} A_n$, we must show that $\mu^*(A) \leq \sum_{n=1}^{\infty} \mu^*(A_n)$, and we may assume that $\mu^*(A_n) < \infty$ for all n. Given $\varepsilon > 0$, we may choose sets $E_{nk} \in \mathscr{F}_0$ with $A_n \subset \bigcup_k E_{nk}$ and $\sum_k \mu(E_{nk}) \leq \mu^*(A_n) + \varepsilon 2^{-n}$. Then $A \subset \bigcup_{n,k} E_{nk}$ and $\sum_{n,k} \mu(E_{nk}) \leq \sum_n \mu^*(A_n) + \varepsilon$. Thus $\mu^*(A) \leq \sum_n \mu^*(A_n) + \varepsilon$, ε arbitrary.

Now if $A \in \mathscr{F}_0$, then $\mu^*(A) \leq \mu(A)$ by definition of μ^*, and if $A \subset \bigcup_n E_n$, $E_n \in \mathscr{F}_0$, then $\mu(A) \leq \sum_n \mu(E_n)$ by 1.2.5 and 1.3.1. Take the infimum over all such coverings of A to obtain $\mu(A) \leq \mu^*(A)$; hence $\mu^* = \mu$ on $\mathscr{F}_0$.

(c) If $F \in \mathscr{F}_0$, $A \subset \Omega$, we must show that $\mu^*(A) \geq \mu^*(A \cap F) + \mu^*(A \cap F^c)$; we may assume $\mu^*(A) < \infty$. Given $\varepsilon > 0$, there are sets $E_n \in \mathscr{F}_0$ with $A \subset \bigcup_n E_n$ and $\sum_{n=1}^{\infty} \mu(E_n) \leq \mu^*(A) + \varepsilon$. Now

$$\mu^*(A \cap F) \leq \mu^*\Big(\bigcup_n (E_n \cap F)\Big) \quad \text{by monotonicity}$$

$$\leq \sum_n \mu(E_n \cap F)$$

since μ^* is countably subadditive and $\mu^* = \mu$ on $\mathscr{F}_0$.

Similarly,

$$\mu^*(A \cap F^c) \leq \sum_n \mu(E_n \cap F^c).$$

Thus

$$\mu^*(A \cap F) + \mu^*(A \cap F^c) \leq \sum_n \mu(E_n) \leq \mu^*(A) + \varepsilon,$$

and the result follows.

(d) If $A = B \cup N$, where $B \in \sigma(\mathscr{F}_0)$, $N \subset M \in \sigma(\mathscr{F}_0)$, $\mu^*(M) = 0$, then $B \in \mathscr{M}$ [note $\mathscr{F}_0 \subset \mathscr{M}$ and $\mathscr{M}$ is a σ-field, so $\sigma(\mathscr{F}_0) \subset \mathscr{M}$]. Also, any set C with $\mu^*(C) = 0$ belongs to $\mathscr{M}$ by definition of μ^*-measurability; hence $A \in \mathscr{M}$. Therefore the completion of $\sigma(\mathscr{F}_0)$ is included in $\mathscr{M}$.

Now assume μ σ-finite on $\mathscr{F}_0$, and let $A \in \mathscr{M}$. If Ω is the disjoint union of sets $A_n \in \mathscr{F}_0$ with $\mu(A_n) < \infty$, then by definition of μ^*, there is a set $B_n \in \sigma(\mathscr{F}_0)$ such that $A \cap A_n \subset B_n$ and $\mu^*(B_n - (A \cap A_n)) = 0$.

[Note that if $A \notin \mathscr{M}$ we obtain only $\mu^*(B_n) = \mu^*(A \cap A_n)$; however if $A \in \mathscr{M}$ (so that $A \cap A_n$ also belongs to $\mathscr{M}$), we have

$$\mu^*(B_n - (A \cap A_n)) = \mu^*(B_n) - \mu^*(A \cap A_n) = 0.]$$

If $B = \bigcup_n B_n$, then $B \in \sigma(\mathscr{F}_0)$, $A \subset B$, and $\mu^*(B - A) = 0$. This argument applied to A^c yields a set $C \in \sigma(\mathscr{F}_0)$ with $C \subset A$ and $\mu^*(A - C) = 0$. Therefore $A = C \cup (A - C)$ with $C \in \sigma(\mathscr{F}_0)$, $A - C \subset B - C \in \sigma(\mathscr{F}_0)$, and $\mu^*(B - C) = \mu^*(B - A) + \mu^*(A - C) = 0$. Thus A belongs to the completion of $\sigma(\mathscr{F}_0)$ relative to μ^*.

Section 1.4

1. Using the formulas of 1.4.5, the following results are obtained:
   (a) 3, (b) 8.5, (c) 5,
   (d) 7.25, (e) $\mu(\frac{1}{2}, \infty) + \mu(-\infty, -\frac{1}{2}) = 7.25$.
4. Let $C_k = \{x \in R^n\colon -k < x_i \le k, i = 1, \ldots, n\}$. Then μ is finite on C_k; hence the Borel subsets B of C_k such that $\mu(a + B) = \mu(B)$ form a monotone class including the field of finite disjoint unions of right-semiclosed intervals in C_k; hence all Borel subsets of C_k belong to the class (see 1.2.2). If $B \in \mathscr{B}(R^n)$, then $B \cap C_k \uparrow B$; hence $a + (B \cap C_k) \uparrow a + B$, and it follows that $\mu(a + B) = \mu(B)$.

   Now if $B \in \overline{\mathscr{B}}(R^n)$, then $B = A \cup C$, $A \in \mathscr{B}(R^n)$, $C \subset D \in \mathscr{B}(R^n)$, with $\mu(D) = 0$. Thus $a + B = (a + A) \cup (a + C)$, and, by Problem 3, $a + A \in \mathscr{B}(R^n)$, $a + C \subset a + D \in \mathscr{B}(R^n)$. By what we have proved above, $\mu(A + D) = \mu(D) = 0$; hence $a + B \in \overline{\mathscr{B}}(R^n)$ and $\mu(a + B) = \mu(B)$.
5. Let A be the unit cube $\{x \in R^n\colon 0 < x_i \le 1, i = 1, \ldots, n\}$, and let $c = \mu(A)$. For any positive integer, r, we may divide each edge of A into r equal parts, so that A is decomposed into r^n subcubes $A_1, \ldots, A_{r^n}$, each with volume r^{-n}. By translation-invariance, $\mu(A_i)$ is the same for all i, so if λ is Lebesgue measure, we have

   $$\mu(A_i) = r^{-n}\mu(A) = r^{-n}c = r^{-n}c\lambda(A) = c\lambda(A_i), \qquad i = 1, \ldots, r^n.$$

   Now any subinterval I of the unit cube can be expressed as the limit of an increasing sequence of sets B_k, where each B_k is a finite disjoint union of subcubes of the above type. Thus $\mu = c\lambda$ on subintervals of the unit cube, and hence on all Borel subsets of the unit cube by the Carathéodory extension theorem. Since R^n is a countable disjoint union of cubes, it follows that $\mu = c\lambda$ on $\mathscr{B}(R^n)$.
6. (a) If $r + x_1 = s + x_2$, $x_1, x_2 \in A$, then x_1 is equivalent to x_2, so that $x_1 = x_2$ since A was constructed by taking one member from each distinct B_x. Thus $r = s$, a contradiction.

If $x \in R$, then $x \in B_x$; if y is the member of B_x that belongs to A, then $x - y$ is a rational number r, hence $x \in r + A$.

(b) If $0 \leq r \leq 1$, then $r + A \subset [0, 2]$; thus

$$\begin{aligned}\sum\{\mu(r + A): 0 \leq r \leq 1, \quad r \text{ rational}\} \\ = \mu\left(\bigcup\{r + A: 0 \leq r \leq 1, \quad r \text{ rational}\}\right) \qquad \text{by (a)} \\ \leq \mu[0, 2] < \infty.\end{aligned}$$

But $\mu(r + A) = \mu(A)$ by Problem 4; hence $\mu(r + A)$ must be 0 for all r. Since R is a countable union of sets $r + A$ by (a), $\mu(R) = 0$, a contradiction.

8. Let $F(x, y) = 1$ if $x + y \geq 0$; $F(x, y) = 0$ if $x + y < 0$. If $a_1 = 0$, $b_1 = 1$, $a_2 = -1$, $b_2 = 0$, then

$$\begin{aligned}\triangle_{b_1 a_1} \triangle_{b_2 a_2} F(x, y) &= F(b_1, b_2) - F(a_1, b_2) \\ &\quad - F(b_1, a_2) + F(a_1, a_2) \\ &= 1 - 1 - 1 + 0 = -1;\end{aligned}$$

hence F is not a distribution function. Other examples: $F(x, y) = \max(x, y)$, $F(x, y) = [x + y]$, the largest integer less than or equal to $x + y$.

12. (a) First assume that μ is finite. Let $\mathscr{C}$ be the class of subsets of R^n having the desired property; we show that $\mathscr{C}$ is a monotone class. For let $B_n \in \mathscr{C}$, $B_n \uparrow B$. Let K_n be a compact subset of B_n with $\mu(B_n) \leq \mu(K_n) + \varepsilon$, $\varepsilon > 0$ preassigned. By replacing K_n by $\bigcup_{i=1}^n K_i$, we may assume the K_n form an increasing sequence. Then $\mu(B) = \lim_{n\to\infty} \mu(B_n) \leq \lim_{n\to\infty} \mu(K_n) + \varepsilon$, so that

$$\mu(B) = \sup\{\mu(K): \quad K \text{ a compact subset of } B\},$$

and $B \in \mathscr{C}$. If $B_n \in \mathscr{C}$, $B_n \downarrow B$, let K_n be a compact subset of B_n such that $\mu(B_n) \leq \mu(K_n) + \varepsilon 2^{-n}$, and set $K = \bigcap_{n=1}^\infty K_n$. Then

$$\mu(B) - \mu(K) = \mu(B - K) \leq \mu\left(\bigcup_{n=1}^{\infty} (B_n - K_n)\right) \leq \sum_{n=1}^{\infty} \mu(B_n - K_n) \leq \varepsilon;$$

thus $B \in \mathscr{C}$. Therefore $\mathscr{C}$ is a monotone class containing all finite disjoint unions of right-semiclosed intervals (a right-semiclosed interval is the limit of an increasing sequence of compact intervals). Hence $\mathscr{C}$ contains all Borel sets.

If μ is σ-finite, each $B \in \mathscr{B}(R^n)$ is the limit of an increasing sequence of sets B_i of finite measure. Each B_i can be approximated from within by compact sets [apply the previous argument to the measure given by $\mu_i(A) = \mu(A \cap B_i)$, $A \in \mathscr{B}(R^n)$], and the above argument that $\mathscr{C}$ is closed under limits of increasing sequences shows that $B \in \mathscr{C}$.

(b) We have $\mu(B) \le \inf\{\mu(V)\colon V \supset B, \quad V \text{ open}\}$

$$\le \inf\{\mu(W)\colon W \supset B, \quad W = K^c, \quad K \text{ compact}\}.$$

If μ is finite, this equals $\mu(B)$ by (a) applied to B^c, and the result follows.

Now assume μ is an arbitrary Lebesgue–Stieltjes measure, and write $R^n = \bigcup_{k=1}^{\infty} B_k$, where the B_k are disjoint bounded sets; then $B_k \subset C_k$ for some bounded open set C_k. The measure $\mu_k(A) = \mu(A \cap C_k)$, $A \in \mathscr{B}(R^n)$, is finite; hence if B is a Borel subset of B_k and $\varepsilon > 0$, there is an open set $W_k \supset B$ such that $\mu_k(W_k) \le \mu_k(B) + \varepsilon 2^{-k}$. Now $W_k \cap C_k$ is an open set V_k and $B \cap C_k = B$ since $B \subset B_k \subset C_k$; hence $\mu(V_k) \le \mu(B) + \varepsilon 2^{-k}$. For any $A \in \mathscr{B}(R^n)$, let V_k be an open set with $V_k \supset A \cap B_k$ and $\mu(V_k) \le \mu(A \cap B_k) + \varepsilon 2^{-k}$. Then $V = \bigcup_{k=1}^{\infty} V_k$ is open, $V \supset A$, and $\mu(V) \le \sum_{k=1}^{\infty} \mu(V_k) \le \mu(A) + \varepsilon$.

(c) Construct a measure μ on $\mathscr{B}(R)$ as follows. Let μ be concentrated on $S = \{1/n\colon n = 1, 2, \ldots\}$ and take $\mu\{1/n\} = 1/n$ for all n. Since $R = \bigcup_{n=1}^{\infty} \{1/n\} \cup S^c$ and $\mu(S^c) = 0$, μ is σ-finite. Since

$$\mu[0, 1] = \sum_{n=1}^{\infty} \frac{1}{n} = \infty,$$

μ is not a Lebesgue–Stieltjes measure. Now $\mu\{0\} = 0$, but if V is an open set containing 0, we have

$$\begin{aligned} \mu(V) &\ge \mu(-\varepsilon, \varepsilon) && \text{for some } \varepsilon \\ &\ge \sum_{k=r}^{\infty} \frac{1}{k} && \text{for some } r \\ &= \infty. \end{aligned}$$

Thus (b) fails. (Another example: Let $\mu(A)$ be the number of rational points in A.)

Section 1.5

2. If $B \in \mathscr{B}(R)$,

$$\begin{aligned} \{\omega\colon h(\omega) \in B\} &= \{\omega \in A\colon h(\omega) \in B\} \cup \{\omega \in A^c\colon h(\omega) \in B\} \\ &= [A \cap f^{-1}(B)] \cup [A^c \cap g^{-1}(B)] \end{aligned}$$

which belongs to $\mathscr{F}$ since f and g are Borel measurable.

5. (a) $\{x\colon f \text{ is discontinuous at } x\} = \bigcup_{n=1}^{\infty} D_n$, where $D_n = \{x \in R^k\colon$ for all $\delta > 0$, there exist $x_1, x_2 \in R^k$ such that $|x_1 - x| < \delta$ and $|x_2 - x| < \delta$, but $|f(x_1) - f(x_2)| \ge 1/n\}$. We show that the D_n are closed. Let $\{x_\alpha\}$

be a sequence of points in D_n with $x_\alpha \to x$. If $\delta > 0$ and $N = \{y: |y - x| < \delta\}$, then $x_\alpha \in N$ for large α, and since $x_\alpha \in D_n$, there are points $x_{\alpha 1}$ and $x_{\alpha 2} \in N$ such that $|f(x_{\alpha 1}) - f(x_{\alpha 2})| \geq 1/n$. Thus $|x_{\alpha 1} - x| < \delta$, $|x_{\alpha 2} - x| < \delta$, but $|f(x_{\alpha 1}) - f(x_{\alpha 2})| \geq 1/n$, so that $x \in D_n$.

The result is true for a function from an arbitrary topological space S to a metric space (T, d). Take $D_n = \{x \in S$: for every neighborhood N of x, there exist $x_1, x_2 \in N$ such that $d(f(x_1), f(x_2)) \geq 1/n\}$. (the above proof goes through with "sequence" replaced by "net.")

In fact T may be a uniform space (see the appendix on general topology, Section A10). If $\mathscr{D}$ is the associated family of pseudo-metrics, we take $D_n = \{x \in S$: for every neighborhood of N of x, there exist $x_1, x_2 \in N$ and $d \in \mathscr{D}$ such that $d(f(x_1), f(x_2)) \geq 1/n\}$ and proceed as above.

The result is false if no assumptions are made about the topology of the range space. For example, let $\Omega = \{1, 2, 3\}$, with open sets $\varnothing$, Ω, and $\{1\}$. Define $f: \Omega \to \Omega$ by $f(1) = f(3) = 1$, $f(2) = 2$. Then the set of discontinuities is $\{3\}$, which is not an F_σ.

(b) This follows from part (a) because the irrationals I cannot be expressed as a countable union of closed sets C_n. For if this were possible, then each C_n would have empty interior since every nonempty open set contains rational points. But then I is of category 1 in R, and since $Q = R - I$ is of category 1 in R, it follows that R is of category 1 in itself, contradicting the Baire category theorem.

6. By Problem 11, Section 1.4, there are c Borel subsets of R^n; hence there are only c simple functions on R^n. Since a Borel measurable function is the limit of a sequence of simple functions, there are $c^{\aleph_0} = c$ Borel measurable functions from R^n to R. By 1.5.8, there are only c Borel measurable functions from R^n to R^k.

7. (a) Since the P_n are measures, $\sum_k P_n(A_k) = P_n(\Omega) = 1$, and it follows quickly that the a_{nk} satisfy the hypotheses of Steinhaus' lemma. If $\{x_n\}$ is the sequence given by the lemma, let $S = \{k: x_k = 1\}$ and let B be the union of the sets A_k, $k \in S$. Then

$$t_n = \frac{1}{1-\alpha} \sum_{k \in S} [P_n(A_k) - P(A_k)] = \frac{1}{1-\alpha}\left[P_n(B) - \sum_{k \in S} P(A_k)\right],$$

and it follows that t_n converges, a contradiction. Thus P is a probability measure. If $B_k \in \mathscr{F}$, $B_k \downarrow \varnothing$, then given $\varepsilon > 0$, we have $P(A_k) < \varepsilon$ for large k, say for $k \geq k_0$. Thus $P_n(A_{k_0}) < \varepsilon$ for large n, say for $n \geq n_0$. Since the A_k decrease, we have $\sup_{n \geq n_0} P_n(A_k) \leq \varepsilon$ for $k \geq k_0$, and since $A_k \downarrow \varnothing$, there is a k_1 such that for $n = 1, 2, \ldots, n_0 - 1$, $P_n(A_k) < \varepsilon$ for $k \geq k_1$. Thus $\sup_n P_n(A_k) \leq \varepsilon$, $k \geq \max(k_0, k_1)$.

(b) Without loss of generality, assume $P_n(\Omega) \le 1$ for all n. Add a point (call it ∞) to the space and set $P_n\{\infty\} = 1 - P_n(\Omega) \to 1 - P(\Omega) = P\{\infty\}$. The P_n are now probability measures, and the result follows from part (a).

Section 1.6

2. $\int_\Omega \sum_{n=1}^\infty |f_n| \, d\mu = \sum_{n=1}^\infty \int_\Omega |f_n| \, d\mu < \infty$; hence $\sum_{n=1}^\infty |f_n|$ is integrable and therefore finite a.e. Thus $\sum_{n=1}^\infty f_n$ converges a.e. to a finite-valued function g.

   Let $g_n = \sum_{k=1}^n f_k$. Then $|g_n| \le \sum_{k=1}^\infty |f_k|$, an integrable function. By the dominated convergence theorem, $\int_\Omega g_n \, d\mu \to \int_\Omega g \, d\mu$, that is,

$$\sum_{n=1}^\infty \int_\Omega f_n \, d\mu = \int_\Omega \sum_{n=1}^\infty f_n \, d\mu.$$

3. Let $x_0 \in (c, d)$, and let $x_n \to x_0$, $x_n \ne x_0$. Then

$$\frac{1}{(x_n - x_0)} \left[\int_a^b f(x_n, y) \, dy - \int_a^b f(x_0, y) \, dy \right] = \int_a^b \left[\frac{f(x_n, y) - f(x_0, y)}{x_n - x_0} \right] dy.$$

   By the mean value theorem,

$$\frac{f(x_n, y) - f(x_0, y)}{x_n - x_0} = f_1(\lambda_n, y)$$

   for some $\lambda_n = \lambda_n(y)$ between x_n and x_0. By hypothesis, $|f_1(\lambda_n, y)| \le h(y)$, where h is integrable, and the result now follows from the dominated convergence theorem (since $[f(x_n, y) - f(x_0, y)]/[x_n - x_0] \to f_1(x_0, y)$, $f_1(x, \cdot)$ is Borel measurable for each x).

8. Let μ be Lebesgue measure. If f is an indicator I_B, $B \in \mathscr{B}(R)$, the result to be proved states that $\mu(B) = \mu(a + B)$, which holds by translation-invariance of μ (Problem 4, Section 1.4). The passage to nonnegative simple functions, nonnegative measurable functions, and arbitrary measurable functions is done as in 1.6.12.

Section 1.7

2. (a) If f is Riemann–Stieltjes integrable, $\alpha = f = \beta$ a.e. $[\mu]$ as in 1.7.1(a). Thus the set of discontinuities of f is a subset of a set of μ-measure 0, together with the endpoints of the subintervals of the P_k. Take a different sequence of partitions having the original endpoints as

interior points to conclude that f is continuous a.e. $[\mu]$. Conversely, if f is continuous a.e. $[\mu]$, then $\alpha = f = \beta$ a.e. $[\mu]$. [The result that f continuous at x implies $\alpha(x) = f(x) = \beta(x)$ is true even if x is an endpoint.] As in 1.7.1(a), f is Riemann–Stieltjes integrable.

(b) This is done exactly as in 1.7.1(b).

3. (a) By definition of the improper Riemann integral, f must be Riemann integrable (hence continuous a.e.) on each bounded interval, and the result follows. For the counterexample to the converse, take $f(x) = 1$, $n \le x < n + 1$, n an even integer; $f(x) = -1$, $n \le x < n+1$, n an odd integer. Then the limit of $r_{ab}(f)$ does not exist. (Alternatively, take $f(x)$ identically 1; then $r_{ab}(f) \to +\infty$ as $a \to -\infty$, $b \to \infty$.)

(b) Define

$$f_n(x) = \begin{cases} f(x) & \text{if } -n \le x \le n, \\ 0 & \text{elsewhere.} \end{cases}$$

Then $f_n \uparrow f$; hence f is measurable relative to the completed σ-field; also, $\int_\Omega f_n \, d\mu \uparrow \int_\Omega f \, d\mu$ by the monotone convergence theorem. But $\int_\Omega f_n \, d\mu = r_{-n,n}(f)$ by 1.7.1(b), and $r_{-n,n}(f) \to r(f)$ by hypothesis; the result follows.

For the counterexample, take

$$f(x) = \begin{cases} \dfrac{(-1)^n}{n+1}, & n \le x < n+1, \quad n = 0, 1, \ldots, \\ 0, & x < 0. \end{cases}$$

We have $r(f) = 1 - \frac{1}{2} + \frac{1}{3} - \cdots$, but

$$\int_R |f| \, d\mu = 1 + \tfrac{1}{2} + \tfrac{1}{3} + \cdots = \infty,$$

so that f is not Lebesgue integrable on R.

Chapter 2

Section 2.1

2. Let $D = \{\omega : f(\omega) < 0\}$; then $\lambda(A \cap D) \le 0$, $\lambda(A \cap D^c) \ge 0$ for all $A \in \mathscr{F}$. By 2.1.3(d),

$$\lambda^+(A) = \lambda(A \cap D^c) = \int_A f^+ \, d\mu$$

since $f^+ = f$ on D^c and $f^+ = 0$ on D. Similarly,

$$\lambda^-(A) = -\lambda(A \cap D) = \int_A f^- \, d\mu$$

since $f^- = -f$ on D, and $f^- = 0$ on D^c. The result follows.

4. If $E_1, \ldots, E_n$ are disjoint sets in $\mathscr{F}$, with all $E_i \subset A$,

$$\sum_{i=1}^n |\lambda(E_i)| = \sum_{i=1}^n |\lambda^+(E_i) - \lambda^-(E_i)| \le \sum_{i=1}^n [\lambda^+(E_i) + \lambda^-(E_i)]$$

$$= |\lambda|\left(\bigcup_{i=1}^n E_i\right) \le |\lambda|(A).$$

Thus the sup of the terms $\sum_{i=1}^n |\lambda(E_i)|$ is at most $|\lambda|(A)$. But

$$\begin{aligned}|\lambda|(A) &= \lambda^+(A) + \lambda^-(A)\\ &= \lambda(A \cap D^c) - \lambda(A \cap D)\\ &= |\lambda(A \cap D^c)| + |\lambda(A \cap D)|\\ &\qquad \text{since } \lambda(A \cap D^c) \ge 0 \quad \text{and} \quad \lambda(A \cap D) \le 0\\ &= |\lambda(E_1)| + |\lambda(E_2)|\end{aligned}$$

and the result follows.

Section 2.2

2. Let $A_n = \{\omega : |g(\omega)| \ge 1/n\}$, $n = 1, 2, \ldots$, so that $A = \bigcup_{n=1}^\infty A_n$. Now

$$\infty > \int_{A_n} |g| \, d\mu \ge \frac{1}{n} \mu(A_n);$$

hence $\mu(A_n) < \infty$. For the example, let μ be Lebesgue measure on $\mathscr{B}(R)$, and let $g(x)$ be any strictly positive integrable function, such as $g(x) = e^{-|x|}$. In this case, $A = R$, so that $\mu(A) = \infty$.

4. If f is an indicator I_A, the result is true by hypothesis. If f is a nonnegative simple function $\sum_{j=1}^n x_j I_{A_j}$, the A_j disjoint sets in $\mathscr{F}$, then

$$\int_\Omega f \, d\lambda = \sum_{j=1}^n x_j \lambda(A_j) = \sum_{j=1}^n x_j \int_{A_j} g \, d\mu = \sum_{j=1}^n x_j \int_\Omega I_{A_j} g \, d\mu$$

$$= \int_\Omega fg \, d\mu \qquad \text{by the additivity theorem.}$$

If f is a nonnegative Borel measurable function, let $f_1, f_2, \ldots$ be nonnegative simple functions increasing to f. By what we have just proved

$\int_\Omega f_n \, d\lambda = \int_\Omega f_n \, g \, d\mu$; hence $\int_\Omega f \, d\lambda = \int_\Omega fg \, d\mu$ by the monotone convergence theorem. Finally, if f is an arbitrary Borel measurable function, write $f = f^+ - f^-$. By what we have just proved,

$$\int_\Omega f^+ \, d\lambda = \int_\Omega f^+ g \, d\mu, \qquad \int_\Omega f^- \, d\lambda = \int_\Omega f^- g \, d\mu$$

and the result follows from the additivity theorem.

6. (a) In the definition of $|\lambda|$, we may assume without loss of generality that the E_i partition A. If A is the disjoint union of sets $A_1, A_2, \ldots \in \mathscr{F}$, then

$$\sum_{j=1}^{n} |\lambda(E_j)| = \sum_{j=1}^{n} \left| \sum_{i=1}^{\infty} \lambda(E_j \cap A_i) \right| \le \sum_{j=1}^{n} \sum_{i=1}^{\infty} |\lambda(E_j \cap A_i)|$$
$$= \sum_{i=1}^{\infty} \sum_{j=1}^{n} |\lambda(E_j \cap A_i)| \le \sum_{i=1}^{\infty} |\lambda|(A_i).$$

Thus $|\lambda|(A) \le \sum_{i=1}^{\infty} |\lambda|(A_i)$. Now to show the reverse inequality, we may assume $|\lambda|(A) < \infty$; hence $|\lambda|(A_i) \le |\lambda|(A) < \infty$. For each i, there is a partition $\{E_{i1}, \ldots, E_{in_i}\}$ of A_i such that

$$\sum_{j=1}^{n_i} |\lambda(E_{ij})| > |\lambda|(A_i) - \frac{\varepsilon}{2^i}, \qquad \varepsilon > 0 \quad \text{preassigned.}$$

Then for any n,

$$|\lambda|(A) \ge \sum_{i=1}^{n} \sum_{j=1}^{n_i} |\lambda(E_{ij})| \ge \sum_{i=1}^{n} |\lambda|(A_i) - \varepsilon.$$

Since n and ε are arbitrary, the result follows.

(b) If $E_1, \ldots, E_n$ are disjoint measurable subsets of A,

$$\sum_{i=1}^{n} |(\lambda_1 + \lambda_2)(E_i)| \le \sum_{i=1}^{n} |\lambda_1(E_i)| + \sum_{i=1}^{n} |\lambda_2(E_i)|$$
$$\le |\lambda_1|(A) + |\lambda_2|(A),$$

proving $|\lambda_1 + \lambda_2| \le |\lambda_1| + |\lambda_2|$; $|a\lambda| = |a|\,|\lambda|$ is immediate from the definition of total variation.

(c) If $\mu(A_i) = 0$ and $|\lambda_i|(A_i^c) = 0$, $i = 1, 2$, then $\mu(A_1 \cup A_2) = 0$ and by (b), $|\lambda_1 + \lambda_2|(A_1^c \cap A_2^c) \le |\lambda_1|(A_1^c) + |\lambda_2|(A_2^c) = 0$.

(d) This has been established when λ is real (see 2.2.5), so assume λ complex, say, $\lambda = \lambda_1 + i\lambda_2$. If $\mu(A) = 0$, then $\lambda_1(A) = \lambda_2(A) = 0$; hence $\lambda \ll \mu$ implies $\lambda_1 \ll \mu$ and $\lambda_2 \ll \mu$. By 2.2.5(b), $|\lambda_1| \ll \mu$, $|\lambda_2| \ll \mu$; hence by (b), $|\lambda| \ll \mu$. The converse is clear since $|\lambda(A)| \le |\lambda|(A)$.

(e) The proof is the same as in 2.2.5(c).

(f) See 2.2.5(d).

(g) The "if" part is done as in 2.2.5(e); for the "only if" part, let $\mu(A_n) \to 0$. Since $|\lambda| \ll \mu$ by (d), $|\lambda|(A_n) \to 0$ by 2.2.5(e); hence $\lambda(A_n) \to 0$.

Section 2.3

2. We have

$$\int_a^b f'(x)\,dx = \int_a^b \lim_{h\to 0} \left[\frac{f(x+h)-f(x)}{h}\right] dx$$

where we may assume $h > 0$

$$\leq \liminf_{h\to 0} \int_a^b \left[\frac{f(x+h)-f(x)}{h}\right] dx$$

$$= \liminf_{h\to 0} \frac{1}{h}\left[\int_b^{b+h} f(x)\,dx - \int_a^{a+h} f(x)\,dx\right]$$

[define $f(x) = f(b)$, $x > b$; $f(x) = f(a)$, $x < a$]

$$\leq \liminf_{h\to 0} \frac{1}{h}[hf(b+h) - hf(a)] \qquad \text{since } f \text{ is increasing}$$

$$= f(b) - f(a).$$

Alternatively, let μ be the Lebesgue–Stieltjes measure corresponding to f (adjusted so as to be right continuous), and let m be Lebesgue measure. Write $\mu = \mu_1 + \mu_2$, where $\mu_1 \ll m$ and $\mu_2 \perp m$. By 2.3.8 and 2.3.9,

$$\int_a^b f'(x)\,dx = \int_a^b f'\,dm$$

$$= \int_a^b D\mu\,dm$$

$$= \int_a^b \frac{d\mu_1}{dm}\,dm = \mu_1(a, b] \leq \mu(a, b] = f(b) - f(a).$$

6. (a) Since A is linear and m is translation-invariant, so is λ; hence by Problem 5, Section 1.4, $\lambda = c(A)m$ for some constant $c(A)$. Now if A_1 and A_2 are linear transformations on R^k, then

$$m(A_1A_2E) = c(A_1)m(A_2E) = c(A_1)c(A_2)m(E)$$

and

$$m(A_1A_2E) = c(A_1A_2)m(E);$$

hence

$$c(A_1A_2) = c(A_1)c(A_2).$$

Since $\det(A_1A_2) = \det A_1 \det A_2$, it suffices to assume that A corresponds to an elementary row operation. Now if Q is the unit cube $\{x: 0 < x_i \leq 1, i = 1, \ldots, k\}$, then $m(Q) = 1$; hence $c(A) = m(A(Q))$. If $e_1, \ldots, e_k$ is the standard basis for R^k, then A falls into one of the following three categories:

(1) $Ae_i = e_j$, $Ae_j = e_i$, $Ae_k = e_k$, $k \neq i$ or j. Then $c(A) = 1 = |\det A|$.
(2) $Ae_i = ke_i$, $Ae_j = e_j$, $j \neq i$. Then $c(A) = |k| = |\det A|$.
(3) $Ae_i = e_i + e_j$, $Ae_k = e_k$, $k \neq i$. Then $\det A = 1$ and $c(A)$ is 1 also, by the following argument. We may assume $i = 1, j = 2$. Then

$$A(Q) = \left\{A \sum_{i=1}^{k} a_i e_i : 0 < a_i \leq 1, \quad i = 1, \ldots, k\right\}$$
$$= \left\{y = \sum_{i=1}^{k} b_i e_i : 0 < b_i \leq 1, \quad i \neq 2, \quad b_1 < b_2 \leq b_1 + 1\right\}.$$

If $B_1 = \{y \in A(Q): b_2 \leq 1\}$, $B_2 = \{y \in A(Q): b_2 > 1\}$, and $B_2 - e_2 = \{y - e_2 : y \in B_2\}$, then $B_1 \cap (B_2 - e_2) = \varnothing$; for if $y \in B_1$, then $b_1 < b_2$ and if $y + e_2 \in B_2 \subset A(Q)$, then $b_2 + 1 \leq b_1 + 1$. Therefore,

$$c(A) = m(A(Q)) = m(B_1) + m(B_2) = m(B_1) + m(B_2 - e_2)$$
$$= m(B_1 \cup (B_2 - e_2)) = m(Q) = 1.$$

(b) Fix $x \in V$ and define $S(y) = T(x + y) - T(x)$, $y \in \{z - x : z \in V\}$. Then if C is a cube containing 0, we have

$$T(x + C) = \{T(x + y): y \in C\} = T(x) + \{S(y): y \in C\}$$
$$= T(x) + S(C).$$

Therefore, $m(T(x + C)) = m(S(C))$, so that differentiability of T at x is equivalent to differentiability of S at 0, and the derivatives, if they exist, are equal. Also, the Jacobian matrix of S at 0 coincides with the Jacobian matrix of T at x, and the result follows.

(c) Let $A = A(0)$, and define $S(x) = A^{-1}(T(x))$, $x \in V$; the Jacobian matrix of S at 0 is the identity matrix, and $S(0) = A^{-1}(0) = 0$. If we show that the measure given by $m(S(E))$, $E \in \mathscr{B}(V)$, is differentiable at 0 with derivative 1, then

$$\left|\frac{m(S(C))}{m(C)} - 1\right| < \frac{\varepsilon}{|\det A|}$$

for sufficiently small cubes C containing 0. Thus by (a), $m(T(C)) = m(AS(C)) = |\det A| m(S(C))$; hence

$$\left|\frac{m(T(C))}{m(C)} - |\det A|\right| = \left|\frac{m(S(C))}{m(C)} - 1\right| |\det A| < \varepsilon,$$

so that μ is differentiable at 0 with derivative $|\det A|$.

(d) (i) If $x \in \bar{C}$, then $|x| \leq \sqrt{k}\,\beta < \delta$; hence $|T(x) - x| \leq \alpha\beta = \frac{1}{2}(\beta_2 - \beta)$. Therefore $T(x) \in C_2$.

(ii) If $x \in \partial C$, then $|T(x) - x| \leq \alpha\beta$ as above, and $\alpha\beta = \frac{1}{2}(\beta - \beta_1)$. Therefore $T(x) \notin C_1$.

(iii) We have $|T(x) - x| \leq \alpha\beta$, and

$$\alpha\beta = \frac{\alpha\beta_1}{1 - 2\alpha} \leq \frac{\frac{1}{4}}{1 - \frac{1}{2}}\,\beta_1 = \frac{1}{2}\,\beta_1,$$

and the result follows.

(iv) If $y \in C_1 - T(C)$ but $y \in T(\bar{C})$, then $y \in T(\partial C)$, contradicting (ii).

Now C_1 is the disjoint union of the sets $C_1 \cap T(C)$ and $C_1 - T(C)$; the first set is open since T is an open map, and the second set is open by (iv). By (iii), $C_1 \cap T(C) \neq \varnothing$, so by connectedness of C_1, we have $C_1 = C_1 \cap T(C)$, that is, $C_1 \subset T(C)$. Also, $T(C) \subset C_2$ by (i).

Therefore $m(C_1) \leq m(T(C)) \leq m(C_2)$, that is,

$$(1 - 2\alpha)^k \beta^k \leq m(T(C)) \leq (1 + 2\alpha)^k \beta^k.$$

Thus $1 - \varepsilon < (1 - 2\alpha)^k < m(T(C))/m(C) \leq (1 + 2\alpha)^k < 1 + \varepsilon$ as desired.

(e) Assume $m(E) = 0$ and $\lambda(E) > 0$. By Problem 12, Section 1.4, and the fact that

$$E = \bigcup_{n,\, j=1}^{\infty} E \bigcap \left\{ \sup_{C_r,\, r < 1/j} \frac{\lambda(C_r)}{m(C_r)} < n \right\},$$

we can find a compact set K and positive integers n and j such that $m(K) = 0$, $\lambda(K) > 0$, and $\lambda(C) < nm(C)$ for all open cubes C containing a point of K and having diameter less than $1/j$. If $\varepsilon > 0$, choose an open set $D \subset K$ such that $m(D) < \varepsilon$.

Now partition R^k into disjoint (partially closed) cubes B of diameter less than $1/j$ and small enough so that if $B \cap K \neq \varnothing$, then $B \subset D$. If the cubes that meet K are $B_1, \ldots, B_t$, we may find open cubes $C_i \supset B_i$, $1 \leq i \leq t$, such that $m(C_i) < 2m(B_i)$ and diam $C_i < 1/j$. Then

$$\lambda(K) \leq \sum_{i=1}^{t} \lambda(B_i) \leq \sum_{i=1}^{t} \lambda(C_i) < n \sum_{i=1}^{t} m(C_i) < 2n \sum_{i=1}^{t} m(B_i)$$

$$\leq 2nm(D) < 2n\varepsilon.$$

Since ε is arbitrary, $\lambda(K) = 0$, a contradiction.

(f) If f is an indicator I_B, let $E = T^{-1}(B)$, so that $B = T(E)$. Then

$$\int_W f(y)\, dy = m(T(E)) \qquad \text{and} \qquad \int_V f(T(x))\, |J(x)|\, dx = \int_E |J(x)|\, dx,$$

and the result follows from part (e). The usual passage to simple functions, nonnegative measurable functions and arbitrary measurable functions completes the proof.

Section 2.4

1. (a) If $f = \{a_1, \ldots, a_n, 0, 0, \ldots\}$, then $\int_\Omega f\, d\mu = \sum_{k=1}^{n} a_k$ by definition of the integral of a simple function. If $f = \{a_n, n = 1, 2, \ldots\}$, with all $a_n \geq 0$, $\int_\Omega f\, d\mu = \sum_{n=1}^{\infty} a_n$ by the result for simple functions and the monotone convergence theorem. If the a_n are real numbers, then $\int_\Omega f\, d\mu = \int_\Omega f^+\, d\mu - \int_\Omega f^-\, d\mu = \sum_{n=1}^{\infty} a_n^+ - \sum_{n=1}^{\infty} a_n^-$ if this is not of the form $+\infty - \infty$. Finally, if the a_n are complex,

$$\int_\Omega f\, d\mu = \sum_{n=1}^{\infty} \operatorname{Re} a_n + i \sum_{n=1}^{\infty} \operatorname{Im} a_n$$

provided Re f and Im f are integrable; since $|\operatorname{Re} a_n|$, $|\operatorname{Im} a_n| \leq |a_n| \leq |\operatorname{Re} a_n| + |\operatorname{Im} a_n|$, this is equivalent to $\sum_{n=1}^{\infty} |a_n| < \infty$.

(b) If $f(\alpha) = 0$ except for $\alpha \in F$, F finite, than $\int_\Omega f\, d\mu = \sum_\alpha f(\alpha)$ by definition of the integral of a simple function. If $f \geq 0$, then $\int_\Omega f\, d\mu \geq \int_F f\, d\mu$ for all finite F; hence $\int_\Omega f\, d\mu \geq \sum_\alpha f(\alpha)$. If $f(\alpha) > 0$ for uncountably many α, then $\sum_\alpha f(\alpha) = \infty$; hence $\int_\Omega f\, d\mu = \infty$ also. If $f(\alpha) > 0$ for only countably many α, then $\int_\Omega f\, d\mu = \sum_\alpha f(\alpha)$ by the monotone convergence theorem. The remainder of the proof is as in part (a).

8. Apply Hölder's inequality with $g = 1$, f replaced by $|f|^r$, $p = s/r$, $1/q = 1 - r/s$, to obtain

$$\int_\Omega |f|^r \, d\mu \le \left(\int_\Omega |f|^{rp} \, d\mu\right)^{1/p} \left(\int_\Omega 1 \, d\mu\right)^{1/q}.$$

Therefore $\|f\|_r \le \|f\|_s \, [\mu(\Omega)]^{1/rq}$, as desired.

9. We have $\int_\Omega |f|^p \, d\mu \le \|f\|_\infty^p \mu(\Omega)$, so $\limsup_{p\to\infty} \|f\|_p \le \|f\|_\infty$. Now let $\varepsilon > 0$, $A = \{\omega\colon |f(\omega)| \ge \|f\|_\infty - \varepsilon\}$ (assuming $\|f\|_\infty < \infty$). Then

$$\int_\Omega |f|^p \, d\mu \ge \int_A |f|^p \, d\mu \ge (\|f\|_\infty - \varepsilon)^p \mu(A).$$

Since $\mu(A) > 0$ by definition of $\|f\|_\infty$, we have $\liminf_{p\to\infty} \|f\|_p \ge \|f\|_\infty$. If $\|f\|_\infty = \infty$, let $A = \{\omega\colon |f(\omega)| \ge M\}$ and show that

$$\liminf_{p\to\infty} \|f\|_p \ge M;$$

since M can be arbitrarily large, the result follows.

If $\mu(\Omega) = \infty$, it is still true that $\liminf_{p\to\infty} \|f\|_p \ge \|f\|_\infty$; for if $\mu(A) = \infty$ in the above argument, then $\|f\|_p = \infty$ for all $p < \infty$. However, if μ is Lebesgue measure on $\mathscr{B}(R)$ and $f(x) = 1$ for $n \le x \le n + (1/n)$, $n = 1, 2, \ldots$, and $f(x) = 0$ elsewhere, then $\|f\|_p = \infty$ for $p < \infty$, but $\|f\|_\infty = 1$.

11. (a) We have $|\int_E (f - z) \, d\mu| \le \int_E |f - z| \, d\mu$. But $\omega \in E$ implies $f(\omega) \in D$; hence $|f(\omega) - z| \le r$; and thus $\int_E |f - z| \, d\mu \le r\mu(E)$. If $\mu(E) > 0$, then

$$\left| \frac{1}{\mu(E)} \int_E f \, d\mu - z \right| = \left| \frac{1}{\mu(E)} \int_E (f - z) \, d\mu \right| \le r;$$

hence $[1/\mu(E)] \int_E f \, d\mu \in D \subset S^c$, a contradiction. Therefore $\mu(E) = 0$, that is, $\mu\{\omega\colon f(\omega) \in D\} = 0$. Since $\{\omega\colon f(\omega) \notin S\}$ is a countable union of sets $f^{-1}(D)$, the result follows.

(b) Let $\mu = |\lambda|$; if $E_1, \ldots, E_n$ are disjoint measurable subsets of A_r,

$$\sum_{j=1}^n |\lambda(E_j)| = \sum_{j=1}^n \left| \int_{E_j} h \, d\mu \right| \le \sum_{j=1}^n \int_{E_j} |h| \, d\mu$$

$$\le r \sum_{j=1}^n \mu(E_j) \le r\mu(A_r).$$

Thus $\mu(A_r) \le r\mu(A_r)$, and since $0 < r < 1$, we have $\mu(A_r) = 0$. If $A = \{\omega\colon |h(\omega)| < 1\} = \bigcup \{A_r\colon 0 < r < 1,\ r \text{ rational}\}$, then $\mu(A) = 0$, so that $|h| \ge 1$ a.e.

Now if $\mu(E) > 0$, then $[1/\mu(E)] \int_E h \, d\mu = \lambda(E)/\mu(E) \in S$, where $S = \{z \in C\colon |z| \le 1\}$. By (a), $h(\omega) \in S$ for almost every ω, so $|h| \le 1$ a.e. $[|\lambda|]$.

(c) If $E \in \mathscr{F}$, $\int_\Omega I_E h \, d|\lambda| = \int_E h \, d|\lambda| = \lambda(E)$ by (b); also, $\int_\Omega I_E g \, d\mu = \int_E g \, d\mu = \lambda(E)$ by definition of λ. It follows immediately that $\int_\Omega fh \, d|\lambda| = \int_\Omega fg \, d\mu$ when f is a complex-valued simple function. If f is a bounded, complex-valued Borel measurable function, by 1.5.5(b), there are simple functions $f_n \to f$ with $|f_n| \le |f|$. By the dominated convergence theorem, $\int_\Omega fh \, d|\lambda| = \int_\Omega fg \, d\mu$. If $f = \bar{h} I_E$, we obtain $|\lambda|(E) = \int_E \bar{h} g \, d\mu$.

(d) In (c), $|\lambda|(E) \ge 0$ for all E; hence $\bar{h}g \ge 0$ a.e. $[\mu]$ by 1.6.11. But if $g(\omega) = |g(\omega)| e^{i\theta(\omega)}$ and $h(\omega) = e^{i\varphi(\omega)}$, then $e^{i(\theta - \varphi)} = 1$ a.e. on $\{g \ne 0\}$, so that $\bar{h}g = |g|$ a.e., as desired.

12. If $I_{(a,b)}$ can be approximated in L^∞ by continuous functions, let $0 < \varepsilon < \frac{1}{2}$ and let f be a continuous function such that

$$\|I_{(a,b)} - f\|_\infty \le \varepsilon;$$

hence $|I_{(a,b)} - f| \le \varepsilon$ a.e. For every $\delta > 0$, there are points $x \in (a, a + \delta)$ and $y \in (a - \delta, a)$ such that $|1 - f(x)| \le \varepsilon$ and $|f(y)| \le \varepsilon$. Consequently, $\limsup_{x \to a^+} f(x) \ge 1 - \varepsilon$ and $\liminf_{x \to a^-} f(x) \le \varepsilon$, contradicting continuity of f.

Section 2.5

4. Let $B_{jk\delta} = \{|f_j - f_k| \ge \delta\}$, $B_\delta = \bigcap_{n=1}^\infty \bigcup_{j,k=n}^\infty B_{jk\delta}$. Then

$$\bigcup_{j,k=n}^{\infty} B_{jk\delta} \downarrow B_\delta,$$

and the proof proceeds just as in 2.5.4.

5. Let $\{f_{n_k}\}$ be a subsequence converging a.e., necessarily to f by Problem 1. By 1.6.9, f is μ-integrable. Now if $\int_\Omega f_n \, d\mu \not\to \int_\Omega f \, d\mu$, then for some $\varepsilon > 0$, we have $|\int_\Omega f_n \, d\mu - \int_\Omega f \, d\mu| \ge \varepsilon$ for n in some subsequence $\{m_k\}$. But we may then extract a subsequence $\{f_{r_j}\}$ of $\{f_{m_k}\}$ converging to f a.e., so that $\int_\Omega f_{r_j} \, d\mu \to \int_\Omega f \, d\mu$ by 1.6.9, a contradiction.

Section 2.6

4. By Fubini's theorem,

$$\mu(C) = \iint_\Omega I_C \, d\mu = \int_{\Omega_1} \left[\int_{\Omega_2} I_C \, d\mu_2 \right] d\mu_1 = \int_{\Omega_1} \mu_2(C(\omega_1)) \, d\mu_1(\omega_1).$$

Similarly, $\mu(C) = \int_{\Omega_2} \mu_1(C(\omega_2)) \, d\mu_2(\omega_2)$. The result follows since $f \ge 0$, $\int_\Omega f = 0$ implies $f = 0$ a.e.

7. (a) Let

$$A_{nk} = \left\{x \in \Omega_1 : \frac{k-1}{n} \le f(x) < \frac{k}{n}\right\},$$

$$B_{nk} = \left\{y \in \Omega_2 : \frac{k-1}{n} \le y < \frac{k}{n}\right\}$$

($n = 1, 2, \ldots, k = 1, 2, \ldots, n$; where $k = n$, include the right endpoint as well). Then

$$G = \bigcap_{n=1}^{\infty} \bigcup_{k=1}^{n} (A_{nk} \times B_{nk}) \in \mathscr{F}.$$

If f is only defined on a subset of Ω_1, replace Ω_1 by the domain of f in the definition of A_{nk}.

(b) Assume $B \subset C_1$. Each $x \in \Omega_1$ is countable and $(x, y) \in B$ implies $y \le x$; hence there are only countably many points $y_{x1}, y_{x2}, \ldots \in \Omega_2$ such that $(x, y_{xn}) \in B$. (If there are only finitely many points $y_{x1}, \ldots, y_{xn}$, take $y_{xk} = y_{xn}$ for $k \ge n$.)
Thus $B = \bigcup_{n=1}^{\infty} G_n$, where G_n is the graph of the function f_n defined by $f_n(x) = y_{xn}$. By part (a), $B \in \mathscr{F}$. [Note that $y_{xn} \in \Omega_2$, which may be identified with [0, 1]; thus (a) applies.] If $B \subset C_2$, each $y \in \Omega_2$ is countable, and $(x, y) \in B$ implies $x < y$, so there are only countably many points $x_{yn} \in \Omega_1$ such that $(x_{yn}, y) \in B$. The result follows as above.

(c) If $F \subset \Omega$, then $F = (F \cap C_1) \cup (F \cap C_2)$ and the result follows from part (b).

9. Assuming the continuum hypothesis, we may replace [0, 1] by the first uncountable ordinal β_1. Thus we may take $\Omega_1 = \Omega_2 = \beta_1$, $\mathscr{F}_1 = \mathscr{F}_2 =$ the image of the Borel sets under the correspondence of [0, 1] with β_1, and $\mu_1 = \mu_2 =$ Lebesgue measure. Let $f = I_C$, where $C = \{(x, y) \in \Omega_1 \times \Omega_2 : y \le x\}$ and the ordering "$\le$" is taken in β_1, not [0, 1]. For each x, $\{y : f(x, y) = 1\}$ is countable, and for each y, $\{x : f(x, y) = 0\}$ is countable; it follows that f is measurable in each coordinate separately. Now $\int_0^1 f(x, y)\, dy = 0$ for all x; hence $\int_0^1 \left[\int_0^1 f(x, y)\, dy\right] dx = 0$. But $\int_0^1 [1 - f(x, y)]\, dx = 0$ for all y; hence $\int_0^1 \left[\int_0^1 f(x, y)\, dx\right] dy = 1$. It follows that f is not jointly measurable, for if so, the iterated integrals would be equal by Fubini's theorem.

Section 2.7

1. Let $\mathscr{G}$ be the smallest σ-field containing the measurable rectangles. Then $\mathscr{G} \subset \prod_{j=1}^{\infty} \mathscr{F}_j$ since a measurable rectangle is a measurable cylinder. But the class of sets $A \subset \prod_{j=1}^{n} \Omega_j$ such that $\{\omega \in \Omega : (\omega_1, \ldots, \omega_n) \in A\} \in \mathscr{G}$ is a

σ-field that contains the measurable rectangles of $\prod_{j=1}^{n} \Omega_j$, and hence contains all sets in $\prod_{j=1}^{n} \mathscr{F}_j$. Thus all measurable cylinders belong to $\mathscr{G}$, so $\prod_{j=1}^{\infty} \mathscr{F}_j \subset \mathscr{G}$.

5.

$$\{x \in R^\infty : f(x) = n\} = \left\{x\colon \sum_{i=1}^{k} x_i < 1, \quad k = 1, 2, \ldots, n-1, \sum_{i=1}^{n} x_i \geq 1\right\}$$

$$\text{if} \quad n = 1, 2, \ldots,$$

$$\{x \in R^\infty : f(x) = \infty\} = \left\{x\colon \sum_{i=1}^{n} x_i < 1, \quad n = 1, 2, \ldots\right\}.$$

In each case we have a finite or countable intersection of measurable cylinders.

Chapter 3

Section 3.2

6. (a) Let $z \in K$, a compact subset of U. If $r < d_0$, a standard application of the Cauchy integral formula yields

$$f(z) = \frac{1}{2\pi} \int_0^{2\pi} f(z + re^{i\theta})\, d\theta.$$

Thus if $0 < d < d_0$,

$$f(z)\frac{d^2}{2} = \int_0^d rf(z)\, dr = \frac{1}{2\pi} \int_0^d r\, dr \int_0^{2\pi} f(z + re^{i\theta})\, d\theta,$$

or

$$f(z) = \frac{1}{\pi d^2} \int_{\theta=0}^{2\pi} \int_{r=0}^{d} f(z + re^{i\theta}) r\, dr\, d\theta$$

$$= \frac{1}{\pi d^2} \iint_D f(x + iy)\, dx\, dy$$

where D is the disk with center at z and radius r. An application of the Cauchy–Schwarz inequality to the functions 1 and f shows that

$$|f(z)| \leq \frac{1}{\pi d^2} (\pi d^2)^{1/2} \|f\|,$$

as desired.

(b) If $f_1, f_2, \ldots,$ is a Cauchy sequence in $H(U)$, part (a) shows that f_n converges uniformly on compact subsets to a function f analytic on U. But $H(U) \subset L^2(\Omega, \mathscr{F}, \mu)$ where $\Omega = U$, $\mathscr{F}$ is the class of Borel sets, and μ is Lebesgue measure; hence f_n converges in L^2 to a function $g \in H(U)$. Since a subsequence of $\{f_n\}$ converges to g a.e., we have $f = g$ a.e. Therefore $f \in H(U)$ and $f_n \to f$ in L^2, that is, in the $H(U)$ norm.

7. (a) If $0 \le r < 1$,

$$\frac{1}{2\pi}\int_0^{2\pi} |f(re^{i\theta})|^2 \, d\theta = \frac{1}{2\pi}\int_0^{2\pi} \sum_n a_n r^n e^{in\theta} \sum_m \bar{a}_m r^m e^{-im\theta} \, d\theta$$

$$= \frac{1}{2\pi} \sum_{n,\, m} a_n \bar{a}_m r^{n+m} \int_0^{2\pi} e^{i(n-m)\theta} \, d\theta$$

since the Taylor series converges uniformly on compact subsets of D

$$= \sum_{n=0}^{\infty} |a_n|^2 r^{2n},$$

which increases to $\sum_{n=0}^{\infty} |a_n|^2$ as r increases to 1.

(b) $$\iint_D |f(x+iy)|^2 \, dx\, dy = \int_0^1 r\, dr \int_0^{2\pi} |f(re^{i\theta})|^2 \, d\theta$$

$$\le 2\pi N^2(f) \int_0^1 r\, dr = \pi N^2(f).$$

(c) $$\iint_D |f_n(x+iy|^2 \, dx\, dy = \int_0^{2\pi}\int_0^1 r^{2n} r\, dr\, d\theta = \frac{2\pi}{(2n+2)} \to 0,$$

but

$$\frac{1}{2\pi}\int_0^{2\pi} |f_n(re^{i\theta})|^2 \, d\theta = \frac{1}{2\pi}\int_0^{2\pi} r^{2n} \, d\theta = r^{2n};$$

hence $N(f_n) = 1$ for all n.

(d) If $\{f_n\}$ is a Cauchy sequence in H^2, part (b) shows that $\{f_n\}$ is Cauchy in $H(D)$. By Problem 6, f_n converges uniformly on compact subsets and in $H(D)$ to a function f analytic on D. Now if $0 < r_0 < 1$, $\varepsilon > 0$, the Cauchy property in H^2 gives

$$\frac{1}{2\pi}\int_0^{2\pi} |f_n(re^{i\theta}) - f_m(re^{i\theta})|^2 \, d\theta \le \varepsilon$$

for $r \le r_0$ and n, m exceeding some integer $N(\varepsilon)$. Let $m \to \infty$; since $f_m \to f$ uniformly for $|z| \le r_0$,

$$\frac{1}{2\pi}\int_0^{2\pi} |f_n(re^{i\theta}) - f(re^{i\theta})|^2 \, d\theta \le \varepsilon, \qquad r \le r_0, \quad n \ge N(\varepsilon).$$

Since r_0 may be chosen arbitrarily close to 1, $N^2(f_n - f) \le \varepsilon$ for $n \ge N(\varepsilon)$, proving completeness.

(e) Since e_n corresponds to $(0, \ldots, 0, 1, 0, \ldots)$, with the 1 in position n, in the isometric isomorphism between H^2 and a subspace of l^2, the e_n are orthonormal. Now if $f \in H^2$, with Taylor coefficients a_n, $n = 0, 1, \ldots$, then $\langle f, e_n \rangle = a_n$, again by the isometric isomorphism. Thus

$$N^2(f) = \sum_{n=0}^{\infty} |a_n|^2 = \sum_{n=0}^{\infty} |\langle f, e_n \rangle|^2$$

and the result follows from 3.2.13(f).

(f) $$\langle e_n, e_m \rangle = \iint_D e_n(x+iy)\bar{e}_m(x+iy)\,dx\,dy = \iint_D e_n(re^{i\theta})\bar{e}_m(re^{i\theta})r\,dr\,d\theta$$

$$= [(2n+2)(2m+2)]^{1/2}\left(\frac{1}{2\pi}\right)\int_0^{2\pi}\int_0^1 r^{n+m}e^{i(n-m)\theta}r\,dr\,d\theta$$

$$= \begin{cases} 0, & n \ne m, \\ 1, & n = m. \end{cases}$$

Thus the e_n are orthonormal. Now if $f \in H(D)$ with

$$f(z) = \sum_{n=0}^{\infty} a_n z^n,$$

then

$$\int_0^{2\pi}\int_0^{r_0} f(re^{i\theta})\bar{e}_m(re^{i\theta})r\,dr\,d\theta$$

$$= \sum_{n=0}^{\infty} a_n \int_0^{2\pi}\int_0^{r_0} r^n e^{in\theta}\left(\frac{2m+2}{2\pi}\right)^{1/2} r^m e^{-im\theta} r\,dr\,d\theta$$

since the Taylor series converges uniformly on compact subsets of D

$$= a_m\left(\frac{2m+2}{2\pi}\right)^{1/2}\frac{r_0^{2m+2}}{2m+2}2\pi = a_m\left(\frac{2\pi}{2m+2}\right)^{1/2} r_0^{2m+2}.$$

Now $f\bar{e}_m$ is integrable on D (by the Cauchy–Schwarz inequality), so we may let $r_0 \to 1$ and invoke the dominated convergence theorem to obtain

$$\langle f, e_m \rangle = a_m \left(\frac{2\pi}{2m+2} \right)^{1/2}.$$

But the same argument with e_m replaced by f shows that

$$\begin{aligned} \|f\|^2_{H(D)} &= \lim_{r_0 \to 1} \sum_{n,m=0}^{\infty} a_n \bar{a}_m \int_0^{2\pi} \int_0^{r_0} r^n e^{in\theta} r^m e^{-im\theta} r \, dr \, d\theta \\ &= \sum_{n=0}^{\infty} \frac{|a_n|^2 2\pi}{2n+2}. \end{aligned}$$

The result now follows from 3.2.13(f).

9. (a) Let g be a continuous complex-valued function on $[0, 2\pi]$ with $g(0) = g(2\pi)$. Then $g(t) = h(e^{it})$, where $h(z)$ is continuous on $\{z \in C\colon |z| = 1\}$. By the Stone–Weierstrass theorem, h can be uniformly approximated by functions of the form $\sum_{k=-n}^{n} c_k z^k$. For the algebra generated by z^n, $n = 0, \pm 1, \pm 2, \ldots$, separates points, contains the constant functions, and contains the complex conjugate of each of its members since $\bar{z} = 1/z$ for $|z| = 1$.

   Thus $g(t)$ can be uniformly approximated (hence approximated in L^2) by trigonometric polynomials $\sum_{k=-n}^{n} c_k e^{ikt}$. Since any continuous function on $[0, 2\pi]$ can be approximated in L^2 by a continuous function with $g(0) = g(2\pi)$, and the continuous functions are dense in L^2, the trigonometric polynomials are dense in L^2.

   (b) By 3.2.6, $\int_0^{2\pi} |f(t) - \sum_{k=-n}^{n} c_k e^{ikt}|^2 \, dt$ is minimized when $c_k = a_k = (1/2\pi) \int_0^{2\pi} f(t) e^{-ikt} \, dt$. Furthermore, *some* sequence of trigonometric polynomials converges to f in L^2 since the trigonometric polynomials are dense. The result follows.

   (c) This follows from part (a) and 3.2.13(c), or, equally well, from part (b) and 3.2.13(d).

10. (a) Let $\{e_n\}$ be an infinite orthonormal subset of H. Take $M = \{x_1, x_2, \ldots\}$, where $x_n = (1 + 1/n)e_n$, $n = 1, 2, \ldots$. To show that M is closed, we compute, for $n \neq m$,

$$\begin{aligned} \|x_n - x_m\|^2 &= \left\| \left(1 + \frac{1}{n}\right) e_n - \left(1 + \frac{1}{m}\right) e_m \right\|^2 \\ &= \left(1 + \frac{1}{n}\right)^2 + \left(1 + \frac{1}{m}\right)^2 \geq 2. \end{aligned}$$

Thus if $y_n \in M$, $y_n \to y$, then $y_n = y$ eventually, so $y \in M$. Since $\|x_n\|^2 = 1 + (1/n)$, M has no element of minimum norm.

(b) Let M be a nonempty closed subset of the finite-dimensional space H. If $x \in H$ and $N = M \cap \{y: \|x - y\| \le n\}$, then $N \ne \varnothing$ for some n. Since $y \to \|x - y\|$, $y \in N$, is continuous and N is compact, $\inf\{\|x - y\| : y \in N\} = \|x - y_0\|$ for some $y_0 \in N \subset M$. But the inf over N is the same as the inf over M; for if $y \in M$, $y \notin N$, then $\|x - y_0\| \le n < \|x - y\|$. Note that y_0 need not be unique; for example, let $H = R$, $M = \{-1, 1\}$, $x = 0$.

Section 3.3

3. Since $\int_a^b |K(s, t)|\, dt$ is continuous in s, it assumes a maximum at some point $u \in [a, b]$. If $K(u, t) = r(t)e^{i\theta(t)}$, $r(t) \ge 0$, let $z(t) = e^{-i\theta(t)}$. Let $x_1, x_2, \ldots$ be a sequence in $C[a, b]$ such that $\int_a^b |x_n(t) - z(t)|\, dt \to 0$; we may assume that $|x_n(t)| \le 1$ for all n and t (see 2.4.14). Since K is bounded,

$$\left| \int_a^b K(s, t)z(t)\, dt \right| = \lim_{n\to\infty} \left| \int_a^b K(s, t)x_n(t)\, dt \right| = \lim_{n\to\infty} |(Ax_n)(s)| \le \|A\|.$$

Set $s = u$ to obtain

$$\int_a^b |K(u, t)|\, dt \le \|A\|,$$

as desired.

7. (a) If $x \in L$, then

$$\|x\|_1 = \left\| \sum_{i=1}^n x_i e_i \right\|_1 \le \sum_{i=1}^n |x_i|\ \|e_i\|_1 \le \left(\max_i \|e_i\|_1 \right) \sum_{i=1}^n |x_i|.$$

But $\sum_{i=1}^n |x_i| \le \sqrt{n}\|x\|_2$, so we may take $k = \sqrt{n} \max_i \|e_i\|_1$.

(b) Let $S = \{x: \|x\|_2 = 1\}$. Since $(L, \| \ \|_2)$ is isometrically isomorphic to C^n, S is compact in the norm $\| \ \|_2$. Now the map $x \to \|x\|_1$ is a continuous real-valued function on $(L, \| \ \|_1)$, and by part (a), the topology induced by $\| \ \|_1$ is weaker than the topology induced by $\| \ \|_2$. Thus the map is continuous on $(L, \| \ \|_2)$; hence it attains a minimum on S, necessarily positive since $x \in S$ implies $x \ne 0$.

(c) If $x \in L$, $x \ne 0$, let $y = x/\|x\|_2$; then $\|y\|_1 \ge m\|y\|_2$ by (b); hence $\|x\|_1 \ge m\|x\|_2$. By (a) and Problem 6(b), $\| \ \|_1$ and $\| \ \|_2$ induce the same topology.

(d) By the above results, the map $T: \sum_{i=1}^n x_i e_i \to (x_1, \ldots, x_n)$ is a one-to-one onto, linear, bicontinuous map of L and C^n [note that $\|\sum_{i=1}^n x_i\|_2$ is the Euclidean norm of $(x_1, \ldots, x_n)$ in C^n]. If $y_j \in L$,

$y_j \to y \in M$, then $y_j - y_k \to 0$ as $j, k \to \infty$; hence $T(y_j - y_k) = Ty_j - Ty_k \to 0$. Thus $\{Ty_j\}$ is a Cauchy sequence in C^n. If $Ty_j \to z \in C^n$, then $y_j \to T^{-1}z \in L$.

9. For (a) implies (b), see Problem 7; if (c) holds, then $\{x\colon \|x\| \le \varepsilon\}$ is compact for small enough $\varepsilon > 0$; hence every closed ball is compact (note that the map $x \to kx$ is a homeomorphism). But any closed bounded set is a subset of a closed ball, and hence is compact.

   To prove that (f) implies (a), choose $x_1 \in L$ such that $\|x_1\| = 1$. Suppose we have chosen $x_1, \ldots, x_k \in L$ such that $\|x_i\| = 1$ and $\|x_i - x_j\| \ge \frac{1}{2}$ for $i, j = 1, \ldots, k$, $i \ne j$. If L is not finite-dimensional, then $S\{x_1, \ldots, x_k\}$ is a proper subspace of L, necessarily closed, by Problem 7(d). By Problem 8, we can find $x_{k+1} \in L$ with $\|x_{k+1}\| = 1$ and $\|x_i - x_{k+1}\| \ge \frac{1}{2}$, $i = 1, \ldots, k$. The sequence $x_1, x_2, \ldots$ satisfies $\|x_n\| = 1$ for all n, but $\|x_n - x_m\| \ge \frac{1}{2}$ for $n \ne m$; hence the unit sphere cannot possibly be covered by a finite number of balls of radius less than $\frac{1}{4}$.

11. (a) Define $\lambda(A) = f(I_A)$, $A \in \mathscr{F}$. If $A_1, A_2, \ldots$ are disjoint sets in $\mathscr{F}$ whose union is A, then $\lambda(A) = \sum_{i=1}^{\infty} f(I_{A_i})$ since f is continuous and $\sum_{i=1}^{n} I_{A_i} \xrightarrow{L^p} I_A$. [Note that

$$\int_\Omega |\sum_{i=1}^{n} I_{A_i} - I_A|^p \, d\mu = \sum_{i=n+1}^{\infty} \mu(A_i) \to 0$$

by finiteness of μ.] Thus λ is a complex measure on $\mathscr{F}$. If $\mu(A) = 0$, then $I_A = 0$ a.e. $[\mu]$, so we may write $I_A \xrightarrow{L^p} 0$ and use the continuity of f to obtain $\lambda(A) = 0$. By the Radon–Nikodym theorem, we have $\lambda(A) = \int_A y \, d\mu$ for some μ-integrable y. Thus $f(x) = \int_\Omega xy \, d\mu$ when x is an indicator; hence when x is a simple function. Since f is continuous, y is μ-integrable, and the finite-valued simple functions are dense in L^p, the result holds when x is a bounded Borel measurable function.

Now let $y_1, y_2, \ldots$ be nonnegative, finite-valued, simple functions increasing to $|y|$. Then

$$\begin{aligned}
\|y_n\|_q^q &= \int_\Omega y_n{}^q \, d\mu \le \int_\Omega y_n^{q-1} |y| \, d\mu = \int_\Omega y_n^{q-1} e^{-i \arg y} y \, d\mu \\
&= f(y_n^{q-1} e^{-i \arg y}) \qquad \text{since} \quad y_n^{q-1} e^{-i \arg y} \quad \text{is bounded} \\
&\le \|f\| \; \|y_n^{q-1}\|_p = \|f\| \; \|y_n\|_q^{q/p} \qquad \text{since} \quad (q-1)p = q.
\end{aligned}$$

Thus $\|y_n\|_q \le \|f\|$; hence by the monotone convergence theorem, $\|y\|_q \le \|f\|$; in particular, $y \in L^q$. But now Hölder's inequality and the fact that finite-valued simple functions are dense in L^p yield $f(x) = \int_\Omega xy \, d\mu$ for all $x \in L^p$. Hölder's inequality also gives

$\|f\| \le \|y\|_q$; hence $\|f\| = \|y\|_q$. If y_1 is another such function, then $g(x) = \int_\Omega x(y - y_1)\, d\mu = 0$ for all $x \in L^p$. By the above argument, $\|y - y_1\|_q = 0$, so $y = y_1$ a.e. $[\mu]$.

(b) (i) If, say, $y_A - y_B > 0$ on the set $F \subset A \cap B$, let $x = I_F$; then $xI_A = xI_B$; hence $\int_\Omega x(y_A - y_B)\, d\mu = 0$, that is,

$$\int_F (y_A - y_B)\, d\mu = 0.$$

But then $\mu(F) = 0$.

(ii) Since $y_{A_n \cup A_m} = y_{A_n}$ a.e. on A_n we have

$$\|y_{A_n}\|_q^q \le \|y_{A_n \cup A_m}\|_q^q = \|y_{A_n}\|_q^q + \int_{A_m - A_n} |y_{A_m}|^q\, d\mu.$$

Since $\|y_{A_n}\|_q^q$ approaches k^q as $n \to \infty$, so does $\|y_{A_n \cup A_m}\|_q^q$, and it follows that $\int_{A_m - A_n} |y_{A_m}|^q\, d\mu \to 0$ as $n \to \infty$. By symmetry, we may interchange m and n to obtain

$$\int_\Omega |y_{A_m} - y_{A_n}|^q\, d\mu = \int_{A_m - A_n} |y_{A_m}|^q\, d\mu + \int_{A_n - A_m} |y_{A_n}|^q\, d\mu \to 0$$

as $n, m \to \infty$. Thus y_{A_n} converges in L^q to a limit y, and since $\|y_{A_n}\|_q \le \|f\|$ for all n, $\|y\|_q \le \|f\|$. If $\{B_n\}$ is another sequence of sets with $\|y_{B_n}\|_q \to k$, the above argument with A_m replaced by B_n shows that $\|y_{A_n} - y_{B_n}\|_q \to 0$; hence $y_{B_n} \to y$ also.

(iii) Let $A \in \mathscr{F}$, $\mu(A) < \infty$. In (ii) we may take all $A_n \supset A$, so that $y_{A_n} = y_A$ a.e. on A; hence $y = y_A$ a.e. on A. Thus if $x = I_A$, then $f(x) = f(xI_A) = \int_\Omega xy_A\, d\mu = \int_\Omega xy\, d\mu$. It follows that $f(x) = \int_\Omega xy\, d\mu$ if x is simple. [If $\mu(B) = \infty$, then x must be 0 on B since $x \in L^p$.] Since $y \in L^q$, the continuity of f and Hölder's inequality extend this result to all $x \in L^p$.

(c) The argument of (a) yields a μ-integrable y such that $f(x) = \int_\Omega xy\, d\mu$ for all bounded Borel measurable x. Let $B = \{\omega\colon |y(\omega)| \ge k\}$; then

$$k\mu(B) \le \int_B |y|\, d\mu = \int_\Omega I_B e^{-i \arg y} y\, d\mu$$
$$= f(I_B e^{-i \arg y}) \le \|f\|\, \|I_B\|_1 = \|f\| \mu(B).$$

Thus if $k > \|f\|$, we have $\mu(B) = 0$, proving that $y \in L^\infty$ and $\|y\|_\infty \le \|f\|$. As in (a), we obtain $f(x) = \int_\Omega xy\, d\mu$ for all $x \in L^1$, $\|f\| = \|y\|_\infty$, and y is essentially unique.

(d) Part (i) of (b) holds, with the same proof. Now if Ω is the union of disjoint sets A_n, with $\mu(A_n) < \infty$, define y on Ω by taking $y = y_{A_n}$ on A_n. Since $\|y_{A_n}\|_\infty \le \|f\|$ for all n, we have $y \in L^\infty$ and $\|y\|_\infty \le$

$\|f\|$. If $x \in L^1$, then $\sum_{i=1}^{n} xI_{A_i} \xrightarrow{L^1} x$; hence

$$f(x) = \sum_{n=1}^{\infty} f(xI_{A_n}) = \sum_{n=1}^{\infty} \int_{\Omega} xy_{A_n}\, d\mu = \sum_{n=1}^{\infty} \int_{A_n} xy\, d\mu = \int_{\Omega} xy\, d\mu.$$

Since $\|f\| \le \|y\|_\infty$ by Hölder's inequality (with $p = 1$, $q = \infty$), the result follows.

Section 3.4

3. Let $\{y_n\}$ be a Cauchy sequence in M, and let x_0 be any element of L with norm 1. By 3.4.5 (c), there is an $f \in L^*$ with $\|f\| = 1$ and $f(x_0) = \|x_0\| = 1$; we define $A_n \in [L, M]$ by $A_n x = f(x)y_n$. Then $\|(A_n - A_m)x\| = |f(x)|\, \|y_n - y_m\| \le \|y_n - y_m\|\, \|x\|$, so that $\|A_n - A_m\| \le \|y_n - y_m\| \to 0$. By hypothesis, the A_n converge uniformly to some $A \in [L, M]$; therefore $\|y_n - Ax_0\| = \|A_n x_0 - Ax_0\| \le \|A_n - A\| \to 0$.
7. Let L be the set of all scalar-valued functions on Ω, with sup norm, and let M be the subspace of L consisting of simple functions

$$x = \textstyle\sum_j x_j I_{A_j},$$

where the A_j are disjoint sets in $\mathscr{F}_0$. Define g on M by

$$g(x) = \textstyle\sum_j x_j \mu_0(A_j).$$

Now $|g(x)| \le \max_j |x_j| \sum_j \mu_0(A_j) \le \max_j |x_j| \mu_0(\Omega) = \mu_0(\Omega)\|x\|$; hence $\|g\| \le \mu_0(\Omega) < \infty$. By the Hahn–Banach theorem, g has an extension to a continuous linear functional f on L, with $\|f\| = \|g\|$. Define $\mu(A) = f(I_A)$, $A \subset \Omega$. Since f is linear, μ is finitely additive, and since f is an extension of g, μ is an extension of μ_0. Now if $\mu(A) < 0$, then

$$f(I_{A^c}) = \mu(A^c) = \mu(\Omega) - \mu(A) > \mu(\Omega).$$

But $\|I_{A^c}\| = 1$, so that $\|f\| > \mu(\Omega)$, a contradiction.
8. Since $A_n x \to Ax$ for each x, $\sup_n \|A_n x\| < \infty$. By the uniform boundedness principle, $\sup_n \|A_n\| = M < \infty$; hence

$$\|Ax\| = \lim_{n\to\infty} \|A_n x\| = \liminf_{n\to\infty} \|A_n x\| \le \|x\| \liminf_{n\to\infty} \|A_n\| \le M\|x\|.$$

12. Let L be the set of all complex-valued functions x on $[0, 1]$ with a continuous derivative x', M the set of all continuous complex-valued functions on $[0, 1]$, with the sup norm on L and M. If $Ax = x'$, $x \in L$, then A is a linear map of L onto M, and A is closed. For if $x_n \to x$ and $x_n' \to y$, then since convergence relative to the sup norm is uniform convergence,

$\int_0^t x_n'(s)\,ds \to \int_0^t y(s)\,ds = z(t)$. Thus $x_n(t) - x_n(0) \to z(t)$; hence $x(t) - x(0) = z(t)$. Therefore $x' = z' = y$. But A is unbounded, for if $x_n(t) = \sin nt$, then $\|x_n\| = 1$, $\|Ax_n\| \to \infty$.

13. By the open mapping theorem, A is open; hence $A\{x \in L: \|x\| < 1\}$ is a neighborhood of 0 in M; say $\{y \in M: \|y\| < \delta\} \subset A\{x \in L: \|x\| < 1\}$. If $y \in M$, $y \neq 0$, then $\delta y/2\|y\|$ has norm less than δ, hence equals Az for some $z \in L$, $\|z\| < 1$. If $x = (2\|y\|/\delta)\,z$, then $Ax = y$ and $\|x\| < 2\|y\|/\delta$. Thus we may take $k = 2/\delta$.

Section 3.5

3. Let $x_1, \ldots, x_n$ be a basis for L, and define $A: C^n \to L$ by $A(a_1, \ldots, a_n) = a_1x_1 + \cdots + a_nx_n$. It is immediate that A is one-to-one onto, linear, and continuous; we must also show that A^{-1} is continuous. If L is one-dimensional, then $A^{-1}(ax) = a$, so the null space of A^{-1} is $\{0\}$, which is closed by the Hausdorff hypothesis. By Problem 2, A^{-1} is continuous. If $\dim L = n + 1$ and the result holds up to dimension n, let f be a nontrivial linear functional on L. Then [see 3.3.4(a)] $N = f^{-1}\{0\}$ is a maximal proper subspace of L; hence by the induction hypothesis, it is topologically isomorphic to C^n. It follows that N is closed in L; the argument is the same as in Problem 7(d), Section 3.3.

   By Problem 2, every linear functional on L is continuous, so if p_i is is the projection of C^n on the ith coordinate space, $p_i \circ A^{-1}$ is continuous for each i, hence A^{-1} is continuous.

5. If L were metrizable, it would have a countable base at 0, which can be assumed to be of the form

$$U_n = \{f \in L: |f(x)| < \delta_n \quad \text{for all} \quad x \in A_n\},$$

   where $\delta_n > 0$ and A_n is a finite subset of $[0, 1]$, $n = 1, 2, \ldots$. Let y be a point in $[0, 1]$ not belonging to $\bigcup_{n=1}^{\infty} A_n$, and let $U = \{f \in L: |f(y)| < 1\}$. Then U is a neighborhood of 0 in L but no $U_n \subset U$, for there are (continuous) functions f with $f = 0$ on A_n but $f(y) = 1$.

6. (a) implies (b): The sets $\{x: d(x, 0) < r\}$, r rational, form a base at 0.
   (b) implies (c): The base at 0 may be assumed to consist of circled convex sets, and the proof of 3.5.7 shows that the corresponding Minkowski functionals generate the topology of L.
   (c) implies (a): If $p_1, p_2, \ldots$ are seminorms generating the topology of L, define

$$d(x, y) = \sum_{n=1}^{\infty} \frac{1}{2^n} \frac{p_n(x - y)}{1 + p_n(x - y)}.$$

It is easily checked that d is a pseudometric, and $d(x_i, x) \to 0$ iff $p_n(x_i - x) \to 0$ for each n, in other words, iff $x_i \to x$ in the topology of L.

7. Let $K_n = \{z: |z| \le 1 - 1/n\}$, $n = 1, 2, \dots$. If K is any compact subset of D, $K \subset K_n$ for large n; hence $\sup\{|f(z)|: z \in K\} \le \sup\{|f(z)|: z \in K_n\} = p_n(f)$. Thus the sets $U_n = \{f \in L: p_n(f) \le 1/n\}$, $n = 1, 2, \dots$, form a countable base at 0, so by Problem 6, L is metrizable.

   Suppose a single norm $\| \;\|$ induces the topology of L. By continuity of p_n, for each n there is a $\delta_n > 0$ such that $\|f\| < \delta_n$ implies $p_n(f) < 1$; hence $\|f\| < 1$ implies $p_n(f) < 1/\delta_n$. Therefore $W = \{f \in L: p_n(f) < 1/\delta_n$ for all $n\}$ is an overneighborhood of 0.

   For each n let z_n be a point in K_{n+1} but not in K_n, and let $f_n \in U_n$ such that $|f_n(z_n)| > 1/\delta_{n+1}$; a choice of f_n is possible because, for example, if

$$1 - \frac{1}{n} < |z_0| < |z_n| \le 1 - \frac{1}{n+1},$$

   then $(z/z_0)^k$ approaches 0 uniformly on K_n as $k \to \infty$, but approaches ∞ at z_n.

   Now $p_{n+1}(f_n) \ge |f_n(z_n)| > 1/\delta_{n+1}$, and since the U_n decrease with n and $f_n \in U_n$, we have, for any k, $f_n \in U_k$ for all $n \ge k$; consequently $f_n \to 0$. But then $f_n \in W$ for large n; hence $p_{n+1}(f_n) < 1/\delta_{n+1}$, a contradiction.

8. (a) We may write $x = xI_{[0,r)} + xI_{[r,1]} = y + z$, and

$$d(x, 0) = \int_0^1 |x(t)|^p \, dt = d(y, 0) + d(z, 0)$$

   since $[0, r) \cap [r, 1] = \varnothing$. Now $d(y, 0) = \int_0^r |x|^p \, dt$, which is continuous in r, approaches 0 as $r \to 0$, and approaches $d(x, 0)$ as $r \to 1$. By the intermediate value theorem, $d(y, 0) = d(z, 0) = \frac{1}{2} d(x, 0)$ for some choice of r.

   (b) By part (a) we can find y_1 with $d(y_1, 0) = \frac{1}{2} d(x, 0)$ and $|f(y_1)| \ge \frac{1}{2}$. Let $x_1 = 2y_1$; then $|f(x_1)| \ge 1$ and $d(x_1, 0) = 2^p d(y_1, 0) = 2^{p-1} d(x, 0)$. Having chosen $x_1, \dots, x_n$ with $|f(x_i)| \ge 1$ and $d(x_i, 0) = 2^{i(p-1)} d(x, 0)$, $i = 1, \dots, n$, apply part (a) to x_n to obtain x_{n+1} with $|f(x_{n+1})| \ge 1$ and $d(x_{n+1}, 0) = 2^{p-1} d(x_n, 0) = 2^{(n+1)(p-1)} d(x, 0)$. Since $p < 1$, $d(x_n, 0) \to 0$ as $n \to \infty$, as desired.

13. (a) Let i be the identity map from $(L, \mathscr{T}_2)$ to $(L, \mathscr{T}_1)$. Since $\mathscr{T}_1 \subset \mathscr{T}_2$, i is continuous, and by the open mapping theorem, $\mathscr{T}_2 \subset \mathscr{T}_1$.

   (b) If $\mathscr{T}_j$ is the topology induced by $\| \;\|_j$, $j = 1, 2$, then $\mathscr{T}_1 \subset \mathscr{T}_2$ by hypothesis, and the result follows from part (a).

15. Define $T: L \to C^n$ by $T(x) = (f_1(x), \dots, f_n(x))$ and define $h: T(L) \to C^n$ by $h(Tx) = g(x)$. Then h is well-defined; for if $Tx_1 = Tx_2$, then $f_i(x_1) = f_i(x_2)$ for all i, so $g(x_1) = g(x_2)$ by hypothesis. Since h is linear on

$T(L)$, it may be extended to a linear functional on C^n, necessarily of the form $h(y_1, \ldots, y_n) = c_1 y_1 + \cdots + c_n y_n$. Thus $g(x) = h(Tx) = c_1 f_1(x) + \cdots + c_n f_n(x)$ for all $x \in L$.

16. (a) Assume $f_i(x) = 0$ for all i. If k is any real number, $f_i(kx) = 0$ for all i, hence $\|kx\| < 1$. Since k is arbitrary, x must be 0.

(b) Since the y_j are linearly independent and T^{-1} is one-to-one, the x_j are linearly independent. If $x \in L$, then $Tx = \sum_{j=1}^{k} c_j y_j$ for some $c_1, \ldots, c_k$; hence $x = \sum_{j=1}^{k} c_j x_j$.

Chapter 4

Section 4.2

2. Let Ω be the first uncountable ordinal, with $\mathscr{F}$ the class of all subsets of Ω, and $\mu(A) = 0$ if A is countable, $\mu(A) = \infty$ if A is uncountable. Define, for each $\alpha \in \Omega$, $f_\alpha(\omega) = 1$ if $\omega \le \alpha$; $f_\alpha(\omega) = 0$ if $\omega > \alpha$. Then $f_\alpha \uparrow f$, where $f \equiv 1$, and $\int_\Omega f_\alpha \, d\mu = 0$ for all α since f_α is the indicator of a countable set. But $\int_\Omega f \, d\mu = \infty$, so the monotone convergence theorem fails.

Section 4.3

4. (a) First note that H is a vector space. For if g is continuous, $\{f \in H: f + g \in H\}$ contains the continuous functions and is closed under pointwise limits of monotone sequences, and hence coincides with H. Thus if $f \in H$ and g is continuous, then $f + g \in H$. A repetition of this argument (notice the bootstrapping technique) shows that if $g \in H$, then $\{f \in H: f + g \in H\} = H$; hence H is closed under addition. Now if a is real, then $\{f \in H: af \in H\} = H$ by the same argument; hence H is closed under scalar multiplication.

Let $\mathscr{S}$ be the open F_σ sets. Then $\mathscr{S}$ is closed under finite intersection, and $I_A \in H$ for all $A \in \mathscr{S}$. (By 4.3.5, I_A is the limit of an increasing sequence of continuous functions.) By 4.1.4, $I_A \in H$ for all $A \in \sigma(\mathscr{S})$ $[= \mathscr{A}(\Omega)$ by 4.3.2]. The usual passage to nonnegative simple functions, nonnegative measurable functions, and arbitrary measurable functions shows that all Baire measurable functions belong to H. But the class of Baire measurable functions contains the continuous functions and is closed under pointwise limits of monotone sequences; hence H is the class of Baire measurable functions.

(b) All functions in H are Baire measurable by part (a), hence $\sigma(H) \subset \mathscr{A}$. But if $A \in \mathscr{A}$, then I_A is Baire measurable, so $I_A \in H$ by part (a).

But then I_A is $\sigma(H)$-measurable by definition of $\sigma(H)$, hence $A \in \sigma(H)$.

5. Let H be the class of Baire measurable functions. Then $K_0 \subset H$, and if $K_\beta \subset H$ for all $\beta < \alpha$, then $K_\alpha \subset H$ since H is closed under monotone limits. Thus $K \subset H$. But K contains the continuous functions and is closed under pointwise limits of monotone sequences. For if $f_1, f_2, \ldots \in K$ and $f_n \to f$, say $f_n \in K_{\alpha_n}$, $n = 1, 2, \ldots$, where $\alpha_n < \beta_1$. If $\alpha = \sup_n \alpha_n < \beta_1$, then $f_n \in K_\alpha$ for all n; hence $f \in K_{\alpha+1} \subset K$. By minimality of H (Problem 4), $H \subset K$.
6. Existence of μ and λ follows from 4.3.13 applied to the real and imaginary parts of E. Now if $\int_\Omega f\, d\mu = 0$ for all $f \in C(\Omega)$, then $\int_\Omega g\, d\mu_1 = \int_\Omega g\, d\mu_2 = 0$ for all real-valued continuous functions on Ω; hence $\mu_1 = \mu_2 = 0$ by 4.3.13. This proves uniqueness of μ (and similarly, uniqueness of λ).

   Now if $f \in C(\Omega)$, then $\int_\Omega f\, d\lambda = \int_\Omega fh\, d|\lambda|$ for some complex-valued Borel measurable h of absolute value 1; hence $|E(f)| \le \int_\Omega |f|\, d|\lambda|$; therefore $\|E\| \le |\lambda|(\Omega)$. On the other hand, let $E_1, \ldots, E_n$ be disjoint Borel subsets of Ω, with $\lambda(E_j) = r_j e^{i\theta_j}$, $j = 1, \ldots, n$. Define $f = \sum_{j=1}^n e^{-i\theta_j} I_{E_j}$ and, for a given $\delta > 0$, let g be a continuous complex-valued function on Ω such that $|g| \le 1$ and $\int_\Omega |f - g|\, d|\lambda| < \delta$ (see 4.3.14). Now

$$\int_\Omega f\, d\lambda = \sum_{j=1}^n r_j = \sum_{j=1}^n |\lambda(E_j)|,$$

   hence

$$|E(g)| = \left| \int_\Omega g\, d\lambda \right| \ge \left| \int_\Omega f\, d\lambda \right| - \delta = \sum_{j=1}^n |\lambda(E_j)| - \delta.$$

   Since δ is arbitrary, it follows that $\|E\| \ge |\lambda|(\Omega)$. But just as in 4.3.13, λ is the unique extension of μ, so $|\lambda|(\Omega) = |\mu|(\Omega)$, completing the proof.

Section 4.4

2. Since there are c Borel subsets of R^n, there are only c measurable rectangles in $\mathscr{F}$. Thus (Problem 11, Section 1.2) card $\mathscr{F} \le c$. But card $\mathscr{F} \ge c$ (consider $\{\omega\colon \omega_t = a\}$, $a \in R$); hence card $\mathscr{F} = c$.
3. Let $\mathscr{C}$ be the class of sets in $\mathscr{F}$ for which the conclusion holds. If $A \in \mathscr{C}$, then $\omega \in A^c$ iff $\omega_{T_0} \notin B_0$, that is, iff $\omega_{T_0} \in B_0{}^c$; thus $A^c \in \mathscr{C}$. If $A_n \in \mathscr{C}$, $n = 1, 2, \ldots$, with $\omega \in A_n$ iff $\omega_{T_n} \in B_{T_n}$, let $T_0 = \bigcup_{n=1}^\infty T_n$, a countable subset of T. Then $\omega \in \bigcup_{n=1}^\infty A_n$ iff $\omega_{T_0} \in B_{T_0}$, where $B_{T_0} = \{y \in \Omega_{T_0}\colon y_{T_n} \in B_{T_n}$ for some $n\}$. Now $\{y \in \Omega_{T_0} : y_{T_n} \in B_{T_n}\} = p_n^{-1}(B_{T_n})$, where p_n is the projection of Ω_{T_0} onto Ω_{T_n}. Thus $p_n\colon (\Omega_{T_0}, \mathscr{F}_{T_0}) \to (\Omega_{T_n}, \mathscr{F}_{T_n})$, so

that $B_{T_0} \in \mathscr{F}_{T_0}$, and consequently $\bigcup_{n=1}^{\infty} A_n \in \mathscr{C}$. Therefore $\mathscr{C}$ is a σ-field containing the measurable cylinders, so $\mathscr{C} = \mathscr{F}$.

4. (a) If the set A in question belongs to $\mathscr{F}$ let T_0 and B_0 be as in Problem 3; assume without loss of generality that $t_0 \in T_0$. Since T_0 is countable, there is a sequence $t_n \to t_0$ with $t_n \notin T_0$. Define $x(t) = 0$, $t \in T_0$; $x(t) = 1$, $t \notin T_0$. Now if $\omega(t) = 0$ for all t, then $\omega \in A$, hence $\omega_{T_0} \in B_0$. But $\omega_{T_0} = x_{T_0}$, so that $x \in A$. Since x is discontinuous at t_0, this is a contradiction.

   (b) Again assume the set A belongs to $\mathscr{F}$ and let T_0 and B_0 be as in Problem 3. Since T_0 is countable, we can find $s \in T$ with $s \notin T_0$. Define $x(t) = c - 1$, $t \neq s$; $x(s) = c + 1$, and let $\omega(t) = c - 1$ for all t. Then $\omega \in A$ and $\omega(t) = x(t)$ for $t \in T_0$; hence $x \in A$, a contradiction.

5. (a) Let $F = \bigcap_{n=1}^{\infty} V_n$, V_n open. Let g_n be a continuous function from Ω_t to $[0, 1]$ with $g_n = 0$ on F, $g_n = 1$ on V_n^c. Let $g = \sum_{n=1}^{\infty} 2^{-n} g_n$. Then g is continuous and $g^{-1}\{0\} = F$. Define $f: \Omega \to R$ by $f(\omega) = g(\omega(t))$. Then $f \in C(\Omega)$ and $\{\omega : f(\omega) = 0\} = A$; therefore $A \in \mathscr{A}(\Omega)$. But if $\mathscr{C}$ is the class of subsets B of Ω_t such that $\{\omega \in \Omega: \ \omega(t) \in B\} \in \mathscr{A}(\Omega)$, then $\mathscr{C}$ is a σ-field containing the closed sets; hence $\mathscr{C} = \mathscr{F}_t$. It follows that all measurable rectangles belong to $\mathscr{A}(\Omega)$; hence $\mathscr{F} \subset \mathscr{A}(\Omega)$.

   (b) If $x, y \in \Omega$, $x \neq y$, then for some t, we have $x(t) \neq y(t)$. Let g be a continuous function from Ω_t to R with $g(x(t)) \neq g(y(t))$. Define $f(\omega) = g(\omega(t))$, $\omega \in \Omega$. Then $f \in C(\Omega)$, f depends on only one coordinate, and $f(x) \neq f(y)$. Since constant functions trivially depend on only one coordinate, the Stone–Weierstrass theorem implies that the algebra A generated by the functions in $C(\Omega)$ depending on one coordinate is dense in $C(\Omega)$. [The algebra A may be described explicitly as the collection of finite sums of finite products of functions depending on one coordinate.]

   Now if $f(\omega) = g(\omega(t))$, $g \in C(\Omega_t)$, then $f = g \circ p_t$, where p_t is the projection of Ω onto Ω_t. Thus every function in A is measurable relative to $\mathscr{F}$ and $\mathscr{B}(R)$. But each function in $C(\Omega)$ is a uniform (hence pointwise) limit of a sequence of functions in A, so that $\mathscr{F}$ makes every function in $C(\Omega)$ measurable. Therefore $\mathscr{A}(\Omega) \subset \mathscr{F}$.

8. (a) Since d is continuous, so is f. If $\omega \neq \omega'$, let $\{\omega_n^*\}$ be a sequence in D with $\omega_n^* \to \omega$. Eventually $d(\omega, \omega_n^*) \neq d(\omega', \omega_n^*)$; hence f is one-to-one. If $f(\omega^n) \to f(\omega)$, but $d(\omega^n, \omega) \geq \delta > 0$ for all n, let ω^* be an element of D such that $d(\omega, \omega^*) < \delta/2$. Since $f(\omega^n) \to f(\omega)$, we have $d(\omega^n, \omega^*) \to d(\omega, \omega^*)$ as $n \to \infty$, and therefore $d(\omega^n, \omega) \leq d(\omega^n, \omega^*) + d(\omega^*, \omega) < \delta$ for large n, a contradiction. Thus f has a continuous inverse.

(b) By part (a), f is a homeomorphism of Ω and $f(\Omega) \subset [0, 1]^\infty$. Since Ω is complete, it is a G_δ in any space in which it is topologically embedded (see the appendix on general topology, Theorem A9.11). Thus $f(\Omega)$ is a G_δ in $[0, 1]^\infty$, in particular a Borel set.

(c) Let $r \in [0, 1)$ with binary expansion $0.\ a_1 a_2 \cdots$ (to avoid ambiguity, do not use expansions that terminate in all ones). Define $g(r) = (a_1, a_2, \ldots)$; g is then one-to-one. If $k/2^n$ is a dyadic rational number with binary expansion $0.a_1 \cdots a_n\, 0\, 0 \cdots$, then

$$g\left[\frac{k}{2^n}, \frac{k+1}{2^n}\right) = \{y \in S^\infty : y_1 = a_1, \ldots, y_n = a_n\}.$$

Thus g maps finite disjoint unions of dyadic rational intervals onto measurable cylinders, and since $g[0, 1) = S^\infty - \{y \in S^\infty : y_n$ is eventually $1\} = S^\infty -$ a countable set, we have $g[0, 1) \in \mathscr{S}^\infty$ and g and g^{-1} are measurable.

(d) Let $A = \{y \in S^\infty : y_n$ is eventually $1\}$, $B = \{y \in S^\infty : y \in g[0, 1)$ and $g^{-1}(y)$ is rational$\}$, where g is the map of part (c). Let q be a one-to-one correspondence of the rationals in $[0, 1]$ and $A \cup B$. Define $h\colon [0, 1] \to S^\infty$ by

$$h(x) = \begin{cases} g(x) & \text{if } x \text{ is irrational,} \\ q(x) & \text{if } x \text{ is rational.} \end{cases}$$

Then h has the desired properties.

(e) Define $h\colon \prod_n \Omega_n \to \prod_n S_n$ by $h(\omega_1, \omega_2, \ldots) = (h_1(\omega_1), h_2(\omega_2), \ldots)$. The mapping h yields the desired equivalence.

Section 4.5

3. Assume $\mu_n \xrightarrow{w} \mu$, and let A be a bounded Borel set whose boundary has μ-measure 0. Let V be a bounded open set with $V \supset \bar{A}$, and let $G_j = \{x \in V : d(x, \bar{A}) < 1/j\}$. If f_j is a continuous map from Ω to $[0, 1]$ such that $f_j = 1$ on $\bar{A}$ and $f_j = 0$ off G_j [see A5.13(b)], then

$$|\mu_n(A) - \mu(A)| \le \left| \int_\Omega (I_A - f_j)\, d\mu_n \right| + \left| \int_\Omega f_j\, d\mu_n - \int_\Omega f_j\, d\mu \right| + \left| \int_\Omega (f_j - I_A)\, d\mu \right|. \tag{1}$$

Since $f_j - I_A$ is 0 on A and also on $G_j{}^c$, the third term on the right side of (1) is bounded by $\mu(G_j - A) \le \mu(\bar{G}_j - A^0)$. Similarly, the first term is bounded by $\mu_n(\bar{G}_j - A^0)$. The second term approaches 0 as $n \to \infty$ since the support of f_j is a subset of the compact set $\bar{G}_j$.

Now let g_{jk} be a continuous function from Ω to $[0, 1]$, with supp $g_{jk} \subset \overline{V}$, such that as $k \to \infty$,

$$g_{jk} \downarrow I_{\{\overline{G}_j - A^0\}} .$$

To verify that the g_{jk} exist, note that $\overline{G}_j - A^0 = \bigcap_{n=1}^{\infty} U_n$, where $U_n = \{x \in \overline{V}: d(x, \overline{G}_j - A^0) < 1/n\}$. Let g'_{jn} be a continuous function from Ω to $[0, 1]$ such that $g'_{jn} = 1$ on $\overline{G}_j - A^0$, $g'_{jn} = 0$ off U_n, and set $g_{jk} = \min(g'_{j1}, \ldots, g'_{jk})$. Now

$$\mu_n(\overline{G}_j - A^0) = \int_{\Omega} I_{\overline{G}_j - A^0} \, d\mu_n \le \int_{\Omega} g_{jk} \, d\mu_n \to \int_{\Omega} g_{jk} \, d\mu \qquad \text{as} \quad n \to \infty.$$

Thus

$$\limsup_{n \to \infty} \mu_n(\overline{G}_j - A^0) \le \int_{\Omega} g_{jk} \, d\mu.$$

Since supp $g_{jk} \subset \overline{V}$ and $\mu(\overline{V}) < \infty$, we may let $k \to \infty$ and invoke the monotone convergence theorem to obtain

$$\limsup_{n \to \infty} \mu_n(\overline{G}_j - A^0) \le \mu(\overline{G}_j - A^0).$$

It follows from (1) and the accompanying remarks that

$$\limsup_{n \to \infty} |\mu_n(A) - \mu(A)| \le 2\mu(\overline{G}_j - A^0).$$

As $j \to \infty$ we have $\overline{G}_j - A^0 \downarrow \overline{A} - A^0 = \partial A$, and since $\mu(\partial A) = 0$ we conclude that $\mu_n(A) \to \mu(A)$, proving that (a) implies (b).

Assume $\mu_n(A) \to \mu(A)$ for all bounded Borel sets with $\mu(\partial A) = 0$, and let f be a bounded function from Ω to R, continuous a.e. $[\mu]$ with supp$f \subset K$, K compact. Let V and W be bounded open sets such that $K \subset V \subset \overline{V} \subset W$ [see A5.12(c)]. Now

$$\overline{V} = \bigcap_{\delta > 0} \{x \in W: d(x, \overline{V}) < \delta\} = \bigcap_{\delta > 0} W_\delta;$$

the W_δ are open and

$$\partial W_\delta \subset \{x \in \overline{W}: d(x, \overline{V}) = \delta\}.$$

Thus the W_δ have disjoint boundaries and $\mu(\overline{W}) < \infty$, so $\mu(\partial W_\delta) = 0$ for some δ. Therefore we may assume without loss of generality that we have $K \subset V$, with V a bounded open set and $\mu(\partial V) = 0$.

Now if $A \subset V$, the interior of A is the same relative to V as to the entire space Ω since V is open. The closure of A relative to V is given by $\overline{A}_V = \overline{A} \cap V$; hence the boundary of A relative to V is $\partial_V A = (\partial A) \cap V$.

If A is a Borel subset of V and $\mu(\partial_V A) = 0$, then $\mu[(\partial A) \cap V] = 0$; also

$$(\partial A) \cap V^c \subset \bar{A} \cap V^c \subset \bar{V} - V = \bar{V} - V^0 = \partial V;$$

hence $\mu[(\partial A) \cap V^c] = 0$, so that $\mu(\partial A) = 0$. Thus by hypothesis, $\mu_n(A) \to \mu(A)$. By 4.5.1, if μ_n' and μ' denote the restrictions of μ_n and μ to V, we have $\mu_n' \xrightarrow{w} \mu'$. Since f restricted to V is still bounded and continuous a.e. $[\mu]$, we have $\int_V f \, d\mu_n' \to \int_V f \, d\mu'$, that is, $\int_\Omega f \, d\mu_n \to \int_\Omega f \, d\mu$. This proves that (b) implies (c); (c) implies (a) is immediate.

Subject Index

A

B

C

D